Nonequilibrium Systems
in Natural Water Chemistry

Nonequilibrium Systems in Natural Water Chemistry

A symposium sponsored by
the Division of Water, Air,
and Waste Chemistry of the
American Chemical Society
at Houston, Texas,
February 24-25, 1970.

J. D. Hem

Symposium Chairman

ADVANCES IN CHEMISTRY SERIES **106**

AMERICAN CHEMICAL SOCIETY

WASHINGTON, D. C.   1971

00017846

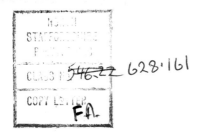

Coden: ADCSHA

Library of Congress Catalog Card 76–170252

ISBN 8412–0127–7

PRINTED IN THE UNITED STATES OF AMERICA

Advances in Chemistry Series

Robert F. Gould, *Editor*

FOREWORD

Advances in Chemistry Series was founded in 1949 by the American Chemical Society as an outlet for symposia and collections of data in special areas of topical interest that could not be accommodated in the Society's journals. It provides a medium for symposia that would otherwise be fragmented, their papers distributed among several journals or not published at all. Papers are refereed critically according to ACS editorial standards and receive the careful attention and processing characteristic of ACS publications. Papers published in Advances in Chemistry Series are original contributions not published elsewhere in whole or major part and include reports of research as well as reviews since symposia may embrace both types of presentation.

CONTENTS

PREFACE

The hydrologist ordinarily needs no reminder that fresh water in natural environments is in constant circulation and that his branch of science requires methods of study appropriate to dynamic systems. The chemist who is concerned with water quality, however, may tend to underemphasize the importance of the motion of water in understanding its chemistry. Certainly he most often sees only small segments of the system, held captive and immobile in water sample bottles. Although there is no reason chemical equilibrium cannot be established in dynamic systems as well as in static ones, an equilibrium or even a steady-state condition does require a certain length of time to become established. Before an intelligent decision can be made as to whether a chemical system is likely to be near equilibrium, one needs to know, at least approximately, both the rates of the chemical reactions involved and the rate of movement of water with respect to other materials participating in the reaction. From the latter rate, the length of time available for equilibration to occur can be estimated and compared with the time required for a reasonably close approach to equilibrium.

Although the need for knowing something about these kinds of rates seems obvious, the amounts of directly applicable information on reaction kinetics is meager and very few chemical studies have taken water circulation rates into careful consideration. Surely a stronger emphasis on such work will be needed in order to understand the chemistry of rivers, lakes, and underground water. One of the objectives of the symposium from which the papers in this volume come was to emphasize these points.

In general, the papers which make up this volume represent work that is being done toward understanding the chemistry of natural aqueous systems without trying to fit them to simple equilibrium models. A broad spectrum of subject matter is covered. Some papers concern field study and others present results of laboratory investigation of synthetic solutions. A few of the papers emphasize theoretical aspects almost exclusively but most are based on current research and demonstrate techniques and approaches that should aid other practitioners in the field of water chemistry.

It is certainly trite to point out the interdisciplinary nature of scientific studies of natural water. One purpose of the symposium where these papers were presented, however, was to bring together some of these different specialists and exchange points of view on shared concerns with the aquatic environment.

The reader of this volume may possibly draw the conclusion that progress to date in our application of nonequilibrium approaches has been dismally small. There is considerable justification for such a conclusion, but progress is being made, nonetheless. Although much can still be learned by application of equilibrium models, one must be wary of oversimplification of the complexities of the real world. It will be necessary eventually to accept the existence of the complications and prepare to deal with them. These papers should help put into proper levels of importance such factors as the slow and unknown rates of many important chemical reactions and the influence of water movement, mixing, and circulation on the composition of water bodies. Instances of progress in study of details of the chemistry of certain specific simple systems which are included in some of the papers may stimulate wider use of nonequilibrium approaches. In future symposia like this one, one may hope real progress will be documented.

The assistance of the authors whose papers are included and of others who have aided in their preparation and publication is gratefully recognized.

John D. Hem

U.S. Department of the Interior
Geological Survey
Menlo Park, California 94025
October 1970

Chemostasis and Homeostasis in Aquatic Ecosystems; Principles of Water Pollution Control

WERNER STUMM[1] and ELISABETH STUMM-ZOLLINGER[2]

Laboratories of Applied Chemistry and Applied Biology, respectively, Harvard University, Cambridge, Mass.

In view of man's inability to adapt to major environmental changes, pollution is equated with disturbance of ecological balance and loss of stability. Increasing the chemical diversity (number of components and phases) makes an equilibrium system more resistant toward external influences imposed on the system. In an ecosystem, its members are interlocked by various feedback loops (homeostasis) and thus adapted to coexistence for mutual advantage; increased diversity makes the system less subject to perturbations and enhances its survival. Because various kinds of disturbance cause similar patterns of change in aquatic ecosystems and affect their stability in a predictable way, general measures of pollution control beyond those of waste treatment can be outlined which mitigate the conflict between resource exploitation and protection of natural waters.

An understanding of the chemistry and biology of natural waters is a prerequisite for an understanding of the ways the environment is affected by man's pollution. In a broad sense, pollution has been characterized as an alteration of man's surroundings in such a way that they become unfavorable to him and to his life. This characterization implies that pollution is not solely caused by contaminants or pollutants added to the environment but can also result from other direct or indirect consequences of man's action.

[1] Present address: Swiss Federal Institute of Technology, (ETH), CH Zürich, Switzerland.
[2] Present address: CH 8700 Küsnacht, Switzerland.

MAN AGAINST NATURE. Man is an integral part of the ecosystem; despite localized large population densities, man as the human animal plays a relatively minor role in the physiology of the ecosphere. Domestic waste and garbage represent a very small fraction of the total detritus produced by organisms. Within the biosphere, the energy involved in man's metabolism (2×10^{15} Kcal per year) may be compared with primary productivity—i.e., the energy fixed by all the plants ($\sim 10^{18}$ Kcal per year). These estimates are based on a daily per capita consumption of 2000 Kcal and a primary productivity of 10^{16} moles year^{-1} of carbon (1). If evenly distributed over the world, man's wastes have a negligible effect on the energy transfer of the ecosphere. Domestic wastes cause localized or temporary unfavorable environmental alteration only where they are discharged in high concentration. On the other hand, man, as an inventive intellectual being, with his capacity of manipulation and dominance dissipates 10 to 20 times (in the USA, 50 to 100 times) as much energy as he requires for his metabolism. The stress imposed upon the environment as a direct or indirect result of this energy dissipation outweighs by far the disturbances caused by the disposal of domestic wastes.

In what way does energy dissipation cause pollution? Obviously, smoke, sulfur dioxide, excess heat and water loss by evaporation, spillage of oil, pesticides, and other petrochemicals into fresh water and oceans, and the leakage of fertilizers from land into the water are some of the byproducts of power consumption and cultural development. Agriculture, forestry, geological exploitation, construction of dams, manipulations of the landscape, urban construction, and other means of civilization counteract the forces of natural selection; they affect the so-called balance of nature and interfere with biological relationships. Most of the energy utilized by our industrial society for its own advantage ultimately causes a simplification of the ecosystem—specifically, a reduction of the food web and a shortening of the food chain (2). The less complex a natural ecosystem, the less stable and the more liable it is to perturbations and to catastrophe. Most of our concern, thus, should be with this simplification of the ecosystem and with the concomitant lack of balance and stability.

Instability as a Measure of Pollution. Man's ability to adapt to a changing environment is very limited because the range of physiological adaptation is narrow and evolutionary adaptation is slow. When man evolved, he found a stable environment capable of resisting change and perturbation. The chemical compositions of the various oceans are quite similar and have probably been essentially constant for the last 100 million years. Similarly, the composition of the atmosphere has remained unchanged, and climatic variations have been extremely slow. In the integrated global ecological system, we have a remarkably well-established balance of production and destruction of organic material as well as of production and consumption of O_2, providing a constant surplus of O_2 in the atmosphere. In view of man's inability to adapt to major environmental changes, pollution may be interpreted as a disturbance in the ecological balance causing loss of stability of the environment.

OBJECTIVES. It is the objective of this presentation to review some of the chemical and biological factors that regulate the composition of natural waters, to illustrate the variables and modes by which stability is imparted to natural systems, to interpret pollution in terms of disturbance of ecological balances and mitigation of ecosystem stability, and to discuss some means of water pollution control beyond those of waste treatment.

Chemical Factors Regulating the Composition of Natural Waters

Terrestrial waters vary in chemical composition; these variations can be understood, at least partially, in terms of the different histories of the waters. Appreciation of some of the pertinent reactions by which natural waters acquire their characteristics can be obtained by carrying out some simple imaginary experiments (Figure 1a). Minerals are mixed with distilled water and exposed to an atmosphere containing CO_2. Congruent

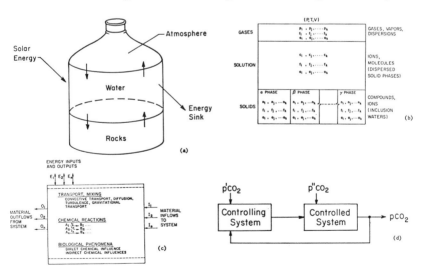

Figure 1.   Generalized models for description of natural water systems

(a) *Mixing rocks, water, and atmosphere*
(b) *Equilibrium models establish boundary conditions toward which aquatic environments must proceed, however slowly*
(c) *Steady state model permits the description of the time-invariant conditions of dynamic and open systems*
(d) *Living systems are controlled by negative feedback (homeostasis)*

and incongruent dissolution reactions (weathering reactions) take place because many constituents of the earth's crust are thermodynamically unstable in the presence of water and the atmosphere; for example:

$$CaCO_3(s) + H_2O = Ca^{2+} + HCO_3^- + OH^-$$
Calcite

$$CaCO_3(s) + H_2CO_3^* = Ca^{2+} + 2\ HCO_3^-$$

$$NaAlSi_3O_8(s) + 5\tfrac{1}{2}\ H_2O$$
$$= Na^+ + OH^- + 2\ H_4SiO_4 + \tfrac{1}{2}\ Al_2Si_2O_5(OH)_4(s)$$
Albite   Kaolinite

$$NaAlSi_3O_8(s) + H_2CO_3^* + 4\tfrac{1}{2}\ H_2O$$
$$= Na^+ + HCO_3^- + 2\ H_4SiO_4 + \tfrac{1}{2}\ Al_2Si_2O_5(OH)_4(s)$$

$$CaAl_2Si_2O_8(s) + 3\ H_2O = Ca^{2+} + 2\ OH^- + Al_2Si_2O_5(OH)_4(s)$$
Anorthite   Kaolinite

$$CaAl_2Si_2O_8(s) + 2\ H_2CO_3^* + H_2O$$
$$= Ca^{2+} + 2\ HCO_3^- + Al_2Si_2O_5(OH)_4(s)$$

$$3\ Ca_{0.33}Al_{4.67}Si_{7.33}(OH)_4(s) + 2\ H_2CO_3^*$$
$$= Ca^{2+} + 2\ HCO_3^- + 8\ H_4SiO_4 + 7\ Al_2Si_2O_5(OH)_4(s)$$
Montmorillonite   Kaolinite

Recognition of the chemical processes involved permits identification of the variables and mechanisms that regulate and control the mineral composition of natural waters. With the help of equilibrium constants for the pertinent reactions, boundary conditions towards which aquatic environments must proceed can be established.

We can also carry out our imaginary experiment by mixing rocks with water in a closed bottle, where we leave a little space at the top for the atmosphere. As epitomized by Sillén (3) and many other researchers (4), such a closed rock–water–atmosphere system constitutes a simplified but representative model of what has fittingly been called "spaceship earth." In an equilibrium system, the concentration of inorganic solutes (ocean) and the CO_2 pressure in the gas phase (atmosphere) are primarily regulated by the heterogeneous reactions involving carbonates and various aluminum silicates, thus illustrating plausibly that the CO_2 content of the atmosphere is regulated at the sea–sediment interface. The volume proportions in this model (Figure 1a) appear unrelated to the real system, but metaphorically the idea of a "gas bubble" is reflected in the mass proportion of CO_2 in the geosphere; for every C atom in the atmospace, there are about 60 C atoms (mostly as HCO_3^-) in the hydrospace and about 40,000 C atoms (largely as CO_3^{2-}) in the sediments (4).

Buffering. Heterogeneous dissolution and precipitation reactions are the principal pH buffer mechanisms in natural waters. It has been shown that the buffer intensity of heterogeneous systems is much larger than that of a homogeneous solution; for example, the buffer intensities at pH = 8 of anorthite–kaolinite and of $CaCO_3$–CO_2 ($10^{-3.5}$ atm) suspensions at hypothetical equilibrium are, respectively, 3000 and 30 times larger than that of a 10^{-3} M HCO_3^- solution (4, 5). In a similar way, as [H^+] is kept constant by heterogeneous equilibrium, the concentrations of other cations and anions in natural waters are buffered by heterogeneous reactions. A water that is in equilibrium with solid $CaCO_3$ will tend to maintain a rather constant pCa even if Ca^{2+} is introduced to the water from external sources.

At equilibrium, the Gibbs phase rule restricts the number of independent variables. It is the basis for organizing and interpreting equilibrium models. A few simple equilibrium systems (Figure 1b) are considered in Table I. They are constructed by incorporating the specific components into closed systems and by specifying the number of phases to be included. The phase rule restricts the number of independent variables (degrees of freedom), F, to which one can assign values on the basis of the number of components, C, and of phases, P:

$$F = C + 2 - P. \qquad (2)$$

In Table Ib, models containing identical components but differing with respect to the number of phases are compared with each other. An increase in P must be accompanied by a decrease in F. The activities in the system, such as number 3 or 5, remain constant and independent of the concentration of the components as long as the phases coexist in equilibrium. In Model 6, only one degree of freedom remains for the given number of components and phases; then P_{CO_2} in the gas phase of the model will be determined by the equilibrium and cannot be varied (manostat). The models given in Table I can be enlarged; the addition of each additional component to an equilibrium system must result in either a new phase or an additional degree of freedom. Sillén (3), who demonstrated the relevance of equilibrium models, has proposed equilibrium systems of different complexity as models for the composition of the ocean and the atmosphere.

MINIMIZING EXTERNAL DISTURBANCE. The displacement of a chemical equilibrium by a change of the parameters (activity, pressure, temperature) on which equilibrium depends is independent of the path of the change, but thermodynamically one can predict the sign of the displacement. The principle of LeChatelier has been expressed qualitatively as follows: "A system tends to change so as to minimize the external stress." As we have seen, for a given number of components the number of independent variables is smaller, the larger the number of coexisting

Table I.   Equilibrium Models;

a: CO₂ and CaCO₃ Solubility Models

Phases	1 Aqueous Solution $CO_2(g)$	2 Aqueous Solution Calcite(s)[d]	3 Aqueous Solution $CO_2(g)$ Calcite(s)
P Components	2 H_2O, CO_2	2 H_2O, CO_2, CaO	3 H_2O, CO_2, CaO
C F Variables[c]	2 2 $t = 25°C$ $-\log p_{co_2} = 3.5$	3 3 $t = 25°C$ $-\log p = 0$[b] $[Ca^{2+}] = C_T$[d]	3 2 $t = 25°C$ $-\log p_{co_2} = 3.5$
Composition pH pHCO₃⁻ pCa pH₄SiO₄	5.7 5.7	9.9[e] 4.1 3.9	8.3 3.0 3.3

[a] From Stumm and Morgan (1).
[b] H_2CO_3* is treated as a nonvolatile acid. The system is under a total pressure of 1 atm.
[c] By specifying p_{CO_2}, the total pressure p is determined ($P = p_{CO_2} + p_{H_2O}$). For the calculation, constants valid at $P = 1$ atm were used.

phases. The simple equilibrium system $CaCO_3$, H_2O, CO_2 with three phases (No. 3, Table I) has an infinite buffer intensity with regard to dilution (H_2O) and to the addition of the base $Ca(OH)_2$ or the acid CO_2; *i.e.*, the system (as long as the three phases coexist in equilibrium) resists attempts to perturbation caused by the addition (or withdrawal) of components of the system. Hence, increasing the number of components and phases—*i.e.*, increasing the chemical diversity—makes the system more resistant toward a larger number of external influences imposed on the system and hence less subject to perturbations resulting from external stresses.

Steady State. In contrast to the models discussed above, natural waters are systems open to their environment, and much of their chemistry depends on the kinetics of physical and chemical processes. If, in such a system, input is balanced by output, a steady state condition is attained, and the system remains unchanged in time. Such a time-invariant con-

Application of Phase Rule[a]

b: Aluminum Silicates and CaCO₃

4	5	6
		Aqueous Solution
	Aqueous Solution	$CO_2(g)$
Aqueous Solution	$CO_2(g)$	*Kaolinite*
$CO_2(g)$	*Kaolinite*	*Ca-montmorillonite*
Kaolinite	*Ca-montmorillonite*	*Calcite*
Ca-montmorillonite	*Calcite*	*Ca-feldspar*
4	5	6
H_2O, CO_2, CaO		
Al_2O_3, SiO_2		
5	5	5
3	2	1
$t = 25°C$	$t = 25°C$	$t = 25°C$
$-\log p_{co_2} = 3.5$	$-\log p_{co_2} = 3.5$	
$8[Ca^{2+}] = [H_4SiO_4]$[d]		
7.4	8.3	9.0
3.9	3.0	3.4
4.2	3.3	3.7
3.2	3.6	3.7
		$-\log p_{co_2} = 4.5$

[d] This additional constraint is necessary for defining the system; other conditions could be specified.
[e] $pCO_3^{2-} = 4.4$.

dition of a chemical reaction system represents a convenient idealized model of a natural water system. While an equlilibrium system at constant temperature and pressure is characterized by a minimum in the Gibbs free energy, energy is required for the maintenance of the steady state (6, 7).

Because the sea has remained constant in its composition for the recent geologic past, it has been described plausibly in terms of a steady state model. For a steady state ocean, for each element, E, the equation $(d[E]/dt)_{input} = (d[E]/dt)_{sedimentation}$ can be written. Because the rate of sedimentation is controlled largely by the rate at which an element is converted (precipitation, ion exchange, biological activity) into an insoluble and settleable form, the residence time is affected by the readiness of the elements to react. Hence, elements that are highly oversaturated (*e.g.*, Al, Fe) have detention times that correspond to the time necessary for ocean mixing ($\sim 10^3$ years). On the other hand, elements with low reactivity such as Na or Li have very long residence times ($\sim 10^8$ years) that are perhaps within one or two orders of magnitude of the age of the ocean.

Steady state models can also aid in understanding fresh water systems (Figure 1c). A presupposition of steady state frequently permits the quantitative evaluation of processes such as exchange reactions between atmosphere and ground waters, mixing relations, limnological transformations of constituents (*see* Figure 4), and local hydrological cycles.

Even highly dynamic natural water systems may be at equilibrium with respect to certain processes; this depends on the time scale of the process. Hence, there always exist in natural waters regions or environments that are locally at equilibrium, even though gradients exist throughout the system as a whole.

Regulation in Ecosystems

Returning to our imaginary closed-bottle experiment, we may expose the bottle in which we mixed rocks with water to some light. There will now be a flow of energy through the system. If the bottle contains organisms, our model (Figure 1a) becomes a microcosmos; a small portion of the light energy is used in algal photosynthesis and becomes stored in the form of organic material. Some of the organic matter becomes oxidized, liberating energy in order to support the life processes (assim-

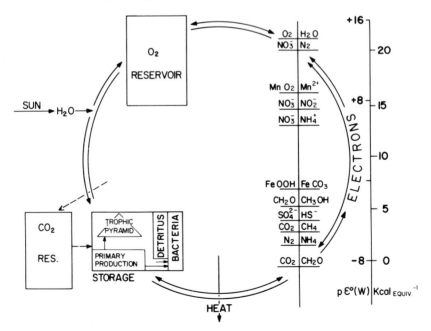

Figure 2.   Photosynthesis and biochemical cycle

ilation) of heterotrophic organisms (consumers and decomposers) (Figure 2). Organisms and their abiotic environment are interrelated and interact upon each other.

The maintenance of life resulting from solar energy (photosynthesis) is the main cause for nonequilibrium conditions (Figure 2). Photosynthesis may be conceived as a process producing localized centers of highly negative p_ϵ ($p_\epsilon = -\log$ electron activity) and oxygen (high p_ϵ). The reduced components (organic compounds) and the equivalent oxidation products (O_2) become partially stored—*e.g.*, in the sediments and in the atmosphere–hydrosphere, respectively. The nonphotosynthetic organisms tend to restore equilibrium by catalytically decomposing the unstable products of photosynthesis through energy-yielding redox reactions, thereby obtaining a source of energy for their metabolic needs. The sequence of redox reactions observed in an aqueous system as a function of p_ϵ values (pH = 7) is also indicated in this figure.

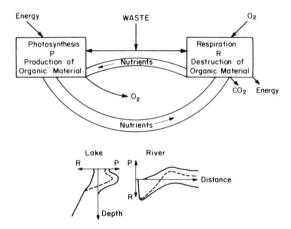

Figure 3.   Balance between photosynthesis and respiration

A disturbance of the P–R (photosynthesis–respiration) balance results from vertical (lakes) or longitudinal (rivers) separation of P and R organisms. An unbalance between P and R functions leads to pollutional effects of one kind or another: depletion of O_2 if P < R or mass development of algae if production rates become larger than the rates of algal destruction by consumer and decomposer organisms (R < P).

Because of the energy flow through the bottle, its contents cannot be in equilibrium, but upon continued exposure to solar energy, eventually a steady state balance between production and destruction of organic material as well as production and consumption of O_2 will be attained (Figure 3). At steady state, a constant surplus of O_2 (equivalent to the reduced organic matter present) prevails in the gas and solution phase. We recognize from our experiment that a system containing living things extracts energy from the stream of radiation and uses this energy to organize the system—*i.e.*, the input of solar energy is necessary to maintain life—and that the flux of energy through the system is accompanied by cycles of water, nutrients, and of other elements (hydrogeochemical cycles) and by cycles of life through different trophic levels. Thus, an ecological system may be defined (8) as a unit of the environment that contains a biological community (primary producers, various trophic levels of consumers and decomposers) in which the flow of energy is reflected in the trophic structure and in material cycles.

In an ecosystem, the energy flow from a source to a sink may lead to an entropy decrease in the intermediate system. In a statistical sense, entropy is a measure of disorder. The second law of thermodynamics demands that any spontaneous process be accompanied by an increase in entropy—*i.e.*, dS(source,sink) + dS(ecosystem) $\geq$ 0. But because dS(source,sink) > 0, the entropy of the ecosystem may decrease: $-$dS(ecosystem) $\leq$ dS(source,sink). This decrease is reflected in the ordering of the ecosystem and the presence of such highly improbable aggregations of energy as living beings (9). Their organization has been acquired at the expense of an increase in entropy of the environment. Thus, the ecosystem may be regarded as an "entropy pump" which employs high-grade solar energy to dissipate excess entropy, thereby maintaining its physical integrity (10).

Interaction Between Organisms and Abiotic Environment. Steady state models may be applied to recognize and evaluate factors that regulate the interaction between biotic and abiotic variables. For example, the growth rate of organisms (*e.g.*, bacteria $dB/dt = \mu B$, where μ is the net growth rate constant (time^{-1}), is determined in a completely mixed system by the hydraulic detention time, $\Gamma_{H_2O} = \mu^{-1}$, because at steady state, the growth of the organisms, μB, is equal to the outflow of organisms B/Γ_{H_2O}.

Another example is illustrated in Figure 4, where some of the important steps in the limnological transformation of phosphorus in a lake are characterized in terms of a steady state model (4). The model simulates a real system by giving a hypothetical balance of the abundance of P in various forms and of the exchange rates. The cycle of phosphorus is determined largely by regeneration of P from biota. Primary production depends to a large extent on the supply of P to the trophogenic layer. For deeper lakes, the rate of supply from sediments is small in comparison with the supply by the hypolimnion and by the introduction of P from waste and drainage. A significant fraction of P introduced into the lake is irretrievably lost to the sediments.

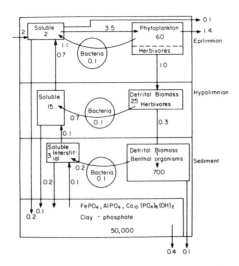

Figure 4.   Simplified steady state model describing important steps in the limnological transformations of P in a lake

The numbers in the boxes are μg P per liter lake volume and the numbers on the arrows give exchange rates in μg P per liter lake volume per day

The dynamics of the transformations, especially the rate of regeneration of nutrients from phytoplankton, detritus, and sediments and the rate of supply of soluble phosphorus to the algae, are often more important in determining productivity than the concentration of soluble P or the quantity of total P present. Values given in Figure 4 for abundance of P in various forms, and for exchange rates, have been chosen so as to be compatible with real systems as reported by various investigators; residence times, $[P]/(d[P]/dt)$, for soluble P and for P in organisms are within the range of observed values. At best, this model simulates a real system to an approximate degree; its main purpose is to offer insight into some attributes and dominant regulatory forces operating in the real system. Manipulating the values (*e.g.*, waste input) within the proper constraints helps illustrate how real systems behave.

A fanciful analogy of the earth's surface geochemical system to a giant chemical engineering plant has been suggested by Lotka (*11*) and, quite recently, by Siever (*12*) and others. Figure 5 gives Siever's recent summary of the main regulatory processes. As he points out, the provocative feature of this hypothetical chemical engineering plant that describes the surface of the earth is that one can readily see the multiplicity

of valves and switches that control the system and the ease with which
this system may be subject to some kind of perturbation if there are vio-
lent movements of any one of the switches. Of particular concern are the
gas regulators of the CO_2 and O_2 tanks. The O_2 tank regulator apparently is governed dominantly by photosynthesis and only secondarily by
weathering; the CO_2-regulating system probably is controlled much more
by the weathering system than it is by photosynthesis.

Homeostasis. Anyone who tries to regulate a chemical reaction system by a multitude of valves or switches (Figure 5) soon becomes frus-
trated with the instability of his experimental system and appreciative of
automatic control devices (servo systems). For example, for the external
control of preselected pH and CO_2 activity, an automatic titrator (pH —
stat) can be used to dose continuously and automatically the quantity
of CO_2 which is necessary to maintain constant pH. Feedback is an
essential feature of such a control system; there are essentially two major
components, a controlling system (error detector) and a controlled system
(*13*) (Figure 1d).

From the controlled system, a signal is supplied *via* feedback loop to
the error detector which generates an error signal which in turn adjusts
the value that controls the CO_2 supply to the reactor. The same kind of
cybernetic mechanism prevails at the level of the individual organism

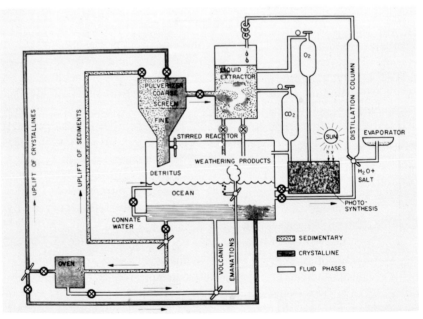

Sedimentology

Figure 5.   Siever's (12) steady state model of the ocean

as well as in groups of organisms and in entire ecological systems. The word "homeostatic" has been used to indicate constancy maintained by negative feedback. A living species adapts to diverse and various environments by two methods: genetic specialization and adaptive plasticity of the phenotype (*14*). This latter type of adaptation, whereby a species adjusts successfully to a spectrum of environments, occurs by homeostatic mode. Information concerning the effects of an organism's own actions on a variable of the system is perceived and fed back to the organism, thereby altering its subsequent performance. Population homeostasis requires a similar feedback of information on whether the system is in balance and, if not, how far it is away from the balance point (*15*).

At the cellular level, mechanisms that specifically regulate the dissimilatory activities of heterotrophic microorganisms include enzyme repression and enzyme inhibition. In enzyme repression, a substrate or a metabolic intermediate related to the substrate represses further synthesis of an unrelated dissimilatory enzyme, whereas the term inhibition applies if a metabolic intermediate inhibits an existing unrelated dissimilatory enzyme (*16*).

The members of a typical ecological community are interdependent by various interrelated homeostatic mechanisms. The community of organisms involved in an anaerobic methane fermentation is a simple example of a system where several types of organisms coexist through interdependency. The transformation of fatty acids to CH_4 is a multistep process requiring several species of methanogenic bacteria. Interaction between species—*e.g.*, between predator and prey—can be considered similarly in terms of negative feedback regulation (*17*). An excessive number of offspring is produced by prey. This number is reduced to a lower level through destruction by the predator. As pointed out by Margalef and others, such destruction is density-dependent, because the existing number of individual predators in turn was determined by the number of prey at a previous time. More commonly, a predator may eat a variety of animals, but since the most abundant kinds would be caught most often, there is a sort of automatic limit on abundance for any given species (*2*).

Our bottle microcosmos will eventually reach a fairly steady state when, despite a constant turnover of individuals in the community, mortality is balanced by reproduction (*18, 19, 20*). The composition of the community of microorganisms may then be described in terms of macroscopic parameters. Experimentally, we can now test to what extent this microcosmos tends to maintain its structure despite environmental changes and shocks.

Ecological Diversity and Stability. The operation of a well-regulated synthetic system in a variable environment is difficult. If we subject a microcosmos—*e.g.*, our bottle—to an external stress (change in light energy, temperature, biotic or abiotic composition), we can test the extent of disturbance by measuring the shift in environmental variables and its parameters. Qualitatively, the following observations are representative. Stress tends to reduce diversity by eliminating the stress-sensitive species and in some cases in turn by decreasing the degree of organization. Despite the reduction in diversity, the total number of organisms of a certain species, usually a generalist, may increase because of the lessened competition by the specialists (*21*). Although with different types of disturbances different microscopic effects are observed, macroscopically the pattern of loss of structure is quite general (*21*). Most important in our experimental system is the observation that the higher the biotic and chemical diversity of the system, the smaller the disturbing effect of the stress.

In a complex ecological community, all its members play their part in cycling nutrients (*15, 22*). They are interlocked by various feedback loops and are thus adapted for mutual advantage; because of the involved network of checks and balances (switches, valves, error and control signals) and the complicated food web, each species has multiple relationships with other species in the community (*10, 23*). The survival of the system as a whole is promoted. A qualitative condition for stability has been stated by E. P. Odum (*24*): In the passage of energy from the sun to the lowest trophic level and hence up to the highest level of carnivore, the degree of choice in the paths that the energy can follow through the food web is a measure of the stability of the community. High chemical buffer intensity and extensive chemostasis together with biotic diversity and extensive homeostasis reflect stability in the ecosystem.

Stream Pollution as a Disturbance of Ecological Balance

Although experience with various kinds of pollution is extensive, characterization of the kind of detrimental effect that results from it may not have been generalized sufficiently. Alteration of the environment does not necessarily constitute pollution; nor does any departure from a condition of positive purity. When is the alteration of the environment unfavorable? There are essentially three categories: (1) pollution by toxicity—*i.e.*, by agents that directly or after incorporation into the food chain represent a direct hazard to man (*e.g.*, poisons, toxic substances, ionizing radiation); (2) pollution resulting from a departure of a balance between photosynthesis and respiration; this type of disturbance is usually caused by organic wastes or by algal nutrients (usually nitrogen and phosphorus compounds) and may also occasionally result from thermal pollution; and (3) pollution as a direct or indirect impairment of the integrity of the diverse ecosystems that together constitute the biosphere; simplification and decrease in stability of an ecosystem tend to reduce its survival value.

We will consider here primarily categories (2) and (3) and will illustrate with a few representative examples how water pollution may be interpreted first as a disturbance in the ecological balance between autotrophic and heterotrophic organisms and then more generally as a loss in stability.

Balance Between Photosynthesis and Respiration. We may consider a stationary state which involves photosynthetic production, $P = dp/dt$ (rate of production of organic material) and heterotrophic respiration, R (rate of destruction of organic material) (Figure 3). We can characterize this steady state chemically by a simple stoichiometry:

$$106\ CO_2 + 16\ NO_3^- + HPO_4^{2-} + 122\ H_2O + 18\ H^+ \text{ (+ Trace elements; energy)}$$

$$P, \quad \Big\downarrow \quad \Big\uparrow R$$

$$\left(\begin{matrix} C_{106}H_{263}O_{110}N_{16}P_1 \\ \text{Algal Protoplasma} \end{matrix} \right) + 138\ O_2 \tag{2}$$

A temporally or spatially localized disturbance of the balance between photosynthesis and respiration (Figure 3b,c) leads to chemical and biological changes which constitute pollution (5, 25, 26). For an open system, the steady state balance is characterized by

$$I + P \approx R + E, \tag{3}$$

where I and E are rate of import and export, respectively, of organic matter. The $P–R$ balance becomes upset when a natural water receives an excess of organic heterotrophic nutrients or an excess of inorganic algal nutrients. This balance may also be disturbed by a physical separation of the P and R organisms. The condition $P > R + E - I$ is characterized by a progressive accumulation of algae, which ultimately leads to an organic overloading of the receiving waters. When $R > P + I - E$, the dissolved oxygen may be exhausted (biochemical oxygen demand) and ultimately NO_3^-, SO_4^{2-}, and CO_2 may become reduced to $N_2(g)$, NH_4^+, HS^-, and $CH_4(g)$. The balance between P and R is necessary to maintain a water in an aesthetically pleasing condition: When $P \approx R$, the organic material is decomposed by respiratory (heterotrophic) activity as fast as it is produced photosynthetically; O_2 produced by photosynthesis can be used for the respiration (Figure 3). In a stratified lake, a vertical separation of P and R results from the fact that the algae remain photosynthetically active only in the euphotic upper layers; algae that have settled under the influence of gravitation serve as food for the

R organisms in the deeper layers of the lake (Figure 3b). Organic algal material that has been synthesized with excessive CO_2 or HCO_3^- in the upper layers of the lake becomes biochemically oxidized in the deeper layers; most of the photosynthetic oxygen escapes to the atmosphere and does not become available to heterotrophs in the deeper water layers. An excessive production on the surface of the lake ($P \gg R$) is paralleled by anaerobic conditions at the bottom of the lake ($R \gg P$).

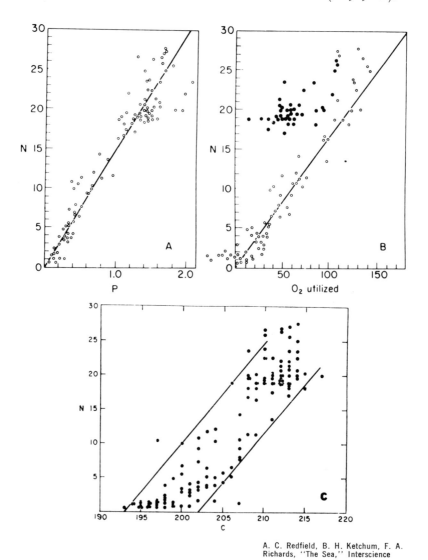

A. C. Redfield, B. H. Ketchum, F. A. Richards, "The Sea," Interscience

Figure 6.   Stoichiometric correlations between soluble nitrate, phosphate, oxygen, and carbonate carbon in the western Atlantic (Ref. 27)

The balance between P and R is responsible for regulating the concentration of O_2 in the oceans and the atmosphere (27). Thus, biological processes in the sea (probably more than half of the total global photosynthesis takes place in the sea) regulate p_{O_2} and in turn the redox intensity of the interface between atmosphere and hydrosphere.

How organisms can influence the composition of natural waters is shown in Figure 6A, which gives a correlation between dissolved nitrate and dissolved phosphate for the waters of the western Atlantic. The mole ratio is $\Delta N/\Delta P \approx 16$. A similar constant correlation is found in other seas and in lakes. The slope of the correlation curves in Figure 6 is readily understandable if we consider that the deviations in the concentrations of phosphate and nitrate result from deviations in the steady state of P and R. As is suggested by the stoichiometry of Reaction 2, in the photosynthetic zone phosphate and nitrate are eliminated from the water for algal growth in a ratio of 1:16. In the deeper water layers, settled algae become mineralized, whereby phosphate and nitrate are released into solution again in a ratio of 1:16. Because O_2 participates in the photosynthesis and respiration reactions, corresponding correlations are obtained between phosphate and O_2 ($\Delta O_2/\Delta P \approx 138$) and between nitrate and O_2 ($\Delta O_2/\Delta N \approx 9$) (Figure 6B). The correlations found are in accord with the stoichiometric composition of plankton. Figure 6C shows the correlation between nitrate nitrogen and carbonate carbon in the waters of the western Atlantic.

It is remarkable that the summation of the complicated processes of the P–R dynamics, in which so many different organisms participate, results in such simple stoichiometry. The stoichiometric formulation of Reaction 2 reflects in a simple way Liebig's Law of the Minimum. It follows from Figure 6 that, as a result of photosynthetic assimilation, sea water becomes exhausted simultaneously in dissolved phosphorus and nitrogen. Not considering temporary and local deviations, one infers from Figures 6 and 7 that nitrogen and phosphorus together determine the extent of organic production in the sea, while phosphorus appears to be the limiting factor in the productivity of the two lakes from which data are represented in Figure 7. For a and b (Lake Zurich), data collected at various depths during the summer stagnation period have been plotted. For b, only results from the deeper water layers were considered. Figure 7c (Lake Norrviken) plots analytical results obtained during winter stagnation (the lake was covered with ice).

The introduction of organic and inorganic nutrients into a stream affect P and R, but need not necessarily disturb the P–R balance. Sewage ponds and algae ponds are examples of systems with very high rates of

assimilation and respiration; such ponds do not produce disorders with
$P \approx R$—*i.e.*, as long as the organic material synthesized is decomposed as
fast as it is produced. A balance between P and R is a necessary but not
sufficient attribute of a nonpolluted and aesthetically pleasing condition.
We may compare the algae pond with similarly productive coral reefs
(28). While in algae ponds we find a simple and short producer–consumer food chain of great instability, coral reefs are stable systems made
up of a complex community with a large number of energy pathways.

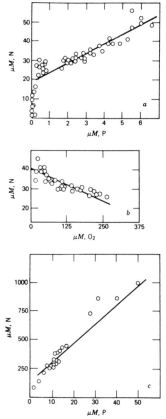

Figure 7.   Correlation be-
tween concentration of ni-
trate, phosphate, and oxygen
in Lake Zurich (a *and* b) *and*
Lake Norrviken (c)

Ecological Succession and Stability. As this comparison between
algae pond and coral reef epitomizes for two ecosystems of comparable
primary plant productivity, the more diversified system is more resistant
to external perturbations. Although the cause and effect relationship

between diversity and stability is not clear (8), it has been shown experimentally that diversity indeed accompanies physical stability.

Our microcosmos bottle, as all ecological systems, whether man-made or natural, must in the long run achieve a steady state. Thermodynamics does not indicate how open systems become adjusted; it merely predicts that natural spontaneous processes are directed toward an entropy increase (free energy decrease). Entropy has been called (29) "time's arrow." In going from an initial to a stationary state, the rate of entropy production decreases. Prigogine (7, 30) has described the steady state as one of minimum entropy production (least free energy dissipation) compatible with external constraints upon the system (e.g., fixed concentrations or affinities in the environment). This interpretation may be considered an extension of LeChatelier's principle to nonequilibrium systems. According to Prigogine's analysis, such a stationary state has a well-known stability against external perturbations because a state of minimum entropy production cannot leave this state by a spontaneous irreversible change. If as a result of some fluctuation it deviates slightly from this state, internal changes will take place and bring back the system to its stable state.

Prigogine's theorem is only valid when the system is fairly close to equilibrium, and thus its validity for ecological systems cannot simply be assumed (6, 31). But even if we keep its limitations in mind, we may consider the Prigogine theorem as a phenomenological principle, because it fits in excellently with many of the observed characteristics of ecological systems and because it formulates in a most concise way observed ecosystem development. Although the general validity of the principle awaits further elucidation, it is in accord with the fact that organisms actually show a decrease of entropy production during development toward the stationary state, and similarly that the succession of ecosystems is directed toward the state with the least production of entropy (per time and mass unit) (30). It also agrees with the principle that the ecosystem organization generally increases during succession, a fact that appears to accord well with descriptive studies on natural systems (8, 17, 25, 32). At stationary state, the system does not come to a halt; although biological and chemical reactions take place, the system is merely characterized by a zero macroscopic change.

The bioenergetic basis of ecosystem development has been documented by many ecologists and has recently been reviewed by Margalef (17) and E. P. Odum (8). H. T. Odum and Pinkerton (29), building on Lotka's law of maximum power in biological systems (11), pointed out that succession involves a fundamental shift of the energy flow in the direction toward a steady state (referred to by ecologists as climax) as increasing energy is being used for maintenance. At steady state, an opti-

mum in metabolic efficiency is attained. In accord with these fundamental considerations, ecological succession has been described by E. P.
Odum (8) as follows: (1) Succession is an orderly process of community
development that is reasonably directional and therefore predictable.
(2) It results from modification of the physical environment by the community; *i.e.*, succession is community-controlled even though the physical
environment determines the pattern, the rate of change, and often sets
limits as to how far development can go. (3) It culminates in a stabilized
ecosystem in which maximum biomass (or high information content) and
symbiotic function between organisms are maintained per unit of available energy flow.

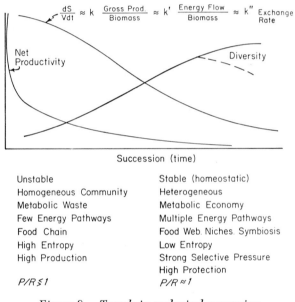

$$\frac{dS}{Vdt} \approx k \; \frac{\text{Gross Prod.}}{\text{Biomass}} \approx k' \; \frac{\text{Energy Flow}}{\text{Biomass}} \approx k'' \; \text{Exchange Rate}$$

Net Productivity

Diversity

Succession (time)

Unstable	Stable (homeostatic)
Homogeneous Community	Heterogeneous
Metabolic Waste	Metabolic Economy
Few Energy Pathways	Multiple Energy Pathways
Food Chain	Food Web. Niches. Symbiosis
High Entropy	Low Entropy
High Production	Strong Selective Pressure
	High Protection
$P/R \lesssim 1$	$P/R \approx 1$

Figure 8.   Trends in ecological succession

*Ecological systems tend to proceed in the direction of a sta-
tionary state and become progressively more self-regener-
ating with regard to energy and materials*

Figure 8 displays qualitatively the development of ecosystems; the
attributes used to describe the developmental stages are those of Odum
(8). Common trends, abstracted from observations, are summarized in
this figure. The time scale for the process of succession is extremely
variable and depends on the type of ecosystem under consideration. The
autotrophic succession of an aquatic laboratory microecosystem into a
mature system may take only a few weeks, while a forest succession
involves a time span of 100 years or more.

For a typical aquatic microcosmic system, the sequence of events based on experimental data by Cooke (*19*) [described by E. G. Odum (*8*)] is as follows: During the first 40 to 60 days, daytime net production (*P*) exceeds nighttime respiration (*R*), so that biomass accumulates in the system. After an early "bloom" at about the 30th day, both rates decline to become approximately equal at the 60th to 80th day. At the same time the *B/P* ratio (grams of biomass carbon supported per gram of daily carbon produced) increases from less than 20 to more than 100 as the steady state is reached. Not only are autotrophic and heterotrophic metabolism balanced in the climax, but large organic structure is supported by small daily production and respiration rates. Similar patterns of development, although at different rates, may be observed on land and in bodies of natural waters.

Margalef (*17, 32*) describes succession as a process of self-organization as it occurs in every cybernetic system with the properties of an ecosystem; he sees succession as a process of accumulating information. It results from indefinite iteration with variables tending toward a stationary endstate.

Man's Influence upon Succession. In accord with our previous considerations, an unpolluted ecological system is one in which energy flows are decreasing while increasingly complex checks and balances are being developed, in which instability is giving way to stability, and where, despite possibly increased biomass, net production is small ($P \approx R$). How does man's action affect the general pattern of ecological succession? Figure 8 suggests a few relevant criteria.

Energy transfer, reciprocally interrelated with biological structure, is perhaps the most important attribute of ecological successional stage. The input of energy causes a reduction in maturity and will "drive" an ecosystem into a younger developmental stage of higher net productivity and of decreased structure. As pointed out by H. T. Odum (*33*), man's success in adapting some natural systems to his use has essentially resulted from applying auxiliary work circuits, using fossil and atomic energy, into plant and animal systems. Obviously, we need to exploit ecosystems for food production, but progress in agricultural food production, essentially achieved by pumping more auxiliary energy (mechanical energy, heat, chemical energy in form of organic and inorganic nutrients) through a system, must be paid for by destruction of homeostatic mechanisms and loss of structure. By clearing land, by planting crops, and by controlling weeds, pests, and other competitors, a monoculture of high crop productivity and of high instability is established. It appears well documented that deforestation causes erosion of primary minerals; nutrients originally locked in the biota can no longer be retained by the soil system (*34*). Not only does the soil system become less stable;

the downhill flow of nutrients from soil to water also is enhanced. As pointed out by Leopold (35), soil and water are not two organic systems, but one; a derangement in either affects the health of both. If energy input remains small, complex nutrient circuits caused by longer food chains retard the drainage and erosion of nutrients into the water and enhance their storage in soils. Dredging of sediments, cutting of reeds, and fluctuations in water level, temperature, or velocity gradients decrease the number of ecological niches and reduce the maturity of the system.

Excess autotrophic or heterotrophic activity, and hence water pollution, is caused by acceleration of energy flow and of exchange of rates resulting from direct input of organic and inorganic waste constituents into receiving waters.

STRANGE MATERIALS. The introduction into receiving waters of materials nonindigenous to ecosystems (toxic substances, ionizing radiation, organic substances which are not of recent biological origin, e.g., petrochemicals) leads to a decrease in the number of pathways for energy flow and thus upsets the community structure. The ecological effects caused by such materials or agents follow the same general pattern known from other types of disturbances. Woodwell (21), in generalizing on pollution, has concentrated on some of the most gross changes in the plant communities of terrestrial ecosystems and has illustrated lucidly that changes in ecosystems caused by many different types of disturbances are similar and predictable. A similarity of response has been documented for either chronic irradiation, fire, exposure to sulfur dioxide, or pollution by herbicides and pesticides. In the cases reported (21), the loss of structure typically involved a shift away from complex arrangements of specialized species toward the generalists; away from forests toward shrubs; away from diversity in birds, plants, and fish toward monotony; away from tight nutrient cycles toward loose ones, with terrestrial systems becoming depleted and with aquatic systems becoming overloaded; away from stability toward instability.

Pollution by Petrochemicals. Oil and petrochemical wastes are quite refractory when introduced into natural waters; they foul up beaches, kill water birds, and become incorporated into the food chain, thus possibly imparting odor, tastes, and toxicity to the water and its food products. It is, perhaps, not sufficiently recognized that such substances might also adversely affect ecological interactions between organisms. As shown recently by Blumer (36) and coworkers for aquatic organisms, behavioral patterns such as food finding, avoidance of injury, choosing of a habitat or host, social communications, sexual behavior, and migration or recognition of territory appear to be controlled sensitively and specifically by chemical cues. Pollution products that often differ structurally from these signaling substances may blur chemotactic stimuli or may mislead by mimicking signals. This example illustrates that pollution cannot be fully characterized by simple yardsticks.

Water Pollution Control

Concepts have to be converted into practice. While our ultimate goal is to stabilize the ecosystems of our environment, we find ourselves caught in a conflict between resource exploitation and pollution control. There is a need to feed the human population and a desire to maintain a culturally advanced civilization and a high quality of life. Hence, we have to continue to utilize energy, exploit the earth's crust for fuels and minerals, manipulate the land, and establish monocultures for high agricultural productivity. Despite technological advances, we lack scientific knowledge and understanding of how to resolve this conflict and how to decide which ecological balance is most desirable for man. As pointed out by Dubos (37), "technological fixes are of course needed to alleviate critical situations, but generally they have only temporary usefulness. More lasting solutions must be based on ecological knowledge of the physiochemical and biological factors that maintain the human organization in a viable relationship with the environment."

The Pollutants. Figure 9 gives a diagram of the exponential growth of the human population, of energy consumption, and of fertilizer production (38). Basically, it is this growth which causes progressively rising levels of ecological disorder. Pollution is inextricably interrelated with the use that is made of energy resources (39).

Most of the conscious and reasonably successful efforts of environmental and sanitary engineers have been and will continue to be directed toward the reduction of organic carbon of biological wastes. For many natural waters, however, other pollutants are ecologically more harmful and may become even more so in the future. Inorganic fertilizing elements are often more serious pollutants than biological wastes for most of our lakes, estuaries and near-shore oceanic waters. While biological wastes and inorganic fertilizers cannot yet cause substantial alterations (besides local and temporal transients) in the composition and biology of the oceans, the invasion of the oceans by petrochemical wastes (fossil carbon compounds) may cause substantial modification of ecological interrelations in the ocean. Conventional biological waste treatment efficiently reduces those organic compounds that would be eliminated anyhow by the forces of self-purification. It is relatively inefficient with regard to strange organic chemicals. For many inland waters, more emphasis upon the elimination of refractory carbon compounds is necessary. The instability of some rivers during episodic shocks of a toxic

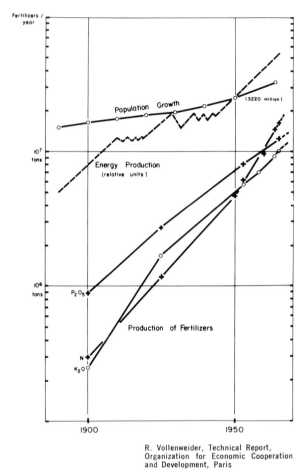

R. Vollenweider, Technical Report,
Organization for Economic Cooperation
and Development, Paris

*Figure 9.   Exponential growth of population, energy
utilization, and fertilizer production (Ref. 38)*

substance is frequently an indication that the river has lost its diversity.
While many refractory chemicals, at the low concentrations encountered,
do not impair the dissolved oxygen balance and do not interfere with the
apparent health and survival of fish, they may prevent some fish from
returning to their streams of birth for breeding (*31*). While one need
not be exceedingly concerned about the disappearance of a species—
i.e., a salmon in a particular stream—such a disappearance signals an
environmental stress and a decrease in community metabolism.

Because of the continuous increase in energy supply, heat will become one of the most relevant water pollutants of the future. Using
ecological criteria, more knowledge has to be developed on how and
where to dispose of heat intelligently.

Practical Problems. As we have seen, many different types of disturbances cause changes in ecosystems that are similar and in broad outline predictable. What has been called (8) the "strategy of ecosystem development" provides a most important basis to anticipate the ecological effects of pollution and to evaluate measures of water quality control. Table II lists some examples of measures and how they grossly affect water quality. This brief list illustrates that water pollution control

Table II.   Some Measures of Water Quality Control

Reducing Ecological Stability	*Promoting Ecological Stability*

Increase of Energy Flow
  by disposing nutrients for
    heterotrophs and autotrophs
  by mixing (destratifying, sediment
    dredging, etc.)
  by heat disposal
  by imposing turbulence
	P–R Balance Restoration
	  by reducing waste input
	  by harvesting or washing out of
	    biomass
	  by reducing relative residence time
	    or by trapping of nutrients
	  by mixing (bringing P and R together
	  by fish management
	  by aeration

Exploitation of Adjacent Soil
  by crop growing, seeding, weeding,
    and grazing
  by fertilizing and irrigating
  by deforestation
  by converting grass land into
    cropland
  by applying herbicides and
    pesticides
	Conservative Land Management
	  by reforestation
	  by restricting monoculture
	    productivity
	  by zoning (maintaining zones
	    adjacent to open waters which
	    are kept free of fertilizers and
	    of low net productivity)
	  by controlling erosion
	  by using detritus agriculture

Reduction of Structure
 	
  by using algicides
  by destruction of niches (removal
    of reeds)
  by episodical physical perturbations (flushouts, temperature
    discharges, shock loadings)
  by excessive harvesting
  by disposal of strange chemicals
  by interfering with chemostasis
	Enhancement of Biological
	Complexity
	  by establishment of ecological
	    niches (zones, waterfront
	    development)
	  by seeding diverse populations and
	    recirculating certain organisms
	  by maintaining relatively high
	    biomass compatible with energy
	    flow
	  by maintining stratification
	  by selective havesting
	  by maintaining high chemical
	    buffer intensity (weathering of
	    rocks)

consists not only of waste treatment; many other physical and biological
means of stream management need consideration. Especially important
is the realization that the coexistence, side by side, of a highly productive
agriculture and of a nonproductive surface water is not possible. The
activities necessary to maintain a soil monoculture (plowing, seeding,
weeding, fertilizing, irrigating, controlling pests, etc.) are incompatible
with and counteract measures designed to keep lakes and bays in a nonpolluted and oligotrophic state. Quite typically, less than half of the
nutrients introduced into a lake originate from municipal wastes (21).
Zoning of land in the watershed—specifically, the restriction of use of
land in the corridor adjoining the surface waters with preference to
diversified ecosystems of long food chain—is among the most powerful
measures for water pollution control.

Some of the measures listed in Table II have dual or multiple effects
and have to be assessed depending on the circumstances. For example,
mixing a body of water—*i.e.*, destratifying a lake—has a variety of effects
and thus may either decrease or increase the lake's ecological stability.
Mixing destroys ecological niches, reduces activity gradients, and shortens
the food chain; hence, the lake becomes more dynamic, with an increased
energy flow per biomass reflected in higher gross productivity and respiration. On the other hand, mixing brings P and R activities into closer contact and better balance, thus reducing net productivity (Figure 10).

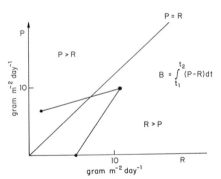

*Figure 10. Mixing of lake may de-
crease net productivity*

Mixing the top layers (photosynthetic functions; $P > R$) with the bottom
layers (heterotrophic functions; $R > P$) increases P and R but may
decrease the standing biomass B. Furthermore, the effective relative
residence time of the nutrients in the lake is reduced, thus decreasing the
total nutrient reserve of the lake. Reducing the residence time of a
nutrient to one half is equivalent to a 50% reduction in the nutrient
input into the lake.

EUTROPHICATION. In stratified waters, the continuous sequence of nutrient assimilation and mineralization of organic matter accompanied by the physical cycle of circulation and stagnation leads to a progressive retention of fertilizing constituents and in turn to a progressive increase in productivity. This aging process from low to high production is often referred to as eutrophication. How does this developmental process enhanced by human influence fit into the pattern of ecological succession? Evolutionary succession in a hypothetical lake to which input of nutrients were curtailed to zero would be in the direction of oligotrophy. Accordingly, as pointed out by Margalef (*17*), oligotrophy as a more mature state should succeed eutrophy, not precede it. The development from an oligotrophic to a eutrophic state—*i.e.*, to a state of lower species diversity and lower maturity—is caused by nutrients that are continuously discharged into these waters by runoff from land and by pollution. This inflow of nutrients into a eutrophic lake keeps it in a state of low maturity.

There is documentation that, in a few instances, lakes that apparently have become sufficiently insulated from nutrient supply have spontaneously developed in the direction of oligotrophy. For example, it is evidenced by the "memory" stored in sediments that Lake Zurich has gone through eutrophic episodes (*40*), a thousand or more years ago, as a result of events not yet known (erosion, deforestation, etc.) from which the lake could recover. This lake has, since the turn of the century, undergone extensive cultural eutrophication (*40*). Hutchinson (*41*) describes the development of a small crater lake between Rome and Siena (Lago di Monterosi). In this case, cultural eutrophication was initiated 2000 years ago not by artificial liberation of nutrients into the water, but by a subtle change in hydrographic regime. Later the lake became rather less eutrophic. Such case histories, together with the changes observed in Lake Washington after sewage had been diverted completely from the lake (*42*), provide important lessons.

Final Remarks

In the last hundred years, man has attempted to master his environment by applying ever increasing power, despite a lack of understanding. Appreciation of the various physical, chemical, and biological factors that regulate the composition of natural waters provides a basis for a more general formulation of pollution. In view of man's inability to adapt to major environmental changes, pollution may be equated with disturbance in ecological balance and loss of stability. Because various kinds of disturbances cause similar and reasonably predictable changes in aquatic ecosystems, various means of water pollution control beyond those of waste treatment can be outlined.

The stress imposed upon the environment results primarily from the fact that the western way of life is dependent on high energy utilization; but continued unlimited growth in energy dissipation is incompatible with maintenance of ecological stability and high quality of life. The ecological constraints demand the alteration of human, social, and economic systems toward a stationary state, where resources and materials have to be recycled as much as possible.

Acknowledgment

Some of the work related to this paper has been supported by grants from the Federal Water Quality Administration, United States Department of the Interior.

Literature Cited

(1) Stumm, W., Morgan, J. J., "Aquatic Chemistry," pp. 340, 405, Wiley, New York, 1970.
(2) Bates, M., "Resources and Man," p. 21, W. H. Freeman, San Francisco, 1969.
(3) Sillén, L. G., "Equilibrium Concepts of Natural Water Systems," ADVAN. CHEM. SER. (1967) 67, 57.
(4) Stumm, W., Leckie, J. O., Intern. Conf. Water Pollution Res., San Francisco, July 1970.
(5) Stumm, W., Stumm-Zollinger, E., Chimia (1968) 22, 325.
(6) Denbigh, K. G., "The Thermodynamics of the Steady State," p. 86, Methuen, London, 1965.
(7) Prigogine, I., "Introduction to the Thermodynamics of Irreversible Processes," p. 91, Wiley-Interscience, New York, 1965.
(8) Odum, E. P., Science (1969) 164, 262.
(9) Morowitz, H. J., "Energy Flow in Biology," p. 18, Academic, New York, 1968.
(10) Patten, B. C., Ecology (1959) 40, 221.
(11) Lotka, A. J., "Elements of Mathematical Biology," p. 227, 294, Dover, New York, 1956.
(12) Siever, R., Sedimentology (1968) 11, 5.
(13) Bayliss, L. E., "Living Control Systems," p. 86, W. H. Freeman, San Francisco, 1966.
(14) Dobzhansky, T., "Evolution Genetics and Man," p. 234, Wiley, New York, 1955.
(15) Wynne-Edwards, V. C., "Regulation and Control in Living Systems," p. 316, H. Calmus, Ed., Wiley, London, 1966.
(16) Stumm-Zollinger, E., Harris, R. H., Proc. 5th Rudolf Research Conf. (1969), in press, M. Dekker, New York, 1971.
(17) Margalef, R., "Perspectives in Ecological Theory," p. 30, University of Chicago, 1968.
(18) Beyers, R. J., Ecol. Monographs (1963) 33, 281.
(19) Cooke, G. D., Bioscience (1967) 17, 717.
(20) Gordon, R. W., Beyers, R. J., Odum, E. P., Eagon, R. G., Ecology (1969) 50, 86.
(21) Woodwell, G. M., Science (1970) 168, 429.
(22) Wynne-Edwards, V. C., Science (1965) 147, 1543.

(23) MacArthur, R. H., *Proc. Natl. Acad. Sci.* (1957) **43**, 294.

(24) Odum, E. P., *Japan. J. Ecol.* (1962) **12**, 108.

(25) Odum, E. P., "Fundamentals of Ecology," p. 43, Saunders, Philadelphia, 1961.

(26) Stumm, W., "Proceedings of the International Conference on Water Pollution Research," Vol. 2, p. 216, Pergamon, New York, 1963.

(27) Redfield, A. C., James Johnson Memorial Volume, p. 177, Liverpool, 1934, *quoted from* Redfield, A. C., Ketchum, B. H., Richards, F. A., "The Sea," M. N. Hill, Ed., Vol. 2, Wiley-Interscience, New York, 1963.

(28) Odum, H. T., *Limnol. Oceanog.* (1956) **1**, 102.

(29) Odum, H. T., Pinkerton, R. C., *Am. Sci.* (1955) **43**, 331.

(30) Prigogine, I., Wiame, J. M., *Experientia* (1946) **15**, 451.

(31) Slobodkin, L. B., *Am. Naturalist* (1960) **876**, 213.

(32) Margalef, R., *Am. Naturalist* (1963) **97**, 357.

(33) Odum, H. T., "Pollution and Marine Ecology," p. 99, T. A. Olson and F. J. Burgess, Eds., Wiley-Interscience, New York, 1967.

(34) Smith, F. E., *Bioscience* (1969) **19**, 317.

(35) Leopold, A., "Symposium on Hydrobiology," p. 17, University of Wisconsin Press, Madison, Wisc., 1941.

(36) Blumer, M., *Oceanus* (1969) **15**, 2.

(37) Dubos, R., "Reason Awake," p. viii, Columbia University Press, New York, 1970.

(38) Vollenweider, R., Technical Report DAS/CSI/68.27, p. 84, Organization for Economic Cooperation and Development, Paris, 1968.

(39) Brown, H., Bonner, J., Weir, J., "The Next Hundred Years," p. 95, Viking, New York, 1963.

(40) Züllig, H., *Schweiz. Z. Hydrol.* (1956) **18**, 6.

(41) Hutchinson, G. E., "Eutrophication," p. 17, National Academy of Sciences, Washington, 1969.

(42) Edmondson, W. T., "Eutrophication," p. 124, National Academy of Sciences, Washington, 1969.

RECEIVED July 13, 1970.

2

Time to Chemical Steady-States in Lakes and Ocean

ABRAHAM LERMAN

Canada Centre for Inland Waters, Burlington, Ontario

In water and sediments, the time to chemical steady-states is controlled by the magnitude of transport mechanisms (diffusion, advection), transport distances, and reaction rates of chemical species. When advection (water flow, rate of sedimentation) is weak, diffusion controls the solute dispersal and, hence, the time to steady-state. Models of transient and stationary states include transport of conservative chemical species in two- and three-layer lakes, transport of salt between brine layers in the Dead Sea, oxygen and radium-226 in the oceanic water column, and reacting and conservative species in sediment.

In natural systems of large dimensions—bodies of water, sediments, atmosphere—many chemical processes are controlled by the transport of reacting species through the system. The distribution of chemical species in natural systems is only too often not homogeneous; concentration gradients and more or less abrupt changes in abundance from one part of an environment to another are commonplace. In general, the nonhomogeneous distributions of chemical species are a combination of (*i*) the geometry of the environment: its shape and location of the "sources" and "sinks" of the chemical species; (*ii*) physics: mechanisms of transport of matter through the system; and (*iii*) chemistry: the nature and rates of the chemical reactions in which the species enter.

Knowledge of these three facets of a natural system is indispensable when we need to understand its present chemical state and also to predict quantitatively the changes in the chemical state and their duration, as would occur when the present characteristics of the system change.

In order to visualize the significance of the geometric, physical, and chemical factors, one might consider a system consisting of a sediment and a water column above it. The geometric factors in this case are the location of the sources of the chemical species—for example, at the sediment–water interface, within the sediment, or distributed throughout the water column—and its sinks, such as removal by a chemical or biochemical reaction occurring throughout the system, removal in outflow, or evaporation.

The relevant chemical aspects of such a system are the concentration or rate of production of the chemical species at the source and the nature and rates of the reactions involving the species. Biological production and consumption of a dissolved substance can in certain cases be treated (1) as if it were a chemical reaction of a simple order.

The transport mechanisms include dispersal by diffusional processes, water flow (advection), settling of biological or detrital particles through the water column, and accumulation of sediment on the floor. In view of the primary significance of diffusion in the transport of dissolved matter in a water column, this mechanism and its bearing on a number of chemical processes will be discussed in detail in this paper.

Different diffusional processes and the magnitude of the characteristic diffusion coefficients are identified in Figure 1. With reference to vertical migration of chemical species through water-filled sediments and water column of lakes and ocean, the relevant diffusional processes are the molecular and eddy diffusivity, respectively. The difference of several orders of magnitude between the molecular and eddy diffusion coefficients reflects the much more rapid dispersal by turbulent eddies in natural bodies of water. The much higher values of the eddy diffusivities in surface waters are owing to the greater effect of the wind-generated turbulence, as compared with the deeper parts of the basin. The values of the diffusion coefficients within a particular type of environment (such as porous media or thermocline layers) may vary by several orders of magnitude, and there is some overlap between different environments (Figure 1). The large variation in the values of the diffusion coefficients reported in the literature for different chemical species in different environments and the laboriousness of their determination in natural environments make it difficult in many cases to obtain accurate estimates of the time required for a certain chemical process to go to completion. However, when the diffusivities are not well known, it is still possible in some systems to choose "reasonable" lower and upper limits of the diffusion coefficients and thereby to bracket the model in short and long time estimates.

The effects of the magnitude of eddy diffusivity on the transport of a dissolved species in a stratified body of water are discussed in some simplified lake models and an example from a real lake in the next two sections.

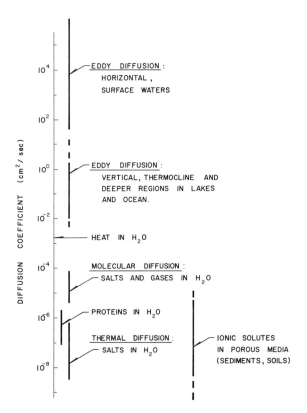

Figure 1. Diffusion coefficients characteristic of various environments; sources of data: Ref. 2, 3, 4, 5, 6, 7, 8, 9, 10, 11, 12, 13, 14, 15, 16

Effect of Eddy Diffusivity on Transport in a Stratified Water Column

An idealized picture of a stratified body of water is a well-mixed layer at the surface, a layer with a more or less pronounced density gradient (pycnocline) below it, and a well-mixed layer below the pycnocline. In many fresh water lakes, the density stratification is thermal in origin, and the concentrations of major dissolved solids are the same in the lighter and denser layer. A difference in concentrations between two layers might arise, for example, when a large influx of warmer water raises the lake level appreciably. A certain amount of mixing is likely to occur in the initial stages of flooding, with the result that a chemical species distributed homogeneously in the original lake retains its homogeneous distribution in the deeper layer, but a concentration gradient comes into being in the mixed layer. Such cases of flooding of a saline layer by a layer of lighter water have been discussed for some Antarctic,

Arctic, and Pacific Coast lakes (*17, 18, 19*). This is shown diagrammatically in the inset of Figure 2; the initial concentration in the upper layer ($C_2°$) is nil. Another possible application of the model is when a dissolved species has been introduced into one of the mixed layers. Diffusion through the pycnocline subsequently establishes a concentration gradient, and the material reaching the other mixed layer is uniformly dispersed within it. The model may apply from the early stages of such a process, after some material has crossed the middle layer, provided the concentration gradient is approximately linear. When a three-layer system remains closed and the dimensions of the water layers do not change, a conservative chemical species in one of the mixed layers would redistribute itself between the two layers because of the diffusional flux down the concentration gradient from one mixed layer into the other.

For a case of transport from the lower into the upper mixed layer, change in the concentration in the upper layer (C_2) as a function of time

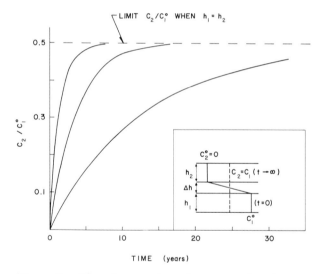

Figure 2. Three-layer water column; transfer of a conservative species from the lower into upper layer

Rate of change in concentration (C_2) in layer h_2. $C_1°$ and $C_2°$ are initial concentrations in layers h_1 and h_2 identified in the inset. Upper and lower layer assumed well mixed; linear concentration gradient in middle layer. Equation 10, with Equations 7 and 9. Thickness of water layers: $h_1 = h_2 = 25$ m, $\Delta h = 10$ m. Concentration–time curves were computed for the following diffusion coefficients, from left to right: $K = 5 \times 10^{-2}$, 1×10^{-2}, and 5×10^{-3} cm$^2 \cdot$ sec^{-1}. Higher value of eddy diffusivity in the middle layer results in a faster attainment of chemical steady-state, indicated by equal concentrations in the three water layers (dashed line in the inset).

is shown in Figure 2; the curves have been calculated for a 60-m-deep water column (lower layer $h_1 = 25$ m, pycnocline $\Delta h = 10$ m, and upper layer $h_2 = 25$ m) for three different eddy diffusion coefficients in the pycnocline (5×10^{-3}, 1×10^{-2}, and 5×10^{-2} cm$^2 \cdot$ sec^{-1}). These values of the eddy diffusion coefficient are in the range reported for pycnoclines

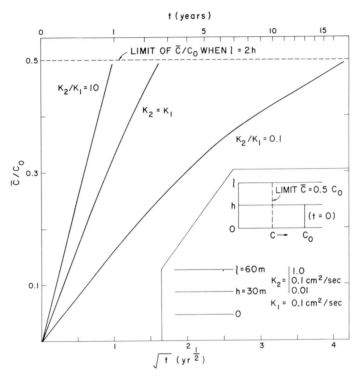

Figure 3. Two-layer water column; transfer of a conservative species from the lower into upper layer

Rate of change in mean concentration $(\overline{C})$ in the upper layer $l - h$. C_o is initial concentration in the lower layer (h). K_1 and K_2 are eddy diffusion coefficients in the lower and upper layer, identified in the inset. Equation 17.

in stratified lakes (Figure 1). The curves in Figure 2 for the rate of increase in concentration show that the concentrations in the two layers would equalize in a relatively short time (10–40 years). The time required to attain equal concentration in such a model lake depends on the eddy diffusivity in the pycnocline and on the vertical dimensions of the individual layers.

Concentration–time curves for a two-layer water column of some-what different physical characteristics are shown in Figure 3. In this model, the two water layers (each 30 m thick, shown in the inset of Figure 3) are not well mixed; each is characterized by a different value of the eddy diffusion coefficient. As in the previous example, the values of the eddy diffusion coefficients were taken to represent the range reported for stratified lakes. Initially, a conservative species (inset of Figure 3) is homogeneously distributed within the lower layer, its concentration being C_o. Migration across the boundary between the two layers and subsequent dispersal within the upper layer would eventually equalize the concentrations. At the limit, the concentration of the species would be homogeneous throughout the two layers and equal to $C_o h/l$.

The conclusion which may be drawn from the curves in Figure 3 is essentially the same as for the three-layer system discussed earlier: the time required to attain equal concentrations is relatively short, 1–20 years.

The preceding discussion of the two simple models suggests that in closed lakes with stationary stratification of the water column, a change in the chemical composition of one of the layers is a transient phenomenon of relatively short duration. In order to maintain a steady concentration gradient in a lake water column, the lake must be open, such that the input of solute is balanced by its removal. Some examples of such cases have been discussed by Hutchinson (7, p. 480 and ff.).

Calculation of Concentration–Time Curves for a Three-Layer Model (Figure 2). The simplest way to estimate how long it would take for the concentrations in the lower and upper layer to become equal is to assume that the flux across the pycnocline is at all times proportional to the concentration difference between the upper and lower layer

$$F = k(C_1 - C_2) \tag{1}$$

where F is the flux of dissolved species (gram $\cdot$ cm^{-2} $\cdot$ sec^{-1}), k is a constant proportionality factor of dimensions cm $\cdot$ sec^{-1}, and C_1 and C_2 are the concentrations in the lower and upper layer, respectively (gram $\cdot$ cm^{-3}).

From the two following equations, C_1 and C_2 may be evaluated.

Rate of concentration change in lower layer: $\dfrac{dC_1}{dt} = -\dfrac{k}{h_1}(C_1 - C_2)$ (2)

Mass balance: $Q_o = C_1 h_1 + C_2 h_2 + (C_1 + C_2)\Delta h/2$ (3)

where Q_o is the total amount of the chemical species (gram $\cdot$ cm^{-2}) in the lake. Q_o may be expressed in terms of the initial concentrations in the two mixed layers and the mean concentration in the pycnocline:

$$Q_o = C_1^\circ h_1 + C_2^\circ h_2 + (C_1^\circ + C_2^\circ)\Delta h/2 \tag{4}$$

Substitution from Equation 3 into Equation 2 and integration give

$$C_1 = \frac{Q_o}{h_1 + h_2 + \Delta h} [1 - e^{-(1+\alpha)kt}] + C_1^\circ e^{-(1+\alpha)kt} \qquad (5)$$

$$C_2 = \frac{Q_o}{h_1 + h_2 + \Delta h} [1 - e^{-(1+\alpha)kt}] + C_2^\circ e^{-(1+\alpha)kt} \qquad (6)$$

where
$$\alpha = \frac{h_1 + \Delta h/2}{h_1(h_2 + \Delta h/2)} \qquad (7)$$

Constant k may be evaluated as follows. When the flux (F) through the pycnocline is eddy diffusional in nature and the concentration gradient is linear, then

$$F = K \frac{\Delta C}{\Delta h} = \frac{K}{\Delta h} (C_1 - C_2) \qquad (8)$$

where K is the eddy diffusion coefficient in the pycnocline layer ($cm^2 \cdot sec^{-1}$) and ΔC is the difference between the concentrations at the pycnocline boundaries. For the case when eddy diffusivity in the pycnocline is constant, comparing Equations 8 and 1 gives

$$k = K/\Delta h \qquad (9)$$

A similar derivation has been given in Ref. 20.

For the model shown in Figure 2, the thickness of the pycnocline $\Delta h = 10$ m, and eddy diffusion coefficient in the pycnocline was given the values of $K = 5 \times 10^{-3}, 1 \times 10^{-2}$, and 5×10^{-2} cm² · sec⁻¹. From Equation 9, the values of k are 1.58, 3.16, and 15.8 m · yr⁻¹. When the initial concentration in the upper layer $C_2^\circ = 0$, Equation 6 can be written in the following form.

$$\frac{C_2}{C_1^\circ} = \frac{h_1 + \Delta h/2}{h_1 + h_2 + \Delta h} [1 - e^{-(1+\alpha)kt}] \qquad (10)$$

The concentration–time curves shown in Figure 2 were calculated using Relationship 10 with the values of the layers' thickness $h_1 = h_2 = 25$ m, $\Delta h = 10$ m, and the values of k derived as explained above.

Constant k defined in Equation 2 is identical with the concept of entrainment velocity (U_e) which has been studied by Turner (21) in experiments on the transport of salt and heat across the interface of a density-stratified two-layer water column. The definition of k in this section also applies to a two-layer model with a stationary interface; in a two-layer water column, $\Delta h = 0$ in Equations 3–7. Then, however, the relationship between k and the eddy diffusion coefficient K in Equation 9 becomes invalid.

Calculation of Concentration–Time Curves for a Two-Layer Model (Figure 3). In a two-layer system, when the diffusion coefficients in the two layers are equal, the concentration of a dissolved substance originally confined to one layer is given by the following relationship (*22*, p. 15)

$$C = \tfrac{1}{2}C_o \sum_{n=-\infty}^{\infty} \left\{ \operatorname{erf} \frac{h + 2nl - z}{2\sqrt{(Kt)}} + \operatorname{erf} \frac{h - 2nl + z}{2\sqrt{(Kt)}} \right\} \tag{11}$$

where C_o is the initial concentration in one layer ($0 < z < h$), h and l are the boundaries of the two layers (shown in the inset of Figure 3), and z is the vertical dimension ($0 \leqslant z \leqslant l$). The significance of the functions erf x (error function) and erfc x (error function complement) which appear in many of the solutions in the text is discussed in the Appendix. The nature of the functions and methods of evaluation, with references, are also summarized in the Appendix.

For the case when the diffusion coefficients in the two layers are not equal, derivation of a closed-form relationship for C is difficult. As an alternative, a simpler method has been used which gives the mean concentration $(\bar{C})$ as a function of time in the upper layer into which the substance diffuses from the lower layer. The method for computing the mean concentration is based on the following: First, it is assumed that the upper layer is semi-infinite, extending from $z = h$ upwards; second, the amount of dissolved matter transported from the lower layer across the plane $z = h$ is evaluated as a function of time; third, the amount of matter transported from the lower into the upper (semi-infinite) layer up to some time t is divided by the height of the upper layer, $l - h$, to obtain the mean concentration $(\bar{C})$ in the upper layer. The mean concentrations computed by this method are within a few percent of the values obtainable by the use of a complete expression for C, such as Relationship 11 (*23*).

The concentration within a semi-infinite medium (C_2) into which the substance diffuses out of the lower layer (initial concentration C_o) may be calculated by the following relationship (*24*, p. 91; coordinates were changed to conform to the coordinates of the model discussed here)

$$C_2 = \tfrac{1}{2}C_o \left\{ -(1 - p) \operatorname{erf} \frac{z - h}{2\sqrt{(K_2 t)}} + (1 - p^2) \sum_{n=1}^{\infty} (-p)^{n-1} \cdot \right.$$

$$\left. \operatorname{erf} \left[\frac{z - h}{2\sqrt{(K_2 t)}} + \frac{nh}{\sqrt{(K_1 t)}} \right] \right\} \tag{12}$$

where K_1 is the diffusion coefficient in the lower layer ($0 < z < h$), K_2 is the diffusion coefficient in the upper layer ($z > h$), and p is a parameter dependent on the densities, specific heats, and diffusion coefficients of the two layers.

$$p = \left(1 - \frac{\rho_1 c_1}{\rho_2 c_2}\sqrt{\frac{K_1}{K_2}}\right)\bigg/\left(1 + \frac{\rho_1 c_1}{\rho_2 c_2}\sqrt{\frac{K_1}{K_2}}\right) \tag{13}$$

In aqueous solutions that are not highly concentrated brines, the product ρc is close to 1 and varies only slightly with concentration. Thus, Equation 13 may be simplified to

$$p = [1 - \sqrt{(K_1/K_2)}]/[1 + \sqrt{(K_1/K_2)}] \tag{14}$$

The flux across the plane $z = h$ may be derived from Equation 12.

$$F_h = -K_2\left(\frac{\partial C_2}{\partial z}\right)_{z=h}$$

$$= \tfrac{1}{2}C_o\sqrt{\frac{K_2}{\pi t}}\left\{1 - p - (1 - p^2)\sum_{n=1}^{\infty}(-p)^{n-1}\exp\left(-\frac{n^2 h^2}{K_1 t}\right)\right\} \tag{15}$$

The amount of matter which has crossed the plane $z = h$ up to time t is obtained by integration of Equation 15.

$$M_t = \int_0^t F_h dt' = C_o\sqrt{\frac{K_2 t}{\pi}}\left\{1 - p - (1 - p^2) \cdot\right.$$

$$\left.\sum_{n=1}^{\infty}(-p)^{n-1}\left[\exp\left(-\frac{n^2 h^2}{K_1 t}\right) - \frac{nh\sqrt{\pi}}{\sqrt{(K_1 t)}}\operatorname{erfc}\frac{nh}{\sqrt{(K_1 t)}}\right]\right\} \tag{16}$$

Erfc is the error function complement, defined as $\operatorname{erfc} x = 1 - \operatorname{erf} x$. On the computation of erfc, see the Appendix. To obtain the mean concentration within the upper layer of thickness $l - h$, the value of M_t (in grams $\cdot$ cm^{-2}) in Relationship 16 is divided by $l - h$.

$$\overline{C}_2 = \frac{C_o}{l - h}\sqrt{\frac{K_2 t}{\pi}}\left\{1 - p - (1 - p^2)\sum_{n=1}^{\infty}(-p)^{n-1} \cdot\right.$$

$$\left.\left[\exp\left(-\frac{n^2 h^2}{K_1 t}\right) - \frac{nh\sqrt{\pi}}{\sqrt{(K_1 t)}}\operatorname{erfc}\frac{nh}{\sqrt{(K_1 t)}}\right]\right\} \tag{17}$$

Relationships 15–17 are physically meaningful for all values of t up to the value at which the mean concentrations in the lower and upper layer become equal and the process of transport across the interface ends. At the end of the process, the concentration of the diffusing substance is the same in the two layers.

$$\overline{C} = C_o h / l \tag{18}$$

This end value of $\overline{C}$ is labelled "limit of $\overline{C}$" in the inset of Figure 3. The concentration–time curves for the upper layer in a two-layer model in Figure 3 were calculated using Relationship 17 with the values of $h = 30$ m, $(l - h) = 30$ m, $K_1 = 0.1$ cm$^2 \cdot$ sec^{-1}, and $K_2 = 0.01$, 0.1, and 1.0 cm$^2 \cdot$ sec^{-1} as shown.

Three-Layer Lake

Diffusion Coefficients in the Pycnocline. Intuitive considerations suggest that in a three-layer lake (well-mixed upper layer, pycnocline, and well-mixed lower layer) the eddy diffusivity in the middle layer should to some extent depend on the steepness of the density gradient in it. When the density gradient is strong, the degree of turbulence and, hence, the magnitude of eddy diffusivity in the pycnocline may be expected to be low, or the gradient would be destroyed by turbulent eddies. The steepness of the density gradient within a water layer determines the stability of stratification: the greater the density gradient, the more stable is the stratification. This relationship between the density gradient and the degree of stability of stratification is commonly expressed in the quantity known as the Brunt–Väisälä stability frequency (25), N

$$N^2 = -g \frac{1}{\rho} \frac{\partial \rho}{\partial z} \tag{19}$$

where g is the acceleration due to gravity, ρ is density, and the depth z is scaled such that increasing depth corresponds to larger negative numbers. On such a scale, the density increasing with depth results in a negative density gradient and N^2 positive. The dimension of N is sec^{-1}. Relationship 19 is valid for a water column up to several hundred meters deep, where the effect of adiabatic compressibility on the density gradient within the water column may be ignored. For a deeper water column, an additional adiabatic term is included in Equation 19. Eckart (25) has discussed the derivation and physical significance of the concept of stability frequency, N. The name stability frequency is borrowed from a mechanical analog of an "isolated" small parcel of water displaced from its hydrostatic equilibrium position in the density gradient. When dis-

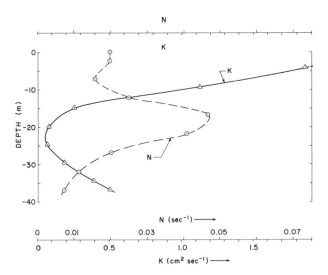

Figure 4. Eddy diffusion coefficients (K) and the Väisälä stability frequency (N) plotted against depth. Points plotted were calculated from temperature–depth profiles for the month of August 1966 in Lake Tiberias. Data sources in Ref. 9.

placed up, the parcel would move down under the force of gravity past its equilibrium position; buoyant force would then displace the parcel upwards, out of the denser liquid and past its equilibrium position. Such mechanical oscillations are mathematically analogous to the concept of stability frequency.

When the density stratification is thermal in nature, the Väisälä stability frequency in Equation 19 may be written in terms of the temperature gradient

$$N^2 = - g\alpha \frac{\partial T}{\partial z} \qquad (20)$$

where α is the coefficient of thermal expansion [$\alpha = (1/\rho)\, \partial\rho/\partial T$]. The values of N^2 are easily calculable from the measured temperature– and/or salinity–depth profiles.

Variation in the stability frequency (N) and eddy diffusion coefficient (K) with depth, determined for one stratified lake (Lake Tiberias), is shown in Figure 4. In the near surface layer of the lake (epilimnion), the turbulence is high and the thermal gradient is very poorly developed; this is reflected in the high value of the eddy diffusion coefficient and low stability frequency. In the thermocline region (depths 15–25 m), the thermal gradient is fairly strong, the stability frequency is high, and the eddy diffusion coefficient is low. The picture reverses itself below the thermocline region where the density gradient is appreciably smaller than in the thermocline; relatively high eddy diffusivity and low stability frequency characterize the deeper region of the water column.

The reciprocal relationship between N and K is apparent from the data plotted in Figure 5: N correlates positively with log $1/K$ (correlation coefficient $r = 0.71$, significant at 0.01 level). The least-squares fit to the 26 points plotted gives

$$\log K = -0.749 - 24.235N \qquad (21)$$

$$K = 0.178 \exp(-55.803N) \qquad (22)$$

The constant coefficient 0.178 in Equation 22 does not imply that the eddy diffusion coefficient approaches the constant value of 0.178 $cm^2 \cdot sec^{-1}$ when the stability frequency N tends to zero. The factor 0.178

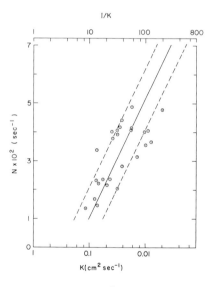

Figure 5. Correlation between eddy diffusion coefficients (K) and stability frequency (N) in the thermocline region of Lake Tiberias (depth 15–25 m)

Solid line: linear least squares fit to the data, Equation 21 in the text. Dashed lines: one standard deviation (±0.168) of the intercept (0.749) of log 1/K. Calculated from April–October 1966, temperature–depth profiles in Lake Tiberias. Sources of data and computation methods discussed in Ref. 9.

is the result of the straight-line fit to the data. In fact, as the stratification is gradually destroyed and the density gradient decreases, the eddy diffusional mixing in the water column is likely to become stronger, until in the absence of a density gradient the eddy diffusion coefficient may become two or more orders of magnitude greater than 0.178 cm^2 · sec^{-1}.

It is not to be expected that the straight-line Equation 21 would apply to the thermocline layers of other lakes. Both N and K were calculated from temperature profiles the shape of which depends in a complex manner on the climate of the area, thermal regime, depth, and volume of the lake. It seems, however, that by arguments presented earlier in this section, an inverse relationship between the stability frequency and eddy diffusion coefficient would, in general, hold in the pycnocline layers of lakes. If such a relationship is established, it would be possible to obtain estimates of K from the values of the stability frequency N, which are much easier to compute.

The highly empirical nature of the correlation between N and K, such as the one shown in Figure 5, is underscored by the theoretical and experimental analysis of the rates of transport across pycnocline which have been reported as dependent on the rate of dissipation of energy in the upper well-mixed layer (21, 26, 27). At present, however, it is difficult to relate the energy dissipation and transport rates in the pycnocline of a real lake.

Thus, with reference to a three-layer lake model in which the eddy diffusion coefficient in the middle layer depends on the concentration (or density) gradient, the eddy diffusion coefficient may be estimated by a relationship of the type of Equation 22 for different values of the gradient and concentration difference between the well-mixed top and bottom layer. A case in which this model might apply is transport of dissolved salts from a saline brine layer on the bottom into a more dilute layer above, when the concentration of total dissolved solids in the pycnocline and surface layer changes continuously in the process. Use of this model to estimate the length of time to a complete mixing of a stratified brine lake, the Dead Sea, will be demonstrated.

Time of Mixing: The Dead Sea. Although it has been stressed earlier that the relationship between N and K established for one lake may not hold for another lake, in the present case the relationship between N and K derived for the thermocline of Lake Tiberias is the only one available and the only one which may be applied to estimate the eddy diffusion coefficients in the pycnocline of the Dead Sea. Implicit is an assumption that the eddy diffusivity of dissolved salts is equal to that of heat.

Both Lake Tiberias and the Dead Sea are located within one rift valley, striking approximately north–south, at a distance of about 100 miles from one another. The stratification of the water column and bottom morphometry of the Dead Sea are shown diagrammatically in Figure 6. The major inflow into the lighter upper water mass is from the River Jordan in the north, two small streams in the east, and a number of springs around the lake. There is no surface outflow from the lake. Calculations of the water budget of the Dead Sea (29) show that the inflow and precipitation are nearly balanced by evaporation from the lake surface. This suggests that there are no other important mechanisms of water removal from the lake, such as underground seepage of brine out of the lake through the sediments. The first record of the chemical

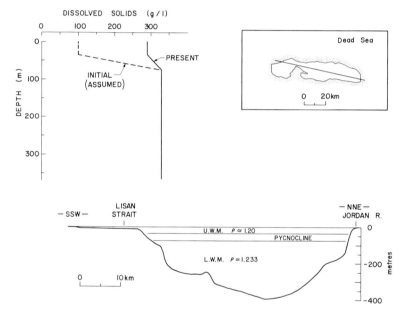

Figure 6. Diagrammatic NNE-SSW cross-section of the Dead Sea drawn along the line shown in the inset; depths from Ref. 28

Densities of upper and lower water mass identified in the cross-section. Diagrammatic concentration–depth profile at present approximated to well-mixed upper and lower layers, with a linear concentration gradient in the pycnocline layer. Initial concentration of dissolved solids 100 grams/liter in the upper layer assumed for the purpose of calculating the age of stratification as discussed in the text.

stratification of the Dead Sea dates back to 1864 (references in Ref. 23), although the stratification is likely to be much older. The stratification, however, cannot be permanent; in a stratified brine lake without outflow, when the lake volume is nearly constant, the dissolved salts brought by the inflow into the upper layer and by transport across the pycnocline continuously increase the concentration in the upper layer so that the concentration eventually becomes homogeneous throughout the water column. A similar, but simpler, process has been discussed for two- and three-layer models in the preceding section.

In order to estimate how long would it take for the Dead Sea water column to become homogeneous, beginning with its present concentration difference between the upper and lower layer, it is necessary to know how the volume of the individual brine layers and rate of salt input vary with time. As these relationships are not known, the following assumptions will be made:

(i) The volume of the Dead Sea and its individual brine layers remains constant.

(ii) The rate at which salts are being added by the inflow remains constant.

(iii) The concentration of salts in the denser lower water mass is constant.

(iv) The relationship between the density gradient in the pycnocline and eddy diffusion coefficient, as in Equations 21 and 22, holds for the Dead Sea pycnocline.

In support of assumption (i), the shore-line records of the Dead Sea (28) indicate that the lake volume did not vary by more than 5% during the last 2000 years. For assumption (iii), a salt layer (NaCl) is exposed on the lake floor, below 40-m depth contour line (28); the lower brine layer is saturated with respect to halite (23), and there are virtually no concentration gradients in it (Figure 6). Thus, if dilution of the lower water mass occurs, the NaCl concentration in the brine would tend to be restored by dissolution of halite on the lake floor until an equilibrium has been reestablished. With regard to assumption (iv), the density gradient in the middle brine layer of the Dead Sea is steeper than in the seasonal pycnocline of the fresh water Lake Tiberias. Hence, the stability frequencies in the present pycnocline of the Dead Sea are higher than the maximum values in Lake Tiberias, as shown in Figures 4 and 5.

The preceding reasoning on the "life expectancy" of the stratification in the Dead Sea may also be applied to its past history, and the length of time needed to attain the present concentration difference starting from some state in the past may be worked out. For this, however, the initial concentration difference between the two layers must be known.

An initial state in the past was assumed as the total dissolved solids concentration in the lower water mass $C_l = 327$ grams/liter, as at present; in the upper water mass it was taken as $C_u = 100$ grams/liter. The time from the initial state to a complete homogenization of the water column was computed as follows.

The rate of concentration increase in the upper water mass owing to eddy diffusional flux across the pycnocline and inflow is (this and subsequent equations are given in the finite difference rather than differential form, insofar as the final working equation arrived at is nonlinear)

$$\frac{\Delta C_u}{\Delta t} = \frac{F}{V_u} + C_J' \text{ (grams} \cdot \text{cm}^{-3} \cdot \text{sec}^{-1}) \tag{23}$$

where C_u is the concentration in the upper water mass (grams $\cdot$ cm^{-3}), F is the eddy diffusional flux from the lower into upper water mass (grams $\cdot$ cm^{-2} $\cdot$ sec^{-1}), V_u is the thickness of the upper water mass (cm^3 $\cdot$ cm^{-2}), assumed constant, and C_J' is the rate of salt addition to the upper water mass by surface inflow per unit volume. C_J' is determined by multiplying the concentration in the inflow (C_J, grams $\cdot$ cm^{-3}) by the rate of inflow (q_J, cm^3 $\cdot$ sec^{-1}) and dividing the product by the volume of the upper water mass V_u: $C_J' = C_J q_J / V_u$ (grams $\cdot$ cm^{-3} $\cdot$ sec^{-1}).

The eddy diffusional flux across the pycnocline is

$$F = K \frac{C_l - C_u}{\Delta z} \text{ (grams} \cdot \text{cm}^{-2} \cdot \text{sec}^{-1}) \tag{24}$$

where K is the eddy diffusion coefficient (cm^2 $\cdot$ sec^{-1}), C_l and C_u are the concentrations in the lower and upper water mass, and Δz is the thickness of the pycnocline layer.

The eddy diffusion coefficient K is, by Relationship 22,

$$K = 0.178 \exp(-55.803N) \text{ (cm}^2 \cdot \text{sec}^{-1}) \tag{25}$$

and the Väisälä stability frequency N, from Equation 19, written in finite difference form is

$$N = \sqrt{\frac{2g(\rho_l - \rho_u)}{\Delta z(\rho_l + \rho_u)}} \text{ (sec}^{-1}) \tag{26}$$

The subscripts l and u denote the density of the lower and upper water mass.

The dependence of the density of the upper water mass on concentration was taken as

$$\rho_u = 1.0057 + 0.6956 C_u \text{ (grams} \cdot \text{cm}^{-3}) \tag{27}$$

which was derived from the density–concentration data for the Dead Sea brines in Neev and Emery (28).

By consecutive substitution from Equation 27 through Equation 23, we obtain

salt from eddy diffusional flux

$$\Delta C_u = \left[\frac{0.178(C_l - C_u)}{V_u \Delta z} \times \right. \qquad \text{salt from inflow}$$

$$\left. \exp\left(-55.803 \sqrt{\frac{2g(\rho_l - 1.0057 - 0.6956 C_u)}{\Delta z(\rho_l + 1.0057 + 0.6956 C_u)}}\right) \right] \Delta t + C_J' \Delta t \tag{28}$$

In Equation 28, ΔC_u (grams · cm^{-3}) is the concentration increment in the upper water mass caused by the eddy diffusional transport across the pycnocline in time interval Δt (first term on the right-hand-side of Equation 28) and addition of salt by inflow in time interval Δt (second term on the right-hand-side).

The constant terms in Equation 28 are the following:

C_l = 0.327 grams · cm^{-3}
V_u = 3.5 × 10^3 cm^3 ·cm^{-2}
Δz = 4.0 × 10^3 cm
g = 980 cm · sec^{-2}
ρ_l = 1.233 grams · cm^{-3}
C_J' = 7.7878 × 10^{-5} grams · cm^{-3} · yr^{-1} (computed from the data on mean concentration of dissolved solids and inflow of the River Jordan, as given in Ref. 28)

Substitution of the constant parameters into Equation 28 and multiplication of the diffusional flux term by 3.156 × 10^7 sec · yr^{-1} gives

$$\Delta C_u = \left[0.40125 \ (0.327 - C_u) \times \right.$$

$$\left. \exp\left(-55.803 \ \sqrt{\frac{0.11138 - 0.34087 C_u}{2.23870 + 0.69565 C_u}}\right) \right] \Delta t + 7.7878 \times 10^{-5} \Delta t \quad (29)$$

where Δt is in years.

Using Equation 29 with some initial value assigned to C_u, the concentration increment ΔC_u can be computed for some short time interval Δt. This value of ΔC_u is added to the initial value of C_u, giving a new concentration C_u. Using the latter value of C_u in Equation 29, the next increment ΔC_u is computed for the next time interval Δt. The entire procedure is repeated until the concentration in the upper water mass equals that in the lower mass, 0.327 grams · cm^{-3}. The time required to attain such a state is the sum of the Δt increments.

In order to minimize the deviations from nonlinearity in the computation of ΔC_u by this method of finite differences, the time interval Δt must be short, such that each successive concentration increment always amounts to no more than a few percent of the concentration in the upper layer C_u. The relative increment $\Delta C_u / C_u$ (in percent) as a function of the dissolved solids concentration in the upper mass C_u is shown in Figure 7 for time increments $\Delta t = 100$ yr and $\Delta t = 50$ yr. For the concentration of the upper water mass in the range from 80 to 327 grams/liter, the concentration increments computed for 50-year time intervals are always under 10% of the concentration in the upper mass. For 100-year intervals, the corresponding increments are twice as large. The slight increase

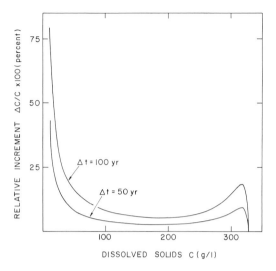

Figure 7. Relative increase in the concentration in the upper brine layer ($\Delta C/C$) owing to river inflow and eddy diffusional transport of salt from the lower brine layer in the Dead Sea. Concentration increment (ΔC) computed from Equation 29 for time steps (Δt) 50 and 100 years. Relative increase shown as a function of increasing concentration in the upper brine layer (C).

in the relative increment $\Delta C_u/C_u$ at high concentration, when the upper and lower water mass are nearly of equal density, is accounted for by the exponential dependence of K in Equation 25 on the density difference between the upper and lower mass.

The concentration in the upper water mass obtained by solving Equation 29 is shown as a function of time in Figure 8. The concentration–time curves were computed for three different values of the inflow term: C_J' as taken for the present mean annual contribution of the River Jordan, $2C_J'$, and $4C_J'$.

The curves in Figure 8 may be read as follows: from some stage in the past when the concentration of dissolved solids in the upper water mass was 100 grams/liter, it would have taken 1700 years to attain the present concentration of 290 grams/liter. In approximately 110 years from the present, the concentrations in the upper and lower mass would become equal. If the rate of addition of salts from the surface inflow were twice the present value, then the present concentration difference between the water masses would have been attained in approximately 1000 years, and there would be 70 years left for the concentrations to become equal. Similar reasoning applies to the curve labelled $4C_J'$.

The calculations and reasoning presented are based on an assumption that all the processes were continuous and characterized by the rate values discussed earlier in this section. If, for example, the salt concentration in the inflow were much lower than at present, the length of time to attain the present concentration in the upper mass would have been correspondingly longer. An opposite effect might have been pronounced if the volume of the upper water mass fluctuated, as in the years when evaporation exceeds inflow. The net effect of such fluctuations would accelerate the rate of the concentration increase in the upper layer. The dissolved solids remain in the brine when more water evaporates, and subsequent flooding introduces additional amounts of dissolved matter. Likewise, possibly stronger evaporation in the shallow southernmost section of the Dead Sea (shown in Figure 6) might have resulted in a concentrated brine flow down the slope and its subsequent mixing with the pycnocline or lower waters. Such a process would also accelerate the destruction of the lake stratification.

Using Equation 29, it is possible to calculate the fraction of the concentration increment in the upper layer resulting from eddy diffusional transport from below, in each 50-year time step.

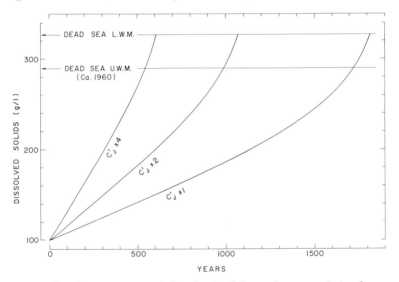

Figure 8.   Concentration of dissolved solids in the upper brine layer (U.W.M.) as a function of time

U.W.M. and L.W.M. are the upper and lower water masses, or brine layers. Computation for a three-layer model assumes a hypothetical concentration of 100 grams/liter in the upper water mass at time $t = 0$. Three concentration–time curves computed for different values of the salt input by surface inflow: Curve C_J' for concentration and discharge as the present mean River Jordan, and curves for twice and four times the present rate of salt input. Equation 29.

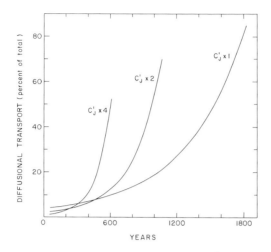

Figure 9.   Fraction of increase in the concentration of the upper water mass (Dead Sea) owing to eddy diffusional transport of salt from the lower brine layer. Computed from Equation 30 for three different values of the surface salt input as identified in Figure 8

Fractional increase owing to diffusion $= \dfrac{\Delta C_u - C_J'\Delta t}{\Delta C_u}$ (30)

The values of the diffusional transport fraction (in percent) obtained from Equation 30 were plotted against time in Figure 9. At the early stages of the process, when the dissolved solids concentration in the upper water mass was relatively low, eddy diffusional transport across the pycnocline amounted only to a small fraction of the concentration increment, the major contribution being attributed to the inflow. As time goes on and the concentration in the upper mass steadily increases, the contribution from the lower water mass gains in importance until it accounts for more than 50% of the dissolved solids being added into the upper mass in a unit of time. For the mean inflow rate as in the River Jordan, the present-day transport from the lower water accounts for 70% of the total salt being added to the upper mass.

It is realized that the premises on which the treatment of a three-layer model of the Dead Sea is based in this section are open to criticisms from different directions. In the absence of any reliable information, however, on the past and future possible changes in the physical and geometric characteristics of the Dead Sea system, the assumption of constant dimensions and rates holding over hundreds or thousands of years is probably admissible insofar as it at least provides a measure of the

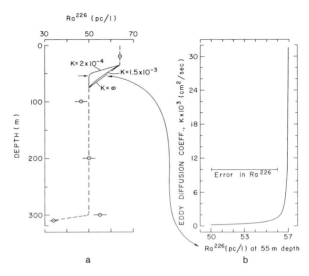

Figure 10. a: *Ra-226 concentrations in the Dead Sea (circles) (31). Solid curves in the pycnocline layer (depth 35–75 m) drawn for a steady-state model and different values of eddy diffusion coefficients (K) in the pycnocline. Equation 31.*

b: *Eddy diffusion coefficient (K) in the pycnocline layer as a function of the shape of concentration– depth profile: K plotted against the Ra-226 concentration at pycnocline midpoint (55 m). Mean error in Ra-226 analyses shows the resulting error in K. K computed from Equation 31 for different values of Ra-226 at pycnocline midpoint.*

life span of a stratified lake. Calculations of the kind presented in this section are likely to be more reliable for shorter periods of time and for lakes of better known behavior.

Reliability of Eddy Diffusion Coefficients. A method for evaluation of the eddy diffusion coefficients in the thermoclines and deeper regions of lakes, based on temperature–depth–time profiles or on the rate of heating of a lake, was originally developed by McEwen (*30*) and further developed and popularized by Hutchinson (*7*). Although in this method routine temperature–depth profiles recorded at short time intervals (several days apart) are fairly convenient to obtain and manipulate, the error in the eddy diffusion coefficients derived from such thermal data is large. It has been shown (*9*) that for a random error in the temperature readings $|\Delta T| = 0.05°C$, an error of 25–50% is associated with the eddy diffusion coefficients (K) in the range 0.05–0.1 cm^2 · sec^{-1}, and it increases with increasing K to approximately 300% when K is in the vicinity of 1.5 cm^2 · sec^{-1}.

When concentration–depth profiles of some dissolved chemical species are used to evaluate the eddy diffusion coefficients of the water column, the rates of reactions removing the species from the water must be known accurately. This knowledge is lacking for most of the common chemical species which are being removed through inorganic or biological reactions in natural waters. The exceptions are those radioactive nuclides which are being removed from the water by radioactive decay only; the decay constants are in general well known.

The nature and accuracy of the chemical data required in order to obtain eddy diffusion coefficients with an accuracy of a factor of 2 may be illustrated in the following example taken from the Ra-226 profile in the Dead Sea (*31*). In Figure 10a is shown the concentration of Ra-226 (in picocurie/liter) at five depths: the concentration in the upper mixed layer (64 pc/liter) is higher than in the lower water mass below the pycnocline. The decrease in the concentration near the bottom suggests that Ra-226 may be migrating from the water column into the sediment. No Ra-226 measurements were taken in the pycnocline (depths 35–75 m). When the transport of Ra-226 from the surface layer down through the pycnocline is effected by eddy diffusion only (*i.e.*, there is neither a vertical nor horizontal flow in the pycnocline), the shape of the Ra-226 concentration–depth profile in the pycnocline is maintained by eddy diffusion and natural decay of the isotope. The limiting shapes of the Ra-226 profile in the pycnocline are (*i*) a linear profile, when the concentration decreases linearly from 64 pc/liter in the upper layer to 50 pc/liter in the lower layer, as shown in Figure 10a, and (*ii*) a strongly curved profile, in which the concentration at the midpoint of the pycnocline (55 m) is equal to that of the lower brine layer, shown near the indicating arrow on the left in Figure 10a. At steady-state, the constant concentration at the pycnocline midpoint and boundaries depends on the eddy diffusion coefficient, decay constant, and the pycnocline thickness by the following relationship (*32*, p. 139)

$$\frac{C_1 + C_3}{2C_2} = \cosh \frac{h}{2} \sqrt{\frac{\lambda}{K}} \tag{31}$$

where C_1 is the Ra-226 concentration at the upper pycnocline boundary ($C_1 = 64$ pc/liter), C_3 is Ra-226 concentration at the lower boundary ($C_3 = 50$ pc/liter), C_2 is the concentration at pycnocline midpoint taken at 55-m depth, h is the pycnocline layer thickness (4×10^3 cm), λ is the Ra-226 decay constant (1.355×10^{-11} sec^{-1}), and K is the eddy diffusion coefficient in the pycnocline. By Relationship 31, K may be evaluated for different values of C_2, the limiting values of which for the case of eddy diffusional transport are 50 pc/liter and $(64 + 50)/2 = 57$ pc/liter, as explained earlier in this section. The eddy diffusion coefficients calculated by Equation 31 are shown in Figure 10b. At the lower limit of C_2, the eddy diffusion coefficient is low, $K = 2 \times 10^{-4}$ cm^2 · sec^{-1}. As C_2 increases and the concentration profile becomes less curved, K also increases. At the upper limit of C_2, when the concentration gradient in the pycnocline is linear, the eddy diffusion coefficient is infinitely large.

The mean error in Ra-226 concentrations plotted in Figure 10a is $\pm 5\%$. Comparing this error with the curve of K values in Figure 10b shows that the error in K computed from some value of C_2 can be as high as several orders of magnitude. The estimates of K are very sensitive to the value of C_2, as may be observed in Figure 10b and in the two profiles close to one another shown in Figure 10a, the one giving $K = 1.5 \times 10^{-3}$ cm^2 · sec^{-1} and the other $K = \infty$.

Approach to Chemical Steady-States

The previous sections dealt with transient and steady chemical states in systems whose geometry was constant and independent of time. In fresh water lakes, however, undergoing periodic seasonal changes from stratification in the summer to complete mixing in the winter, the distribution of the chemical species is greatly affected by the changes in the structure of the water column. A steady-state distribution of a chemical species which has been established during the period of stratification is destroyed when the lake turnover occurs; subsequently, the concentrations in the water column change and tend to a new steady-state which may be established under the new physical conditions.

When there is a constant source of a reacting chemical species in the water column or at its boundaries (e.g., water–air and/or water–sediment interface) then, by a rule of thumb, a steady-state may be attained within a period of time equal to "a few" half-lives of the species. In detail, a steady-state concentration is attained after "infinitely long" time. The time required for the concentration to come close to the steady-state value at any point in the water column depends on its distance from the source, transport properties of the medium (i.e., its diffusivity and distribution of advective velocities), and the rates of the reactions removing the species from the water. A concentration of 95% of a steady-state value may be arbitrarily taken as sufficiently close to a steady-state and indicating that the transient state has effectively come to an end. The time required to attain this concentration level (i.e., when $C = 0.95 C_{ss}$) at some point of a concentration–depth profile will be referred to as the time to steady-state. By way of generalization, a chemical species with a constant half-life would attain a steady-state concentration at any point in the water column sooner when the distance from the source is small and the dispersal within the water column is fast (*i.e.*, eddy diffusion coefficient is large).

If the half-life of a chemical species is much longer than several months, then on the time scale of seasonal turnovers of the lake, the species may be regarded as conservative, and the question whether it may attain a steady-state distribution in the water column depends only on its source and the transport characteristics of the medium.

While in the ocean transient states of chemical species are primarily characteristic of the surface waters where seasonal variations in biological productivity greatly affect the concentrations of dissolved nutrients (33), in lakes subject to seasonal turnovers transient concentrations are probably a rule. The effects exerted by the physical characteristics of the environment on the chances of a reacting chemical species to attain a steady-state concentration profile will be discussed for a lake, ocean, and sediment column.

Lake. A homogeneous water column of height h and eddy diffusion coefficient K will be considered. A chemical species is being continuously introduced into the water column owing to its constant concentration (C_o) at one boundary $(z = h)$, and it is being removed by a first-order reaction (reaction rate constant λ) occurring throughout the water column. This model may, with some imagination, be compared with a case of a chemical species forming at the sediment–water interface and transported upwards by eddy diffusion; the upper boundary of the water column at $z = 0$ (*e.g.*, the water–air interface) is impermeable to the dissolved species, such that its only sink is the chemical reaction.

The concentration (C) as a function of time (t) and position (z) in the water column is given by the following differential equation.

$$\frac{\partial C}{\partial t} = K \frac{\partial^2 C}{\partial z^2} - \lambda C \tag{32}$$

The initial and boundary conditions of this model are

$$\text{Initial conditions: at } t = 0 \text{: } C = 0 \text{ in } z > 0 \tag{33}$$

$$\text{Boundary conditions: at } t > 0 \text{: } C = C_o \text{ at } z = h \tag{34}$$

$$\frac{dC}{dt} = 0 \text{ at } z = 0 \tag{35}$$

The last condition of the zero concentration gradient at the upper boundary $z = 0$ defines its impermeability. The solution is (the Appendix contains a note on the methods of solution of this and similar equations discussed in the text):

$$C = \frac{C_o}{2} \sum_{n=0}^{\infty} (-1)^n \left\{ \exp\left(-[(2n+1)h - z]\sqrt{(\lambda/K)}\right) \right.$$

$$\times \operatorname{erfc}\left[\frac{(2n+1)h - z}{2\sqrt{(Kt)}} - \sqrt{(\lambda t)}\right] +$$

$$+ \exp\left([(2n+1)h - z]\sqrt{(\lambda/K)}\right) \times \operatorname{erfc}\left[\frac{(2n+1)h - z}{2\sqrt{(Kt)}} + \sqrt{(\lambda t)}\right] +$$

$$+ \exp\left(-[(2n+1)h + z]\sqrt{(\lambda/K)}\right) \times \operatorname{erfc}\left[\frac{(2n+1)h + z}{2\sqrt{(Kt)}} - \sqrt{(\lambda t)}\right] +$$

$$+ \exp\left([(2n+1)h + z]\sqrt{(\lambda/K)}\right) \times \operatorname{erfc}\left[\frac{(2n+1)h + z}{2\sqrt{(Kt)}} + \sqrt{(\lambda t)}\right] \right\} \quad (36)$$

A relationship for the stationary concentration C_{ss} may be determined either by substituting $\partial C/\partial t = 0$ into Equation 32 and solving the right-hand side or by evaluating the limit of C in Equation 36 when t tends to infinity. The steady-state solution is

$$C_{ss} = \frac{C_o[e^{z\sqrt{(\lambda/K)}} + e^{-z\sqrt{(\lambda/K)}}]}{e^{h(\sqrt{\lambda/K})} + e^{-h\sqrt{(\lambda/K)}}} \quad (37)$$

or

$$C_{ss} = \frac{C_o \cosh z\sqrt{(\lambda/K)}}{\cosh h\sqrt{(\lambda/K)}} \quad (38)$$

Considering the shape of the stationary concentration–depth profile given by Equation 37 or 38, at the source boundary ($z = h$) $C_{ss} = C_o$, whereas at the other boundary ($z = 0$), the concentration is

$$C_{ss} = \frac{C_o}{\cosh h\sqrt{(\lambda/K)}} \quad (39)$$

The latter relationship shows that when the eddy diffusion coefficient K is large and the cosh term consequently tends to 1, the steady-state concentration is close to C_o throughout the water column.

A concentration at some point of the water column equal to 95% of the steady-state concentration may be regarded as a value reasonably close to, and indicative of, the steady-state attained: $C = 0.95C_{ss}$. As the concentration C given by Relationship 36 depends jointly on h, z, λ, K, and t, a somewhat simpler form is obtained by considering the concentration change at the boundary $z = 0$.

The concentration at $z = 0$ is, from Equation 36

$$C_{z=o} = C_o \sum_{n=0}^{\infty} (-1)^n \left\{ \exp\left[-(2n+1)h\sqrt{(\lambda/K)}\right] \times \right.$$

$$\left. \mathrm{erfc}\left[\frac{(2n+1)h}{2\sqrt{(Kt)}} - \sqrt{(\lambda t)}\right] + \exp\left[(2n+1)h\sqrt{(\lambda/K)}\right] \times \right.$$

$$\left. \mathrm{erfc}\left[\frac{(2n+1)h}{2\sqrt{(Kt)}} + \sqrt{(\lambda t)}\right] \right\} \quad (40)$$

Additional simplification of Equation 40 may be achieved by writing λ and t in terms of the half-life (τ) of the reacting species

$$\lambda = 0.693/\tau \quad (41)$$

$$t = j\tau \quad (42)$$

where j is the number of half-lives. Substitution of Equations 41 and 42 into Equation 40 gives:

$$C_{z=0} = C_o \sum_{n=0}^{\infty} (-1)^n \left\{ \exp\left[-0.833\,(2n+1)h/\sqrt{(K\tau)}\right] \times \right.$$

$$\left. \mathrm{erfc}\left[\frac{(2n+1)h}{2(\sqrt{K\tau j})} - 0.833\sqrt{j}\right] + \exp\left[0.833\,(2n+1)h/\sqrt{(K\tau)}\right] \times \right.$$

$$\left. \mathrm{erfc}\left[\frac{(2n+1)h}{2\sqrt{(K\tau j)}} + 0.833\sqrt{j}\right] \right\} \quad (43)$$

In Equation 43, the terms C_o, h, K, and τ are constants for each particular case, and the only independent variable is the number of half-lives, j. Therefore, when h, K, and τ are taken as a dimensionless parameter $h/\sqrt{(K\tau)}$, it follows from Equation 43 that for any value of $h/\sqrt{(K\tau)}$ there is only one value of j which gives the concentration equal to the 95% of the steady-state concentration (*i.e.*, $C = 0.95C_{ss}$). The values of time when $C = 0.95C_{ss}$ were computed for different values of the parameter $h/\sqrt{(K\tau)}$ and plotted in Figure 11; the larger is the value of $h/\sqrt{(K\tau)}$, the longer it takes to attain the 95% level of the steady-state concentration. The quotient $h/\sqrt{(K\tau)}$ is large when the water column is deep (large h), eddy diffusional transport is slow (small eddy diffusion coefficient K), and the half-life of the reacting species is short. Under such conditions, it would take longer to attain a steady-state concentration than in a shallow water column, or when the eddy diffusivity is large, or when the half-life of the species is long.

As an example of estimating the time required for the concentration to attain the value of 95% of the steady-state concentration, one might

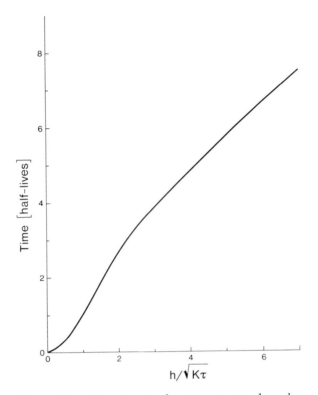

Figure 11. Time to steady-state at one boundary
(z = 0) of a water layer of thickness h

Model: concentration of a chemical species constant (C_o) at
one boundary (z = h). Transport through the water column
by eddy diffusion (K), and removal by first order reaction
(half-life τ). The other boundary of the water column (z =
0) is impermeable to the chemical species. Time to steady-
state at the impermeable boundary shown for different values
of dimensionless quotient h/ $\sqrt{(K\tau)}$. Equations 39, 43.

consider a water column 30 m deep ($h = 3 \times 10^3$ cm), characterized by
a mean eddy diffusion coefficient $K = 1$ cm$^2 \cdot$ sec^{-1}, through which a dissolved species of half-life $\tau = 2$ months ($= 5.184 \times 10^6$ sec) is being
dispersed, when the concentration at one of the boundaries is constant.
In this case $h/\sqrt{(K\tau)} = 1.318$, and from the curve in Figure 11, the
time to steady-state is approximately $t = 1.5$ half-lives or 3 months. Thus,
if the water column retains its characteristics for longer than 3 months, a
stationary concentration may become established.

Extending the analogy to lakes which are mixed during four or five
months of the year, steady-state concentrations may become established
in the water column only in those cases when the value of $h/\sqrt{(K\tau)}$,

characteristic of the water column and chemical species, gives the time to steady-state, as read off the curve in Figure 11, which is less than the length of time the lake water column remains homogeneously mixed.

Ocean. Despite the fact that no changes in the chemical composition of the ocean water have been established in recent time, it is instructive to consider the transient behavior of reacting chemical species in the oceanic water column. The case of the nuclear fallout products transported through the surface and thermocline layer of the ocean is the best known, although not yet completely understood, case of a transient chemical event on a world-wide scale (*34, 35, 36, 37*).

The distribution of a number of dissolved species (O_2, C-14, Ra-226, salinity) in the Central Pacific water column, at depths between 1 and 4 km, has been shown (*11*) to be consistent with a steady-state model of the water column in which the concentration–depth profiles are stationary and the concentrations at the boundaries 1 and 4 km are stipulated at their present values. The physical model of the water column is based on two transport mechanisms: vertical eddy diffusion (eddy diffusion coefficient $K = 1.3$ cm$^2 \cdot$ sec^{-1}) and upwelling of deep water (advection velocity $U = 1.4 \times 10^{-5}$ cm $\cdot$ sec^{-1}, or approximately 1 cm per day) (*11*).

For a water column of such physical characteristics, the ratio K/U, the scale height, is approximately 900 m. For distances much greater than the scale height, the transport resulting from flow is much more important than the transport by eddy diffusional dispersal. Conversely, for distances much shorter than the scale height, the eddy diffusional mode is the main mechanism of transport. In such a water column, the time required for the concentration to attain a steady-state will be calculated for chemical species reacting by two different mechanisms: (*i*) zero-order reaction, when the rate of change in concentration ($\partial c/\partial t$) is constant and independent of the concentration of the species, and (*ii*) first-order reaction, when $\partial c/\partial t$ is proportional to the concentration. Dissolved oxygen consumed in the oxidation of organic matter will be considered as an example of a zero-order reaction, and radioactive decay will be discussed as an example of a first-order reaction.

OXYGEN. The O_2 concentration in the oceanic water column at intermediate depths decreases upwards to the layer of oxygen minimum. The concentration–depth profile for the Central Pacific, shown diagrammatically in Figure 12, is considered stationary and maintained by the supply of oxygen in the upwelling water and its removal in the course of oxidation of the organic matter settling through the water column (*11, 38*). The rate of oxygen consumption in the process of oxidation of organic matter has been reported as approximately constant and independent of the oxygen concentration in the water; the reported values of the consumption rate (v) are in the range 0.0027–0.0053 ml $\cdot$ liter$^{-1} \cdot$ yr^{-1} (*11*).

Exponential decrease with depth in the oxygen consumption rate has been considered in a steady-state distribution by Wyrtki (*39*).

A good fit of the measured oxygen concentrations by a curve calculated from the eddy diffusion–advection model has been obtained by

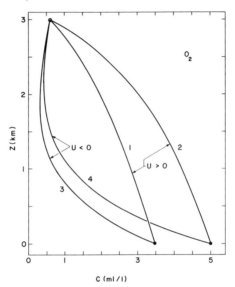

Figure 12. Diagrammatic stationary concentration–depth profiles of dissolved oxygen in a 3-km-thick water layer with fixed concentrations at the boundaries (depths −1 and −4 km). Vertical dimension z = depth + 4 km. Profiles computed for K = 1.2 cm² · sec⁻¹, U = ±1.4 cm · sec⁻¹. Note the difference between profile shapes for advection up (U > 0) and down (U < 0). Equations 47, 51.

Munk (*11*) using the rate of oxygen consumption $\nu = 0.0027$ ml. · liter⁻¹-yr⁻¹, eddy diffusion coefficient $K = 1.3$ cm² · sec⁻¹, and upwelling velocity $U = 1.4 \times 10^{-5}$ cm · sec⁻¹. The oxygen concentration values at the boundaries are stipulated as constant: 3.5 ml/liter at the lower boundary ($z = 0$) and 0.6 ml/liter at the upper bundary ($z = h \equiv 3$ km). The concentration–depth profile is given by the following relationship (*11*)

$$C = C_o + \left(C_1 - C_o + \frac{\nu h}{U} \right) \times \frac{\exp \ (Uz/K) - 1}{\exp \ (Uh/K) - 1} - \frac{\nu z}{U} \tag{44}$$

which is a solution of the differential equation

$$K \frac{\partial^2 C}{\partial z^2} - U \frac{\partial C}{\partial z} - \nu = 0 \tag{45}$$

describing the change in concentration as a function of depth (z), with the boundary conditions $C = C_o$ at $z = 0$ and C_1 at $z = h$.

In order to be able to consider the rate of change in the oxygen concentration and the rate of its approach to a steady-state, new concentra-

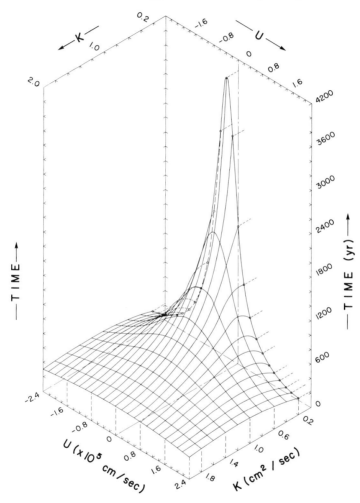

*Figure 13.   Time to steady-state for oxygen concentration at
z = 1 km*

*Computed as explained in the text using initial stationary concentra-
tion $(C_{t=0})$ given by profile 1 in Figure 12 and new steady-state
given by profile 2. Time to steady-state shown for different values of
eddy diffusion coefficient (K) and advective velocity (U). Time to
steady-state defined as the time when the oxygen concentration has
attained 95% of the concentration difference between the new and
initial steady-state profiles: $C - C_{t=0} = 0.95(C_{t=\infty} - C_{t=0})$.
C from Equation 50.*

tions at the boundaries of the water column must be stipulated. It will be assumed that the oxygen concentration at the lower boundary ($z = 0$) has increased "instantaneously" to 5.0 ml/liter, whereas the concentration at the upper boundary remains unchanged at 0.6 ml/liter. For such new boundary conditions, the time-dependent oxygen concentrations may be evaluated by solving the following differential equation

$$\frac{\partial C}{\partial t} = K \frac{\partial^2 C}{\partial z^2} - U \frac{\partial C}{\partial z} - \nu \tag{46}$$

with the following initial and boundary conditions.

Initial conditions: at $t = 0$:

$$C = C_{t=0} \equiv C_o + \left(C_1 - C_o + \frac{\nu h}{U}\right) \times \frac{\exp(Uz/K) - 1}{\exp(Uh/K) - 1} - \frac{\nu z}{U} \tag{47}$$

Boundary conditions: at $t > 0$: $C = C_2$ at $z = 0$ (48)

$$C = C_3 \text{ at } z = h \tag{49}$$

The solution is:

$$C = C_{t=0} + \tfrac{1}{2}(C_2 - C_o) \sum_{n=0}^{\infty} \left\{ \exp(-Uhn/K) \times \right.$$

$$\mathrm{erfc}\left[\frac{2nh + z}{2\sqrt{(Kt)}} - \frac{U}{2}\sqrt{\frac{t}{K}}\right] +$$

$$\exp[U(nh + z)/K] \times \mathrm{erfc}\left[\frac{2nh + z}{2\sqrt{(Kt)}} + \frac{U}{2}\sqrt{\frac{t}{K}}\right] -$$

$$\exp\{-U[(n+1)h - z]/K\} \times \mathrm{erfc}\left[\frac{(2n+2)h - z}{2\sqrt{(Kt)}} - \frac{U}{2}\sqrt{\frac{t}{K}}\right] -$$

$$\left. \exp[Uh(n+1)/K] \times \mathrm{erfc}\left[\frac{(2n+2)h - z}{2\sqrt{(Kt)}} + \frac{U}{2}\sqrt{\frac{t}{K}}\right]\right\} +$$

$$\tfrac{1}{2}(C_3 - C_1) \sum_{n=0}^{\infty} \left\{ \exp(-U[(n+1)h - z]/K) \times \right.$$

$$\mathrm{erfc}\left[\frac{(2n+1)h - z}{2\sqrt{(Kt)}} - \frac{U}{2}\sqrt{\frac{t}{K}}\right] +$$

$$\exp(Uhn/K) \times \mathrm{erfc}\left[\frac{(2n+1)h - z}{2\sqrt{(Kt)}} + \frac{U}{2}\sqrt{\frac{t}{K}}\right] -$$

$$\exp(-Uh(n+1)/K) \times \mathrm{erfc}\left[\frac{(2n+1)h + z}{2\sqrt{(Kt)}} - \frac{U}{2}\sqrt{\frac{t}{K}}\right] -$$

$$\left. \exp(U(nh + z)/K) \times \mathrm{erfc}\left[\frac{(2n+1)h + z}{2\sqrt{(Kt)}} + \frac{U}{2}\sqrt{\frac{t}{K}}\right]\right\} \tag{50}$$

In Equation 50, the term $C_{t=0}$ is the initial steady-state concentration–depth profile given in Equation 47 and shown in curves 1 and 3 in Figure 12 for two values, positive and negative, of advection velocity U. The infinite series in Equation 50 converge rapidly when the values of U, K, h, and z used are those discussed later in this section.

All the time-dependent terms in Equation 50 assembled between the braces are independent of the oxygen consumption rate, v. This means that the rate at which the concentration C at any depth approaches a steady-state value $(\partial C/\partial t)$ is independent of the zero order reaction rate constant. The time-dependent terms contain, however, the eddy diffusion coefficient and advection velocity, and the rate of approach to steady-state is therefore dependent on these two physical characteristics of the environment.

A new steady-state profile with the boundary conditions of Equations 48 and 49 is established when t tends to infinity.

$$C_{t=\infty} = C_2 + (C_3 - C_2 + \frac{vh}{U}) \times \frac{\exp{(Uz/K)} - 1}{\exp{(Uh/K)} - 1} - \frac{vz}{U} \qquad (51)$$

The time to steady-state was calculated for the oxygen concentration change at $z = 1$ km (*i.e.*, depth 3 km; *see* Figure 12) as the time when the concentration referred to the initial steady-state $(C - C_{t=0})$ has attained the value of 95% of the new steady-state, Equation 51, referred to the same initial point: $(C_{t=\infty} - C_{t=0})$.

The problem thus is: what is t when $C - C_{t=0} = 0.95(C_{t=\infty} - C_{t=0})$ at $z = 1$ km?

The values of t were calculated for different diffusion coefficients and advective velocities by substitution of successively increasing values of t into Equation 50 and using Equations 47 and 51 until the value of $(C - C_{t=0})/(C_{t=\infty} - C_{t=0}) = 0.95$ sought was obtained by interpolation between two numbers close to 0.95. The value of the oxygen consumption rate used in the computation of $C_{t=0}$ and $C_{t=\infty}$ was $v = 0.0027$ ml · liter^{-1} · yr^{-1}. Calculation of the time needed to attain 95% of the steady-state concentration was done for the eddy diffusion coefficient (K) in the range 0.2–2.0 cm^2 · sec^{-1} and advective velocity (U) in the range from +2.4 to −2.4 cm · sec^{-1}. The results are plotted in a three-dimensional diagram in Figure 13. The time to steady-state shown in Figure 13 is symmetrical with respect to the positive and negative advective velocities because the concentration at both the lower and upper boundary have been stipulated. The symmetry of the time values means that the time required to attain a steady-state concentration depends on the absolute value of the advective velocity but not on its sign; the same length of time is obtained for upward and downward flow. The concentration values, however, at any depth very much depend on the direction of flow, as shown in the concentration–depth profiles in Figure 12 drawn for the positive (up) and negative (down) velocity. The time–velocity–diffusivity surface in Figure 13 shows that when the eddy diffusion coefficient of the water column is large, the advective velocity has very little effect on the time it takes to attain a steady-state: for $K = 2.0$ cm$^2 \cdot$ sec^{-1}, the time to steady-state is between 300 and 400 years for all the values of U shown. Conversely, in a water column of low eddy diffusivity, the advective velocity is the main controlling factor of the length of time needed to reach a steady-state. For $K = 0.2$ cm$^2 \cdot$ sec^{-1}, the time to steady-state is 4000 years when the advective velocity is near 0, and it decreases to 300 years when the advective velocity approaches $\pm 2.4 \times 10^{-5}$ cm $\cdot$ sec^{-1}.

Under present conditions ($K = 1.3$ cm$^2 \cdot$ sec^{-1}, $U = 1.4 \times 10^{-5}$ cm $\cdot$ sec^{-1}), it would take approximately 500 years for the oxygen concentration at $z = 1$ km to attain a new steady-state, as can be read off Figure 13. This estimate of time to steady-state in a model layer with the concentrations at the boundaries maintained constant may be compared with the time estimate obtained from a different model.

For the same physical conditions, the same initial concentration–depth profile (Figure 12, curve 1), and the same new concentration at the lower boundary $z = 0$ (5.0 ml/liter), transient concentration–depth profiles have been computed for a semi-infinite water layer extending upwards from $z = 0$ (40). Although the computations were done for a semi-infinite layer model, the oxygen concentrations were considered only within the 3-km-thick layer between $z = 0$ and $z = 3$. In the semi-infinite layer model, the oxygen concentration at $z = 1$ km attains the 95% value of the steady-state concentration after approximately 1000 years. After 500 years, the concentration is only 88% of the steady-state value. The difference between the estimates of time to steady-state from the two models is understandably accounted for by a faster approach to steady-state in a water layer the two boundaries of which are maintained at constant concentrations.

RADIUM-226 IN WATER. Ra-226 will be considered as an example of a chemical species which is being removed from the oceanic water column by a first-order chemical reaction, radioactive decay. Other possible mechanisms of removal, such as uptake by detrital silicates and organisms, will not be discussed. The supply of Ra-226, however, from organic matter decomposing in the water column will be considered.

The source of Ra-226 in the deep ocean is the flux of Ra-226 from the ocean floor sediments where it forms by the decay of ionium (Th-230) (5, 41) and the decomposition of organic matter sinking through the water column (42, 43). The present concentration of Ra-226 at inter-

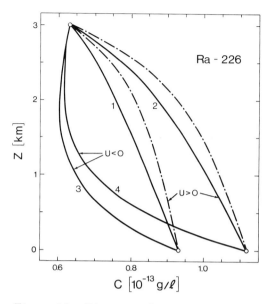

Figure 14.   Diagrammatic stationary concentration–depth profiles of Ra-226 in a water column

Same depths, eddy diffusion coefficient, and advection as in Figure 12. Solid curves: eddy diffusion, advection, and radioactive decay. Equations 52, 59. Dashed curves: eddy diffusion, advection, decay, and production. Equation 62.

mediate depths (1–4 km) decreases from bottom up, and the concentration–depth profile has been explained as maintained by upwelling, eddy diffusivity, and radioactive decay (*11*). The concentration of Ra-226 in Central Pacific decreases from 0.93×10^{-13} grams/liter at 4-km depth to 0.63×10^{-13} grams/liter at 1-km depth (Figure 14). A reasonable compatibility between the observed concentration values and a diffusion–advection model has been obtained by Munk (*11*) using the eddy diffusion coefficient $K = 1.3$ cm$^2 \cdot$ sec^{-1} and upwelling velocity $U = 1.4 \times 10^{-5}$ cm $\cdot$ sec^{-1}, as in the case of the oxygen profile. The Ra-226 concentration between the depths 4 and 1 km is given by the relationship (*11*)

$$C = \frac{C_1 - C_o \exp (R_2 h)}{\exp (R_1 h) - \exp (R_2 h)} \exp (R_1 z) + \frac{C_o \exp (R_1 h) - C_1}{\exp (R_1 h) - \exp (R_2 h)} \exp (R_2 z)$$

(52)

where R_1 and R_2 are constants.

$$R_1 = \frac{U}{2K} + \sqrt{\frac{U^2}{4K^2} + \frac{\lambda}{K}}; \quad R_2 = \frac{U}{2K} - \sqrt{\frac{U^2}{4K^2} + \frac{\lambda}{K}}$$

C_o is the concentration at the lower boundary ($z = 0$), and C_1 is the concentration at the upper boundary ($z = h \equiv 3$ km). Equation 52 is the solution of the following differential equation describing the concentration change at steady-state

$$K \frac{\partial^2 C}{\partial z^2} - U \frac{\partial C}{\partial z} - \lambda C = 0 \tag{53}$$

with the boundary conditions: at $z = 0$, $C = C_o$, and at $z = h$, $C = C_1$. λ is the decay constant of Ra-226 ($\lambda = 4.279 \times 10^{-4}$ yr^{-1}).

A time-dependent concentration of Ra-226 within the same layer of water, when the concentrations at the lower and upper boundary have been stipulated, may be obtained by solving the following differential equation

$$\frac{\partial C}{\partial t} = K \frac{\partial^2 C}{\partial z^2} - U \frac{\partial C}{\partial z} - \lambda C \tag{54}$$

with the initial conditions: at $t = 0$ in $0 < z < h$

$$C = C_{t=0} \equiv \frac{C_1 - C_o \exp (R_2 h)}{\exp (R_1 h) - \exp (R_2 h)} \exp (R_1 z) +$$

$$\frac{C_o \exp (R_1 h) - C_1}{\exp (R_1 h) - \exp (R_2 h)} \exp (R_2 z) \tag{55}$$

where the constants R_1 and R_2 were defined under Equation 52 and the boundary conditions

$$\text{at } t > 0: C = C_2 \text{ at } z = 0 \tag{56}$$

$$C = C_3 \text{ at } z = h \tag{57}$$

The solution of Equation 54 with the initial and boundary conditions of Equations 55–57 is

$$C = C_{t=0} + \frac{1}{2} (C_2 - C_o) \exp (Uz/2K) \times$$

$$\sum_{n=0}^{\infty} \left\{ \exp\left(-(2nh + z) \sqrt{\frac{U^2}{4K^2} + \frac{\lambda}{K}}\right) \times \mathrm{erfc} \left[\frac{2nh + z}{2\sqrt{(Kt)}} - \sqrt{\left(\frac{U^2}{4K} + \lambda\right)t} \right] + \right.$$

$$\exp\left((2nh + z) \sqrt{\frac{U^2}{4K^2} + \frac{\lambda}{K}}\right) \times \mathrm{erfc} \left[\frac{2nh + z}{2\sqrt{(Kt)}} + \sqrt{\left(\frac{U^2}{4K} + \lambda\right)t} \right] -$$

$$\exp\left(-[(2n + 2)h - z] \sqrt{\frac{U^2}{4K^2} + \frac{\lambda}{K}}\right) \times \mathrm{erfc} \left[\frac{(2n + 2)h - z}{2\sqrt{(Kt)}} - \right.$$

$$\sqrt{\left(\frac{U^2}{4K} + \lambda\right)t}\right] - \exp\left([(2n+2)h - z]\sqrt{\frac{U^2}{4K^2} + \frac{\lambda}{K}}\right) \times$$

$$\mathrm{erfc}\left[\frac{(2n+2)h - z}{2\sqrt{(Kt)}} + \sqrt{\left(\frac{U^2}{4K} + \lambda\right)t}\right]\right\} +$$

$$\tfrac{1}{2}(C_3 - C_1)\exp\left(U(z - h)/2K\right) \times$$

$$\sum_{n=0}^{\infty}\left\{\exp\left(-[(2n+1)h - z]\sqrt{\frac{U^2}{4K^2} + \frac{\lambda}{K}}\right) \times \mathrm{erfc}\left[\frac{(2n+1)h - z}{2\sqrt{(Kt)}} -\right.\right.$$

$$\sqrt{\left(\frac{U^2}{4K} + \lambda\right)t}\right] + \exp\left([(2n+1)h - z]\sqrt{\frac{U^2}{4K^2} + \frac{\lambda}{K}}\right) \times$$

$$\mathrm{erfc}\left[\frac{(2n+1)h - z}{2\sqrt{(Kt)}} + \sqrt{\left(\frac{U^2}{4K} + \lambda\right)t}\right] - \exp\left(-[(2n+1)h + z]\right.$$

$$\sqrt{\frac{U^2}{4K^2} + \frac{\lambda}{K}}\right) \times \mathrm{erfc}\left[\frac{(2n+1)h + z}{2\sqrt{(Kt)}} - \sqrt{\left(\frac{U^2}{4K} + \lambda\right)t}\right] -$$

$$\exp\left([(2n+1)h + z]\sqrt{\frac{U^2}{4K^2} + \frac{\lambda}{K}}\right) \times$$

$$\mathrm{erfc}\left[\frac{(2n+1)h + z}{2\sqrt{(Kt)}} + \sqrt{\left(\frac{U^2}{4K} + \lambda\right)t}\right]\right\} \tag{58}$$

The term $C_{t=0}$ is the initial steady-state concentration given by Relationship 55.

In Equation 58, the time-dependent terms between the braces contain the decay constant λ. Therefore, the rate of change in Ra-226 concentration at any depth $(\partial C/\partial t)$ depends on the decay rate constant. Thus, in the case of a first-order reaction (radioactive decay), the rate of change in concentration depends on the reaction rate constant, whereas it has been shown in the preceding section that for a zero-order reaction (oxygen consumption), the rate of change in concentration $(\partial C/\partial t)$ is independent of its rate constant.

A new steady-state concentration $(C_{t=\infty})$ will be attained when t tends to infinity in the time-dependent terms of Equation 58.

$$C_{t=\infty} = \frac{C_3 - C_2 \exp\,(R_2 h)}{\exp\,(R_1 h) - \exp\,(R_2 h)} \exp\,(R_1 z) +$$

$$\frac{C_2 \exp\,(R_1 h) - C_3}{\exp\,(R_1 h) - \exp\,(R_2 h)} \exp\,(R_2 z) \qquad (59)$$

The constants R_1 and R_2 were defined under Equation 52.

The time it takes the Ra-226 concentration to build up to a steady-state may be considered, as before, the time when the concentration has attained the 95% value of the difference between the old and new steady-state concentrations:

$$C - C_{t=0} = 0.95\ (C_{t=\infty} - C_{t=0}) \qquad (60)$$

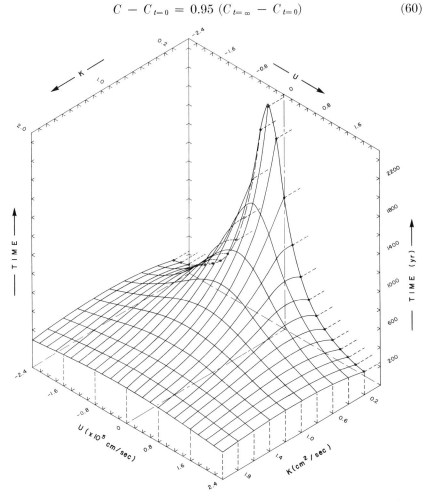

Figure 15. Time to steady-state of Ra-226 concentration at z = 1 km.
Notation as in Figure 13. C from Equation 58.

The values of time which satisfy Equation 60 were calculated for different values of the eddy diffusion coefficient (range 0.2–2.0 cm² · sec⁻¹) and advective velocity (range from +2.4 × 10⁻⁵ to −2.4 × 10⁻⁵ cm · sec⁻¹) and plotted in the t–K–U graph in Figure 15. The curves in Figure 15 are symmetrical about $U = 0$, indicating that the direction of flow (up or down) has no effect on the time it takes to reach a steady-state concentration. When the turbulence in the water column is relatively high (K in the vicinity of 2.0 cm² · sec⁻¹), the advection has little effect on the time to steady-state; the time values are in the range 300–400 years. When turbulence is low (low values of K), then advection dominates the picture. The time to steady-state decreases from approximately 2000 years when U is near 0 to 300 years when the absolute value of U is high.

There are indications that release of Ra-226 by decomposing organic matter is a mechanism of some significance in maintaining the Ra-226 concentrations in ocean water (*42, 43*). If this additional supply of Ra-226 is expressed as a constant production rate Q (grams · liter⁻¹ · yr⁻¹), then a steady-state concentration–depth profile may be obtained from the differential equation

$$K \frac{\partial^2 C}{\partial z^2} - U \frac{\partial C}{\partial z} - \lambda C + Q = 0 \tag{61}$$

the solution of which for constant boundary concentrations (C_o at $z = 0$ and C_1 at $z = h$) is

$$C_{ss} = \frac{Q}{\lambda} + \frac{C_1 - C_o \exp{(R_2 h)} - [1 - \exp{(R_2 h)}]Q/\lambda}{\exp{(R_1 h)} - \exp{(R_2 h)}} \exp{(R_1 z)} +$$

$$\frac{C_o \exp{(R_1 h)} - C_1 + [1 - \exp{(R_1 h)}]Q/\lambda}{\exp{(R_1 h)} - \exp{(R_2 h)}} \exp{(R_2 z)} \tag{62}$$

The constants R_1 and R_2 were defined under Equation 52.

For new concentrations at the boundaries of the water column, C_2 at $z = 0$ and C_3 at $z = h$, transient concentrations may be evaluated from the following equation

$$C = C_{t=0} + \tfrac{1}{2}(C_2 - C_o) \exp{(Uz/2K)} \times [\text{summation terms from equation 58}]$$

$$+ \tfrac{1}{2}(C_3 - C_1) \exp{[U(z - h)/2K]} \times [\text{summation terms from Equation 58}] \tag{63}$$

where the initial distribution $C_{t=0}$ is given by Equation 62. At the new steady-state, the concentrations are given by Equation 62 with C_3 replacing C_1 and C_2 replacing C_0.

Estimates of the production rate of Ra-226 at intermediate depths in the Pacific are in the range 1×10^{-18}–50×10^{-18} grams $\cdot$ liter$^{-1} \cdot$ yr^{-1} (43). Two steady-state profiles computed from Equation 62 using $Q = 21 \times 10^{-18}$ grams $\cdot$ liter$^{-1} \cdot$ yr^{-1} are shown in Figure 14 (dashed curves). The differences between the concentrations shown by the solid curves 1 and 2 in Figure 14 (no production, $Q = 0$) and the corresponding dashed curves are 10% or less. The small difference between the two models shows that the sampling and analytical accuracy must be high if the differences in the Ra-226 production in the water column are to be inferred from observations.

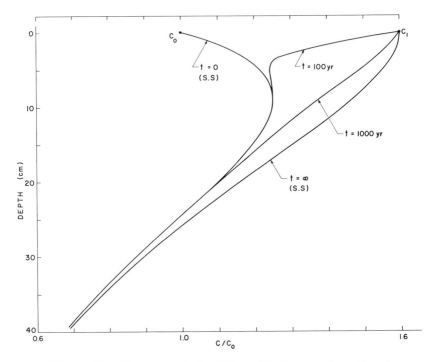

Figure 16.   Diagrammatic Ra-226 profiles in oceanic sediment

Profile at $t = 0$: steady-state, Equation 64, diffusion coefficient of Ra-226 in sediment $K = 1 \times 10^{-9}$ cm$^2 \cdot$ sec^{-1}, sedimentation rate $U = 9.5 \times 10^{-12}$ cm $\cdot$ sec^{-1} (3 mm/1000 yr; (5)). Concentration scale normalized to the value of initial Ra-226 at the sediment–water interface $C_o = 1$. New concentration at the interface $C_1 = 1.6C_o$. Profiles at 100 and 1000 years after the change in boundary concentration computed from Equation 70. New steady-state ($t = \infty$) from Equation 71. Constant Th-230 concentration at the interface taken as $C_T^\circ = 100C_o$.

RA-226 IN SEDIMENT. Migration of Ra-226 in the sediment column in deep ocean has been inferred from the disequilibrium of Ra-226 and its parent ionium (Th-230) detected in a number of sediment cores (5). Whereas Th-230 taken up by the sediment particles from the sea water shows no tendency to migrate nor redistribute itself in the sediment (5, 44), Ra-226 migrates, and this process results in the flux of Ra-226 from the sediment into water column. A generalized concentration–depth profile of Ra-226 in the sediment, adopted from the Indian Ocean data, is shown in Figure 16, labelled $t = 0$. The concentration profile, considered stationary, is maintained by a balance between the supply of Ra-226 from the decay of the parent Th-230 (which is being added to the sediment at a constant rate and constant concentration) and the decay of Ra-226 and its migration through the sediment column. From the steady-state profile of Ra-226 in the Indian Ocean, the diffusion coefficient for Ra-226 in the sediment has been estimated as 1×10^{-9} cm$^2 \cdot$ sec^{-1} (5). This value is approximately three orders of magnitude lower than the values of the diffusion coefficients of ionic solutes in aqueous solutions, and it is also much too low to be accounted for by the tortuosity of the pore space in the sediment. Chemical interaction of Ra-226 with the sediment may be the reason for the low value of the diffusion coefficient obtained (5). A stationary concentration–depth profile of a chemical species in the sediment may be perturbed by any combination of such factors as a change in the rate of deposition, change in the rate of supply, and change in the concentration at the sediment–water interface resulting from external causes.

Such changes, disturbing the existing chemical steady-state, would cause the concentration of the species to vary as a function of time until a new steady-state has been established. The time required to attain a new steady-state for Ra-226 in the sediment will be evaluated for a simple, but hypothetical, case of the Ra-226 concentration at the sediment–water interface increasing by a factor of 1.6. Such an increase would keep the Ra-226/Th-230 atom ratio at the sediment–water interface at the value of 1.6/100, which is still below the equilibrium value of approximately 2/100; the present-day ratio is near 1/100.

The present $(t = 0)$ Ra-226 profile in Figure 16 is given by the following relationship describing a steady-state distribution of a decaying species

$$C = A \exp\left[-z\lambda_T/U\right] + (C_o - A) \exp\left[\left(\frac{U}{2K} - \sqrt{\frac{U^2}{4K^2} + \frac{\lambda_R}{K}}\right)z\right] \quad (64)$$

where A is a constant.

$$A = \frac{\lambda_T C_T{}^\circ}{\lambda_R - \lambda_T - K\lambda_T{}^2/U^2} \quad (65)$$

λ_T is the decay constant of Th-230, $C_T{}^\circ$ is Th-230 concentration at the sediment–water interface, λ_R is the decay constant of Ra-226, C_0 is the Ra-226 concentration at the sediment–water interface, U is the rate of sedimentation, and K is the diffusion coefficient of Ra-226 in the sediment.

A relationship for a non-steady-state concentration of Ra-226 may be obtained by solving the following differential equation

$$\frac{\partial C}{\partial t} = \lambda_T C_T{}^\circ \exp\left(-z\lambda_T/U\right) + K\frac{\partial^2 C}{\partial z^2} - U\frac{\partial C}{\partial z} - \lambda_R C \tag{66}$$

where the term $\lambda_T C_T{}^0 \exp\left(-\lambda_T z/U\right)$ is the rate of production of Ra-226 (grams $\cdot$ cm^{-3} $\cdot$ yr^{-1}) in the sediment owing to decay of Th-230. The concentrations at the interface, rate of sedimentation, and diffusion coefficient are considered constant. Equation 66 is to be solved with the following conditions.

Initial conditions: at $t = 0$:

$$C = C_{t=0} \equiv A\,\exp\left[-z\lambda_T/U\right] + (C_o - A)\,\exp\left[\left(\frac{U}{2K} - \sqrt{\frac{U^2}{4K^2} + \frac{\lambda_R}{K}}\right)z\right] \tag{67}$$

$$\text{Boundary conditions: at } t > 0: C = C_1 \text{ at } z = 0 \tag{68}$$

$$C = 0 \text{ at } z = \infty \tag{69}$$

The solution of Equation 66 is

$$C = C_{t=0} + \tfrac{1}{2}(C_1 - C_o)\,\exp\left[\left(\frac{U}{2K} - \sqrt{\frac{U^2}{4K^2} + \frac{\lambda_R}{K}}\right)z\right] \times$$

$$\left\{\operatorname{erfc}\left[\frac{z}{2\sqrt{(Kt)}} - \sqrt{\left(\frac{U^2}{4K} + \lambda_R\right)t}\right] + \right.$$

$$\left. \exp\left[2z\sqrt{\frac{U^2}{4K^2} + \frac{\lambda_R}{K}}\right] \times \operatorname{erfc}\left[\frac{z}{2\sqrt{(Kt)}} + \sqrt{\left(\frac{U^2}{4K} + \lambda_R\right)t}\right]\right\} \tag{70}$$

When a new steady-state has been attained, the Ra-226 concentration as a function of depth becomes

$$C_{t=\infty} = A\,\exp\left[-z\lambda_T/U\right] + (C_1 - A)\,\exp\left[\left(\frac{U}{2K} - \sqrt{\frac{U^2}{4K^2} + \frac{\lambda_R}{K}}\right)z\right] \tag{71}$$

which is analogous to Relationship 64 for the initial steady-state. The concentration–depth profile for a new steady-state, with the Ra-226 concentration at the sediment–water interface taken as $C_1 = 1.6C_0$, is shown in Figure 16 in the curve labelled $t = \infty$. Two curves for transient con-

centrations at 100 and 1000 years after the change in the boundary concentration, computed using Equation 70, are also in Figure 16. After a
time as short as 1000 years, the concentration–depth profile is already
very close to the new steady-state profile; the differences in concentration
between the two curves are 7% and less. In the initial profile, the Ra-226
concentration increases from the sediment–water interface down. Such a
concentration gradient is a prerequisite condition for maintaining diffusional flux of Ra-226 from the sediment into the overlying water. At
the new steady-state, however, the concentration decreases from the
interface down, which indicates that there would be no Ra-226 flux out
of the sediment. The new steady-state profile would be attained in approximately 3000 years; concentration curve for $t = 3000$ is within 1%
of the steady-state concentration. The time is obviously short when
viewed in perspective of the history of oceanic sediments. It may be
verified from Equation 70 for transient concentrations that, in general,
rapid rates of sedimentation (large U) or high diffusivities within the
sediment (large K) would result in a more rapid attainment of a steady-
state. The half-life of the chemical species (1620 yr in the case of
Ra-226) has relatively little effect on the length of time it takes to establish a new steady-state. Even when the migrating species is a stable
nuclide ($\lambda_R = 0$ in Equations 64, 70, and 71), it would take less than
10,000 years for its concentration to come to within 5% of the steady-state
value in the upper 10–20 cm of the sediment. The generality of the arguments may be stressed by pointing out that the time to steady-state
depends on how fast the time-dependent terms (those between the braces
in Equation 70) tend to their limiting values of 2 and 0 as t tends to
infinity. These terms depend on U, K, and λ_R, but not on the chemical
nature of the parent species (*i.e.*, neither on its concentration $C_T{}^\circ$ nor its
decay constant λ_T).

The shortness of time for transition from one stationary concentration
profile to another demonstrates that even in the slowly deposited deep
oceanic sediments it might be difficult to detect near the sediment–water
interface any changes (if such occurred) in the past chemical history of
the ocean.

Appendix

The second-order partial differential equations given in the text of
the paper contain time derivatives of concentration ($\partial C/\partial t$) and terms
containing $\partial C/\partial z$ and C. The solutions of these equations, unless referred
to a literature source, were obtained by the method of Laplace transformation with the aid of standard tables of Laplace transforms. Good
working summaries of the Laplace transformation method as applied to

solution of problems in heat flow and diffusion are in References 22 and 32. Tables of Laplace transforms in References 22, 32, and 45 are given in the form which is particularly convenient for solving equations with constant coefficients, of the type used in this paper.

The functions erf (error function) and erfc (error function complement) appear in many of the solutions given in the paper. These functions appear in the process of integration of terms containing e^{-y^2} (y is some function of K, U, λ, t, and z) and in the process of inverting (with the aid of the tables) the transformed concentration variable $\widetilde{C}$ back to the original concentration C, to be given in the solution as a function of z and t (and the constants). Discussion and mathematical definitions of the error function are given in many texts and, among those listed in the references of this paper, in the Handbook of Mathematical Functions (46), Carslaw and Jaeger (32), and Crank (22). The error function is defined as

$$\text{erf } x = \frac{2}{\sqrt{\pi}} \int_0^x e^{-y^2} dy \tag{72}$$

where y is integration variable, and it may be a function of x. The error function complement is defined as

$$\text{erfc } x = \frac{2}{\sqrt{\pi}} \int_x^\infty e^{-y^2} dy \tag{73}$$

Erf x and erfc x are interrelated,

$$\text{erfc } x = 1 - \text{erf } x \tag{74}$$

For negative argument,

$$\text{erf } (-x) = - \text{erf } x \tag{75}$$

$$\text{erfc } (-x) = 1 + \text{erf } x$$

$$= 2 - \text{erfc } x \tag{76}$$

In many of the solutions given in the paper, the limiting values of concentration C when either z or t approach 0 or infinity may be verified by substitution of the appropriate values of the functions erf and erfc. The values of these functions when the argument is zero, plus- or minus-infinity are:

erf $(0) = 0$	erfc $(0) = 1$
erf $(\infty) = 1$	erfc $(\infty) = 0$
erf $(-\infty) = -1$	erfc $(-\infty) = 2$

Erfc x is a rapidly decreasing function of x. When x increases indefinitely, erfc x tends to zero faster than e^{x^2} tends to infinity. Owing to this, the product e^{x^2} erfc x and, consequently, the product e^xerfc x tend to zero as x increases indefinitely. Products of exponentials and error functions appear in some of the solutions discussed in the paper.

Tables of erf x are available for values of x between 0 and 2.00, in steps of 0.01 (46). Erf x, erfc x, and several related functions have been tabulated for values of x between 0 and 3.0, in steps of 0.05 and 0.1 (22, 32). References to older tables are in Carslaw and Jaeger (32, p. 482).

In this paper, the values of erf x and erfc x were computed from an approximation for erf x given in Ref. 47 for $0 \leqslant x < 3$.

$$\text{erfc } x = 1/(1 + a_1 x + a_2 x^2 + a_3 x^3 + a_4 a^4 + a_5 x^5)^8 \qquad (77)$$

where $a_1 = 0.14112821$ $\qquad\qquad$ $a_4 = -\ 0.00039446$

$\qquad$ $a_2 = 0.08864027$ $\qquad\qquad$ $a_5 = \qquad 0.00328975$

$\qquad$ $a_3 = 0.02743349$

For values of $x \geqslant 3$, the following series was used (32).

$$\text{erfc } x = \frac{e^{-x^2}}{\sqrt{\pi}} \left(\frac{1}{x} - \frac{1}{2x^3} + \frac{3}{2^2 x^5} - \frac{15}{2^3 x^7} + \frac{105}{2^4 x^9} - \frac{945}{2^5 x^{11}} + \frac{10365}{2^6 x^{13}} \right) \qquad (78)$$

Relationships 77 and 78 are easily programmable for use in a digital computer. In combination with Relationships 75 and 76, they allow computation of erfc x between the limits minus- and plus-infinity.

Other forms of series expansion, rational approximations, and methods of interpolation from tables of erf x and erfc x are given in the Handbook of Mathematical Tables (46, pp. 297–9, 304).

Acknowledgment

The material in the part of this paper dealing with the Dead Sea and Lake Tiberias was prepared in 1969 at the Isotope Department, Weizmann Institute of Science, Rehovot, Israel. At that time, I benefited from constructive discussions with Joel R. Gat and Aaron Nir of the Weizmann Institute, and I thank Dr. Gat for communication of unpublished data on Ra-226 in the Dead Sea. For critical reading and discussion of the paper I am indebted to J. Stewart Turner (Cambridge, U.K., and Woods Hole, Mass.), Harmon Craig (La Jolla, Calif.), and the editorial reviewers, Edward D. Goldberg (La Jolla, Calif.) and Derek W. Spencer (Woods Hole, Mass.).

Literature Cited

(1) Riley, G. A., *Theory of Food-Chain Relations in the Ocean,* "The Sea," M. N. Hill, Ed., Vol. 2, p. 438, Interscience, New York, 1963.
(2) Berner, R. A., "An Idealized Model of Dissolved Sulfate Distribution in Recent Sediments," *Geochim. Cosmochim. Acta* (1964) **28**, 1497–503.
(3) Csanady, G. T., "Turbulent Diffusion in Lake Huron," *J. Fluid Mech.* (1963) **17**, 360–84.
(4) Duursma, E. K., "Molecular Diffusion of Radioisotopes in Interstitial Water of Sediments," *Symp. Disposal of Radioactive Wastes into Seas, Oceans, and Surface Waters,* p. 355–71, International Atomic Energy Agency, Vienna, 1966.
(5) Goldberg, E. D., Koide, M., *Rates of Sediment Accumulation in the Indian Ocean,* "Earth Science and Meteoritics," J. Geiss, E. D. Goldberg, Eds., p. 90, North-Holland, Amsterdam, 1963.
(6) Himmelblau, D. M., "Diffusion of Dissolved Gases in Liquids," *Chem. Rev.* (1964) **64**, 527–50.
(7) Hutchinson, G. E., "Treatise on Limnology," 1, p. 480, Wiley, New York, 1957.
(8) Landolt-Börnstein, "Zahlenwerte und Funktionen. Transportphänomene II." 6 Auflage, Band II, Teil 5, Bandteil b. Table 25425, p. 227, Springer, Berlin, 1968.
(9) Lerman, A., Stiller, M., "Vertical Eddy Diffusion in Lake Tiberias," *Verhandl. Intern. Verein. Limnol.* (1969) **17**, 323–33.
(10) Lerman, A., Weiler, R. R., "Diffusion and Accumulation of Chloride and Sodium in Lake Ontario Sediment," *Earth Planet. Sci. Lett.* (1970) **10**, 150–6.
(11) Munk, W. H., "Abyssal Recipes," *Deep-Sea Res.* (1966) **13**, 707–30.
(12) Murthy, R. J., "An Experimental Study of Horizontal Diffusion in Lake Ontario," *Proc. 13th Conf. Great Lakes Res., Intern. Assoc. Great Lakes Res.* (1970) **1**, 477–89.
(13) Neurath, H., "The Investigation of Proteins by Diffusion Measurements," *Chem. Rev.* (1942) **40**, 357–94.
(14) Okubo, A., "A New Set of Oceanic Diffusion Diagrams," Tech. Rept. 38, Chesapeake Bay Inst., 1968. Clearinghouse Fed. Sci. Tech. Inf., Springfield, Va., Doc. AD 675 269.
(15) Stommel, H., "Horizontal Diffusion Due to Oceanic Turbulence," *J. Marine Res.* (1949) **8**, 199–225.
(16) Van Schaik, J. C., Kemper, W. D., Olsen, S. R., "Contribution of Adsorbed Cations to Diffusion in Clay–Water Systems," *Soil Sci. Soc. Am. Proc.* (1966) **30**, 17–22.
(17) Hattersley-Smith, G., Keys, J. E., Serson, H., Mielke, J. E., "Density Stratified Lakes in Northern Ellesmere Island," *Nature* (1970) **255**, 55–6.
(18) Shirtcliffe, T. G. L., "Lake Bonney, Antarctica: Cause of the Elevated Temperatures," *J. Geophys. Res.* (1964) **69**, 5257–68.
(19) Williams, P. M., Mathews, W. H., Pickard, G. L., "A Lake in British Columbia Containing Old Sea-Water," *Nature* (1961) **191**, 830–2.
(20) Frank-Kamenetskii, D. A., "Diffusion and Heat Transfer in Chemical Kinetics," 2nd ed., p. 27–34, Plenum, New York, 1969.
(21) Turner, J. S., "The Influence of Molecular Diffusivity on Turbulent Entrainment Across a Density Interface," *J. Fluid Mech.* (1968) **33**, 639–56.
(22) Crank, J., "The Mathematics of Diffusion," p. 9–61, 121–46, Oxford University Press, Oxford, U.K., 1956.
(23) Lerman, A., "Chemical Equilibria and Evolution of Chloride Brines," *Mineral. Soc. Am. Spec. Publ.* **3**, 291–306.

(24) Lovering, T. S., "Heat Conduction in Dissimilar Rocks and the Use of Thermal Models," *Bull. Geol. Soc. Am.* (1936) **47**, 87–100.

(25) Eckart, C., "Hydrodynamics of Oceans and Atmospheres," p. 57–71, Pergamon, New York, 1960.

(26) Kato, H., Phillips, O. M., "On the Penetration of Turbulent Layer into Stratified Fluid," *J. Fluid Mech.* (1969) **37**, 643–55.

(27) Turner, J. S., "A Note on Wind Mixing at the Seasonal Thermocline," *Deep-Sea Res.* (1969) *Suppl. to Vol.* **16**, 297–300.

(28) Neev, D., Emery, K. O., "The Dead Sea, Depositional Processes and Environments of Evaporites," *Bull. Israel Geol. Surv.* (1967) **41**, 1–147.

(29) Neumann, J., "Tentative Energy and Water Balances for the Dead Sea," *Bull. Res. Council Israel* (1958) **7G**, 137–63.

(30) McEwen, G. F., "A Mathematical Theory of the Vertical Distribution of Temperature and Salinity in Water Under the Action of Radiation, Conduction, Evaporation, and Mixing Due to the Resulting Convection," *Bull. Scripps Inst. Oceanog.* (1929) *Tech. Ser.* **2**, 197–306.

(31) Gat, J. R., Gilboa, G., Isotope Dept., Weizmann Institute of Science, Rehovot, Israel, "Radium-226 in the Dead Sea," personal communication, 1969.

(32) Carslaw, H. S., Jaeger, J. C., "Conduction of Heat in Solids," 2nd ed., 50–496, Oxford University Press, Oxford, U.K., 1959.

(33) Redfield, A. C., Ketchum, B. H., Richards, F. A., *The Influence of Organisms on the Composition of Sea-Water,* "The Sea," M. N. Hill, Ed., Vol. 2, p. 26, Interscience, New York, 1963.

(34) Bowen, V. T., Noshkin, V. E., Volchok, H. L., Sugihara, T. T., "Strontium-90: Concentrations in Surface Waters of the Atlantic Ocean," *Science* (1969) **164**, 825–7.

(35) Broecker, W. S., "Radioisotopes and the Rate of Mixing Across the Main Thermoclines of the Ocean," *J. Geophys. Res.* (1966) **71**, 5827–36.

(36) Münnich, K.-O., Roether, W., "Transfer of Bomb C-14 and Tritium from the Atmosphere to the Ocean. Internal Mixing of the Ocean on the Basis of Tritium and C-14 Profiles," *Symp. Radioactive Dating and Methods of Low Level Counting,* p. 93, International Atomic Energy Agency, Vienna, 1967.

(37) Rooth, C., Ostlund, H. G., "Tracing the Oceanic Tritium Transient," Tech. Rept., Univ. of Miami, Rosenstiel School of Marine and Atmospheric Sciences, 1970, 1–27.

(38) Sverdrup, H. U., Johnson, M. W., Fleming, R. H., "The Oceans," p. 161, Prentice-Hall, Englewood Cliffs, N. J., 1942.

(39) Wyrtki, K., "The Oxygen Minima in Relation to Ocean Circulation," *Deep-Sea Res.* (1962) **9**, 11–28.

(40) Lerman, A., "Sea Water—Geochemical Balance," R. W. Fairbridge, Ed., "Encyclopedia of Earth Sciences," 4A, Van Nostrand Reinhold, New York, 1971 (in press).

(41) Koczy, F. F., "Natural Radium as a Tracer in the Ocean," *Proc. 2nd U.N. Intern. Conf. Peaceful Uses of Atomic Energy* (1958) **18**, 351–7.

(42) Craig, H., "Dissolved Gases, Deuterium, Oxygen, and Carbon Isotopes in the Ocean," *Bat-Sheva Seminar on Marine Geochemistry,* Intern. Summer School, Weizmann Institute of Science, Rehovot, Israel, June 1969 (unpublished).

(43) Craig, H., Scripps Institution of Oceanography, La Jolla, Calif., "Rates of the Ra-226 Production in the Pacific," personal communication, 1971.

(44) Bernat, M., Goldberg, E. D., "Thorium Isotopes in the Marine Environment," *Earth Planet. Sci. Lett.* (1969) **5**, 308–12.

(45) Nixon, F. E., "Handbook of Laplace Transformations," 2nd ed., Prentice-Hall, Englewood Cliffs, N. J., 1965, 260 pp.

(46) Abramowitz, M., Stegun, I. A., Eds., "Handbook of Mathematical Functions with Formulas, Graphs and Mathematical Tables," p. 279–329, National Bureau of Standards, Washington, D. C., 1966.
(47) Hastings, C., "Approximations for Digital Computers," p. 186, Princeton University Press, Princeton, N. J., 1955 (*also cited in* Ref. *46,* p. 299).

RECEIVED May 27, 1970.

Rates of Physical and Chemical Processes in a Carbonate Aquifer

WILLIAM BACK and BRUCE B. HANSHAW

U. S. Geological Survey, Washington, D. C. 20242

For much of the Tertiary carbonate aquifer system of Florida, the velocity of ground-water flow ranges from 2 to 8 meters per year. Water in the recharge area is undersaturated with respect to both calcite and dolomite. As the water moves downgradient, it attains equilibrium with respect to calcite in about 4000 carbon-14 years and with respect to dolomite in about 15,000 carbon-14 years. Combining the amount of entropy produced from chemical and physical processes with carbon-14 ages provides an approximation of the total entropy (excluding thermal energy from heat flow) production for the system as a function of time and distance. The values range from about −2 to 7 mcal/kg/°K/1000 years for various flow paths of about 100 km.

A general reference base for irreversible processes is provided by entropy production which serves as a unifying concept relating changes in both physical and chemical energy. Distribution of entropy production provides an integrating variable for use in evaluating the relative importance of physical and chemical processes at points within a system or between two hydrologic systems.

Because the concept of entropy is generally not familiar to hydrologists, a brief introduction is probably in order. A thorough and rigorous explanation can be obtained from standard works such as those by Fast (*1*), Fitts (*2*), Katchalsky and Curran (*3*), Klotz (*4*), Lewis and Randall (*5*), and Prigogine (*6*). A statement of the second law of thermodynamics is generally used as a definition of entropy of a system as follows: $dS \geqslant DQ/T$, where dS is an infinitesimal change in entropy for an infinitesimal part of a process carried out reversibly, DQ is the heat absorbed, and T is the absolute temperature at which the heat is absorbed. In one sense, entropy is a mathematical function for the term

DQ/T, which is an exact differential, whereas DQ alone cannot be integrated without having a path specified (4, p. 101). That is, DQ/T is both
an extensive variable and a thermodynamic function and merits a symbol
and name—i.e., entropy, which comes from the Greek word meaning
"evolution."

The second law of thermodynamics is often stated to be the law of
dissipation or degradation of energy; however, this can lead to confusion
because it seems to violate the first law of thermodynamics, a statement
of conservation of energy. When the second law is stated in the above
form, it is really referring to the degradation of the "useable" energy of a
system. Entropy is therefore an indication of the degradation of a system
or an index of the exhaustion of a system (4, p. 130).

It follows logically that the combination of all spontaneous reactions
within a natural system will tend to increase the entropy of that system,
and this is the basis for the statement that entropy of the universe is
striving toward a maximum. Although energy and entropy are expressed
in somewhat similar units, calories per mole for energy and calories per
mole per degree for entropy, confusion arises if they are thought of as
having similar attributes or characteristics. As Klotz (4, p. 129) points
out, one can think of energy as being a kind of material fluid, and hence
it flows from one area to another and is conserved. Entropy, on the other
hand, must be viewed as an index of condition or character rather than
as the measure of content of some imaginary fluid and is the index of
capacity for spontaneous change. Entropy summarizes in a concise form
the possible ways in which the variables of temperature, pressure, and
composition may change in natural processes.

One of the fundamental tasks required to achieve the ultimate goal
of hydrogeology is to understand the controls on energy distribution and
transformation within an aquifer system. If this is accepted, it then becomes the hydrologists' role to bring together into one concept the fluxes
and forces of the chemical reactions, of the hydrodynamic flow paths,
and of heat. This idea was clearly articulated and developed by G. B.
Maxey (7, p. 145), who stated in part:

"Aquifer systems have been studied by three separate methods of
analysis: (1) hydrodynamic, utilizing a distributed potential system; (2)
hydrochemical, using parameters of water quality; and (3) hydrothermal,
using distribution and gradients of temperature. The various approaches
have been dictated largely by the specialized training and experience of
the individual research worker. However, the complexity of present
hydrologic problems now requires bringing together the various aspects
into a single concept of a functioning system."

It follows that one of the fundamental objectives of hydrogeochem-
istry is to evaluate the relative significance of various processes that
control the total energy distribution and energy dissipation within a hy-

drologic system. Classical "thermostatics" can provide only a partial
description of the functioning of a hydrogeochemical system, and it is
necessary to apply the principles of irreversible or nonequilibrium thermodynamics.

In the functioning of a carbonate aquifer, rainfall infiltrates through
the soil zone, becomes charged with carbon dioxide, moves to the water
table, dissolves soluble minerals of the aquifer, increases in chemical
concentration, and continues to move to deeper parts of the aquifer, eventually to discharge to the ocean. All of these chemical and physical processes are irreversible reactions and can be thoroughly understood only
by the application of principles of irreversible thermodynamics. The
processes and reactions could be formulated and expressed in energy
terms, but it intuitively appeared more simple to us to bring together
the products of these processes through the concept of entropy rather
than through an energy function.

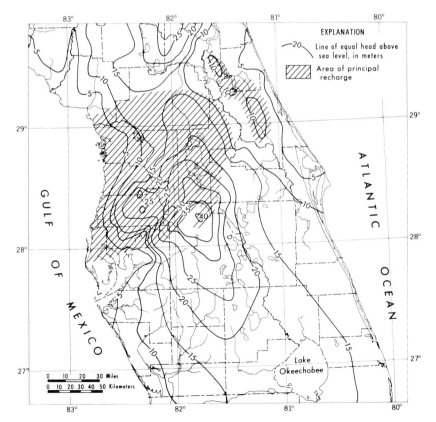

Figure 1.   Principal artesian aquifer of central Florida, showing area of
major recharge (after Ref. 9, plate 12)

Hydrogeology

Certain principles of irreversible thermodynamics can be applied by considering the interrelations between geology, ground-water flow pattern, and chemical character of water in the Floridian peninsula. The principal artesian aquifer of Florida consists chiefly of Tertiary limestone, with minor amounts of dolomite, and ranges in age from middle Eocene to middle Miocene. It is one of the most extensive limestone aquifers in the United States. The Tertiary limestones crop out in north-central Florida and in a broad belt extending from western Florida through southeastern Alabama, Georgia, and southeastern Carolina, approximately paralleling the Fall Line. The Ocala Limestone of late Eocene age is one of the most productive water-bearing formations of the principal aquifer (8, p. 31).

Figure 1 shows the height of the energy surface in meters above sea level. Two mounds tend to dominate the ground-water flow of central Florida: one near the center of the map that is 40 meters above sea level and another smaller one to the west that is about 25 meters above sea level. The general pattern of flow is primarily down the potentiometric gradient and perpendicular to the contours.

Also shown is the area of principal recharge. North of the two mounds, the overlying sediments that form the confining bed are thin to nonexistent, and because of exposed limestone in this area, a large amount of recharge occurs; however, the potentiometric surface is low owing to rapid discharge of the water. This is a region in which a great deal of water is discharged through many springs, such as Silver and Rainbow Springs, that exist in the area of the ground-water saddle formed by the central mound and a potentiometric high north of the study area.

Although the potentiometric surface has essentially the same gradient and shape north and south of the mounds, less recharge occurs in southern parts of the elongated dome than in the northern part because of a thicker confining bed and lower transmissivity of the aquifer to the south. Water that flows southward discharges upward through the confining bed and also to the ocean and gulf. The maximum gradient of the potentiometric surface of central Florida is about 2.5 meters per kilometer with an average gradient of about 1 meter per kilometer.

The ground water of central Florida comprises one major hydrologic system, and it has recently been shown that a geochemical system is coexistent with the hydrologic system (*10, 11*). Depths of wells sampled during this study range from about 100 to 500 meters. A body of salt water that underlies the entire Florida peninsula ranges in depth from near sea level at parts of the shoreline to about 700 meters in central Florida. The interface between the fresh water and salt water forms one of the boundaries of the fresh-water system. Geochemical mapping, including distribution of chloride, sulfate, calcium, magnesium, and carbon-14 concentrations, shows a systematic pattern of increase downgradient. It was concluded that, although the wells have a range of total depths and open intervals, they are sampling parts of the same hydrologically-connected geochemical system (*11*).

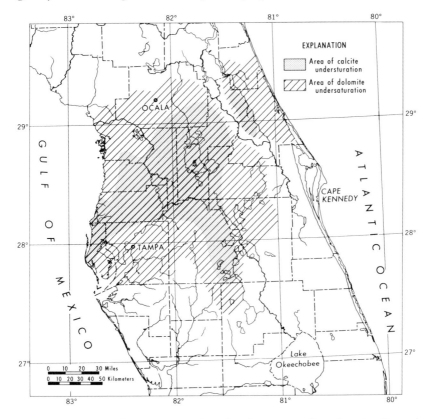

Figure 2. Principal artesian aquifer, showing areas of undersaturation of ground water with respect to calcite and dolomite

Chemical Reactions

In the carbonate aquifer system of central Florida, two major controls on the chemical character of the water are solution of calcite and of dolomite. One way to evaluate the significance of these reactions as chemical controls is to determine the departure from equilibrium of the water with respect to each of the minerals. To calculate departure from equilibrium, solubility products of $10^{-8.35}$ and 2×10^{-17} were used for calcite and dolomite, respectively. The departure from equilibrium with respect to calcite is shown in Figure 2. The area of undersaturation coincides closely with the area of major recharge (Figure 1). The equilibrium boundary outlines the elongated dome on the potentiometric map, which indicates that some water is recharged into the aquifer along the top of the dome and is thereby lowering the amount of saturation in this area.

A preliminary map of departure from equilibrium with respect to dolomite is also shown in Figure 2. The area of undersaturation is larger for dolomite than for calcite. Because the dolomite value is exactly one half that of calcite, it follows that in order for a water that is in equilibrium with calcite to become saturated with respect to dolomite, it is only necessary for the magnesium concentration to equal the calcium concentration (12, 13, 14). The area of recharge and the area of highest potentiometric surface show that the water is undersaturated with respect to dolomite; downgradient, it progressively attains equilibrium with dolomite and eventually becomes supersaturated.

In making thermodynamic calculations to determine departure from equilibrium, thermodynamic data for pure stoichiometric calcite and dolomite were used. However, mineralogic and x-ray examination of aquifer material has shown that the calcite may have several mole percent magnesium; the dolomite that occurs in the system is generally calcium-rich (13). Therefore, both of these minerals in the natural state have a higher free energy and hence a somewhat higher solubility than the pure minerals. Thus, part of the supersaturation that we have calculated may be more apparent than real.

Rates of Flow and Chemical Reactions

For the past several years, we have been working to evaluate the radiocarbon technique for dating ground water; that is, to determine the amount of time the water has been out of contact with the atmosphere (15, 16, 17). This is done by means of the carbon-14 activity of the dissolved carbonate species. Results of part of this work give the age of water as a function of position in the aquifer system. In the recharge area, there are waters of mixed origin, and the age varies according to the amount of mixing of exceedingly young water with somewhat older water. However, downgradient from the area of principal recharge, the age of the water increases in a systematic manner. Thus, by combining results from radiocarbon concentration with changes in the chemical and physical parameters of the system, rates of chemical and physical processes which occur within a system may be derived. Within the recharge area, the maximum apparent age of the mixed water is approximately 5000 years. Downgradient from the recharge area, the water increases to approximately 30,000 years before present, which is the oldest age found in that part of the aquifer system.

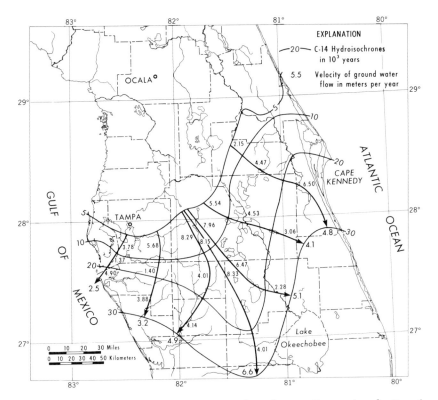

Figure 3. Residence time of water within the aquifer and velocity of ground-water flow; values at the arrow tips are averages for entire flow paths

The values on the flow lines in Figure 3 were calculated from radio-carbon dates and indicate velocities in meters per year for various segments along particular flow paths, which are shown by the heavy lines with arrows. The average values for the entire path range from about 2.5 meters per year to 6.5 meters per year. Along short reaches, the range of velocities is about 1.5 to 8.5 meters per year.

In addition to estimating velocity of ground-water flow, carbon-14 concentrations permit estimation of the rate of solution and precipitation of carbonate minerals. The aquifer is composed of approximately 2/3 calcite and 1/3 dolomite distributed throughout the section. Saturation with respect to calcite occurs rather rapidly, and it is only in areas of principal recharge that undersaturated waters are generally found (Figures 1 and 2). However, the kinetics of dolomite formation and dissolution are quite slow, and the area of undersaturation extends farther downgradient than does the area of calcite undersaturation. By combining age of water from Figure 3 with saturation boundaries of calcite and dolomite from Figure 2, an approximation can be obtained for the time required for water to become saturated with these minerals. The results show that water attains equilibrium with respect to calcite in about 4000 carbon-14 years and with respect to dolomite in about 15,000 carbon-14 years.

Rate of Entropy Production

Thus far in this discussion, these fluid-filled formations of Florida have been considered as part of a geologic system, a hydrologic system, and as a coexisting geochemical system. It would seem desirable to combine the results of the various natural processes into one unifying concept.

The input to the combined system occurs on the potentiometric highs in the form of rainfall containing minor amounts of total dissolved solids. Initial changes in water chemistry occur within the soil zone where the water is charged with large amounts of CO_2 gas. This CO_2-rich water percolates into the ground-water system where the CO_2 attacks the carbonate minerals. This is an irreversible chemical process whereby the CO_2 in the water reacts with the minerals and brings them into solution.

Likewise, simple gravitational movement of water from potentiometric highs to oceanic base level is an irreversible physical process which produces a loss of potential energy. The basis for evaluating energy distribution of a ground-water system is the potential theory best explained in a classical paper by Hubbert (8). Potential is composed of the sum of two terms, a gravitational potential energy and a pressure energy. Potential is equal to the work required to transform a unit of mass of fluid from an arbitrarily chosen standard state to the state at the point under consideration (18, p. 797–8). For the standard state, it is convenient to use an elevation of zero, a pressure of 1 atm, and a velocity of zero. Potential, ϕ, for ground water can be expressed as follows (19, p. 1959)

$$\phi = gz + \frac{P}{\rho} \tag{1}$$

where g is acceleration owing to gravity, z is elevation, P is gage pressure, and ρ is density. In almost all instances, the kinetic energy of flowing ground water is negligible because of the low velocities of flow. "Total head" as used by hydrologists is related to "potential" by the expression $\phi = gh$, where h is head. Although it may intuitively seem that the total potential could include terms other than gravity and pressure to reflect chemical and thermal energy changes, Hubbert's potential concept is restricted to mechanical energy only and is so used in this paper.

Head is an intensive state variable and is independent of the process that produces a change in head. Thus, to calculate energy loss from flow, a known reversible process can replace the unknown irreversible process.

Head loss thus represents useable energy lost from the system and may be thought of as a measure of change in entropy. Likewise, the irreversible process of dissolving minerals in the aquifer system has an entropy change associated with it. One way in which the physical and chemical processes within such a system can be compared is through use of entropy concepts.

Because none of the potential energy of a ground-water system is converted to kinetic energy, all the energy is transformed to heat which is absorbed by the system. Therefore, changes in entropy owing to head loss, which can be treated as a reversible process, are obtained by calculating the changes in potential energy associated with flow through the system. This provides a determination of minimum entropy production caused by change in altitude. When it becomes possible to separate all sources of heat to the system (earth heat flow, solar radiation, heats of solution and precipitation, and frictional heat production), the additional

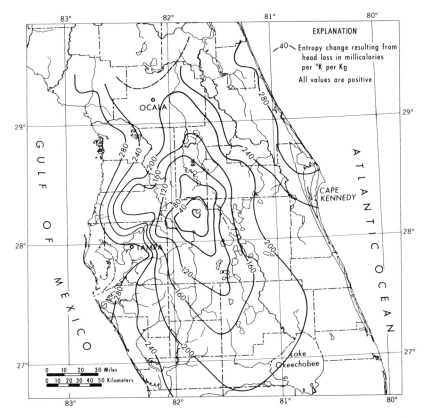

Figure 4.   Distribution of entropy change resulting from head loss within the aquifer

entropy production can be combined with the minimum to obtain total entropy produced from physical processes.

Kilogram-meters can be converted readily to millicalories (mcal) per kilogram as follows

$$1 \text{ kg-meter} = 2.34 \times 10^3 \text{ mcal} \tag{2}$$

To convert to entropy between any two points

$$\text{Entropy [mcal/kg/°K]} = \frac{\Delta h \text{[meters]} \times \dfrac{2.34 \times 10^3 \text{ mcal}}{\text{kg-meters}}}{\text{°K}} \tag{3}$$

where Δh is loss in head between two points. The temperature of ground water in this system ranges from about 23°C in the recharge area to a maximum of about 28°C in the deepest part of the system. A simple sensitivity test can be made as follows: If $T = 298.16° \pm 2°K$ and head $= 30$ m, then for $T = 296.16°K$ (23°C), the calculated entropy is 237.0 mcal; for $T = 298.16°K$ (25°C), the calculated entropy is 235.5 mcal; and for $T = 300.16°K$ (27°C), the calculated entropy is 234.0 mcal. This suggests that over the narrow range of observed temperatures, the entire system may be approximated by assuming an isothermal system at 25°C. For this preliminary study, the assumption of an isothermal system permits neglecting thermal energy transfer from sources mentioned above. This topic will be rigorously evaluated in a subsequent study.

The results of calculating entropy production from head values ranging between elevations of 40 meters to about sea level are shown on Figure 4. Note that the high point on the potentiometric surface is designated as having a zero entropy level. This is the input boundary of the system, and by our definition, the entropy of the water attributed to position is zero at this point. Therefore in order to depict the entropy increase attributed to downgradient flow, the equation was modified to

$$\text{Change in entropy} = (h_{max} - h_i) \frac{2.34 \times 10^3 \text{ mcal}}{\text{°K}} \tag{4}$$

As the water flows down the potentiometric surface, entropy is progressively produced by this physical process to about 300 mcal/kg/°K.

The mineralogy of the Floridian aquifer consists of approximately 65% calcite and 34% dolomite, with minor amounts of gypsum scattered through the formation. Gypsum may be locally abundant in some parts of the aquifer system. Therefore, only three chemical reactions need be considered to describe the major chemical changes in this system. First is the solution of calcite by means of water and soil CO_2 gas; second is the solution of dolomite, also by means of water and soil CO_2 gas; and

Table I.   Standard Entropy Values[a]

	$S°$, $Cal/°K/Mole$
Ca^{2+} aq.	-13.2
Mg^{2+} aq.	-28.2
HCO_3^-	22.7
CO_2 aq.	29.0
H_2O	16.716
SO_4^{2-}	4.1
$CaCO_3$ [calcite]	22.2
$CaMg(CO_3)_2$ [dolomite]	37.09[b]
$CaSO_4 \cdot 2H_2O$ [gypsum]	46.36

[a] Values from Rossini *et al.* (*20*) except where noted.
[b] Value for dolomite from Stout *et al.* (*21*).

third is solution of gypsum ($CaSO_4 \cdot 2H_2O$) to form calcium ions, sulfate ions, and water. Although sulfate reduction occurs within the system, the simplifying assumption has been made that the decrease in sulfate concentration is not significant for these calculations. In order to determine the chemical entropy production of the system, the entropy of each of these three reactions was calculated using values in Table I as follows:

Calcite

$$CaCO_3 + H_2O + CO_{2aq} = Ca^{2+} + 2HCO_3^- \tag{5}$$

$$\Delta S_{reaction} = -35.7 \text{ cal/°K/mole}$$

Dolomite

$$CaMg(CO_3)_2 + 2H_2O + 2CO_{2aq} = Ca^{2+} + Mg^{2+} + 4HCO_3^- \tag{6}$$

$$\Delta S_{reaction} = -79.9 \text{ cal/°K/mole}$$

Gypsum

$$CaSO_4 + 2H_2O = Ca^{2+} + SO_4^{2-} + 2H_2O \tag{7}$$

$$\Delta S_{reaction} = -22.1 \text{ cal/°K/mole}$$

The change in entropy at any point, i, in the system owing to the above three equations is given by the following relationships

$$\Delta S_{i,calcite} = \Delta S_{R,calcite} [m_{Ca} - (m_{Mg} + m_{SO_4})] \tag{8}$$

$$\Delta S_{i,dolomite} = \Delta S_{R,dolomite} \times m_{Mg} \tag{9}$$

$$\Delta S_{i,gypsum} = \Delta S_{R,gypsum} \times m_{SO_4} \tag{10}$$

$$\Delta S_{chem} = \Delta S_{calcite} + \Delta S_{dolomite} + \Delta S_{gypsum} \tag{11}$$

where ΔS is change of entropy, m is molality, subscript i is at any site, subscript R is for reactions identified by subscripts calcite, dolomite, and

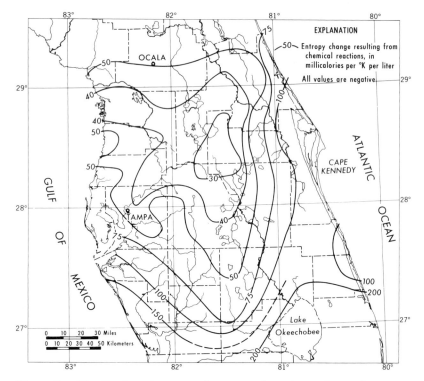

Figure 5.   Distribution of entropy change resulting from chemical reactions within the aquifer

gypsum, ΔS_{chem} is total entropy change resulting from the chemical reactions studied, and

$$\Delta S_{reaction} = \Delta S_{products} - \Delta S_{reactants} \qquad (12)$$

These calculations are based on the approximation that the amount of entropy produced per mole of substance dissolved is linear for concentrations ranging from infinite dilution to saturation with respect to these three phases.

The change in entropy owing to the solution of calcite is given by Equation 8. For these calculations, entropy of the reaction was multiplied by molality of calcium ion less molality of magnesium and sulfate ions. The reason for this subtraction is that, assuming no other source for magnesium, the total amount of dolomite that has gone into solution may be approximated by the amount of magnesium at any point within the aquifer. Therefore, because congruent solution of stoichiometric dolomite means that the amount of magnesium from dolomite equals the amount of calcium from dolomite, the contribution of calcium from dolomite solution can be determined by subtracting magnesium in Equation 8. For nonstoichiometric dolomites, an error is introduced in this calculation equal to the percentage that the dolomite is magnesium-rich or magnesium-deficient. Likewise, the amount of gypsum going into solution may be approximated by the amount of sulfate; subtracting sulfate is equivalent to accounting for the calcium contributed to the aqueous phase by the dissolution of gypsum.

In areas where supersaturation of carbonate minerals exists, it was found that the total amount of magnesium plus sulfate was greater than the amount of calcium. In that case, we concluded that calcite was precipitating, and this equivalent amount of entropy change was added to the total chemical entropy production.

The essence of the above discussion is that, because the change of entropy of a reaction is based on mole concentrations, it is necessary only to determine what fraction of a mole of a particular mineral has dissolved in order to estimate the entropy change for that reaction at any given site. Another point that needs clarification is that the negative entropy change for these reactions would appear to violate the second law in that a natural spontaneous reaction is not tending to maximize entropy. However, the second law refers to the universe, or at least to a total system, and not to a single reaction or process. Therefore, it is acceptable for any individual reaction or process to have a negative entropy change.

In addition, every reaction which produces entropy generates disorder within the reacting substances. In the case under consideration, the entropy change is in the H_2O, in the small number of ions in solution, and in the interactions between H_2O and the ions. H_2O is the dominant component, and because it has a high degree of disorder, solution of carbonate minerals tends to order the H_2O, primarily by hydration of carbonate and calcium ions, thereby providing for a decrease in entropy in that part of the system.

The results of these calculations and their distribution are shown in Figure 5. The change in entropy levels arising from chemical reactions ranges from about -30 to -200 mcal/liter/°K. The spatial distribution of the changes in chemical entropy and mechanical entropy is in excellent agreement, and surprisingly, the values are nearly equal but with opposite signs. The negative sign on the entropy change resulting from chemical reactions indicates that part of the heat absorbed from energy transformation resulting from ground-water flow is available to dissolve carbonate minerals and thereby may be converted to chemical energy.

A summation of the entropy produced from chemical and physical processes is shown in Figure 6 to range from 0 to 200 mcal/kg/°K. In the recharge area, the entropy production from head loss is greater than the negative entropy change resulting from chemical reactions. Far down-gradient where the flow is less vigorous and the water is supersaturated with respect to carbonate minerals, the values are equal and produce a "total" entropy of zero or even of negative values. The negative values indicate an additional source of energy not accounted for in these preliminary calculations. Possibilities include various thermal sources and sinks and inadequate accounting for precipitation of carbonate minerals, which would provide for positive entropy change.

In Figure 7, the hydroisochrones determined from carbon-14 dating (8) and the entropy map (Figure 6) are combined, which permits the calculation of rates of entropy production along particular flow paths within the hydrogeochemical system. The rate of entropy production, in units of mcal/kg/1000 years, ranges from −11 to 25 for various segments and from −2 to 7 for flow-line averages.

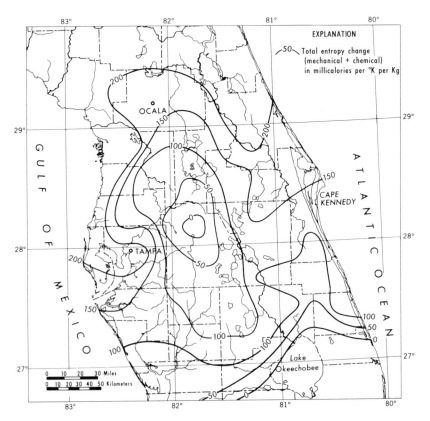

Figure 6. Distribution of entropy change resulting from combination of physical and chemical processes

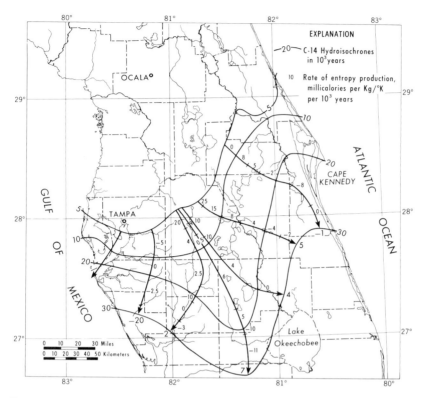

Figure 7. Rate of entropy production resulting from both chemical and physical processes; values at tips of arrows are averages for the entire flow path

Concluding Remarks

We think that the approach discussed here will provide a greater understanding of the combined physical and chemical functioning of an aquifer system and hopefully will have transfer value to other natural systems. We expect that further application and refinement of methods to study the distribution of entropy production will provide an integrating variable for use in evaluating the relative importance of physical and chemical processes at points within a system or between two hydrologic systems. The next step is to develop techniques that will provide a method for predicting chemical as well as physical changes resulting from natural and artificial stresses imposed on an aquifer system. Our working hypothesis is that our data represent the solution to a set of simultaneous transport equations. The next task is to formulate the transport equations in a manner compatible with field- and laboratory-measured parameters as a verification of the model provided by irreversible thermodynamics.

Acknowledgment

We have benefited greatly from discussions with J. D. Bredehoeft, M. K. Hubbert, V. T. Stringfield, B. F. Jones, Thomas Maddock, III, F. J. Pearson, Jr., R. A. Robie, and A. H. Truesdell of the U. S. Geological Survey; J. V. A. Sharp of the University of Nevada; P. A. Domenico of the University of Illinois; and with G. M. Lafon of the State University of New York at Binghamton. We are particularly grateful to Thomas Maddock, III; V. T. Stringfield; J. D. Bredehoeft, and Donald Langmuir, Penn State University, for their reading and criticism of the manuscript. We are deeply indebted to Meyer Rubin for his cooperation in the field on many occasions and for providing carbon-14 analyses used in this study. All the above-named individuals are in no way responsible for whatever omissions or errors may be present nor does this acknowledgment necessarily imply their complete agreement with the approach and procedures used in this study. Part of this work was supported by Army Research Office grant No. EN-5830.

Literature Cited

(1) Fast, J. D., "Entropy," McGraw-Hill, New York, 313 pp., 1962.
(2) Fitts, Donald D., "Nonequilibrium Thermodynamics," McGraw-Hill, New York, 173 pp., 1962.
(3) Katchalsky, A., Curran, Peter F., "Nonequilibrium Thermodynamics in Biophysics," Harvard University Press, Cambridge, Mass., 248 pp., 1967.
(4) Klotz, Irving M., "Chemical Thermodynamics," W. A. Benjamin, New York, 468 pp., 1964.
(5) Lewis, Gilbert Newton, Randall, Merle, "Thermodynamics," revised ed., Kenneth S. Pitzer and Leo Brewer, Eds., McGraw-Hill, New York, 723 pp., 1961.
(6) Prigogine, I., "Introduction to Thermodynamics of Irreversible Processes," Wiley, New York, 147 pp., 1967.
(7) Maxey, George B., "Hydrogeology Today," *Geol. Soc. Am., Annual Meetin, Atlantic City, 1969*, Abstracts, p. 145.
(8) Stringfield, V. T., "Artesian Water in Tertiary Limestone in the Southeastern States," *U.S. Geol. Surv. Profess. Papers* (1966) **517.**
(9) Stringfield, V. T., "Artesian Water in the Florida Peninsula," *U.S. Geol. Surv. Water-Supply Papers* (1936) **773-C.**
(10) Back, William, "Preliminary Results of a Study of Calcium Carbonate Saturation of Ground Water in Central Florida," *Bull. Intern. Assoc. Sci. Hydrology* (1963) **8** (3), 43–51.
(11) Back, William, Hanshaw, Bruce B., "Comparison of Chemical Hydrogeology of the Carbonate Peninsulas of Florida and Yucatan," *J. Hydrology* (1970) **10** (4), 330–68.
(12) Barnes, Ivan, Back, William, "Dolomite Solubility in Ground Water," *U.S. Geol. Surv. Profess. Papers* (1964) **475-D,** D179–80.
(13) Hanshaw, Bruce B., Back, William, Deike, Ruth G., "A Geochemical Hypothesis for Dolomitization by Ground Water," *Soc. Econ. Geol.,* in press.
(14) Hsu, K. J., "Solubility of Dolomite and Composition of Florida Ground Waters," *J. Hydrology* (1963) **1** (4), 288–310.

(15) Hanshaw, Bruce B., Back, William, Rubin, Meyer, "Radiocarbon Determinations for Estimating Ground-Water Flow Velocities in Central Florida," *Science* (1965) **148**, 494–5.

(16) Hanshaw, Bruce B., Back, William, Rubin, Meyer, Friedman, Irving, "The Evaluation and Application of ^{14}C Dating of Ground Water," in preparation.

(17) Pearson, F. J., Jr., White, D. E., "Carbon-14 Ages and Flow Rates of Water in Carrizo Sand, Atascosa County, Texas," *Water Resources Res.* (1967) **3**, 251–61.

(18) Hubbert, M. King, "The Theory of Ground-Water Motion," *J. Geol.* (1940) **49** (8), part 1, 785–944 (reprinted *in* "The Theory of Ground-Water Motion and Related Papers," Hafner Publishing Co., New York, 1969).

(19) Hubbert, M. King, "Entrapment of Petroleum under Hydrodynamic Conditions," *Bull. Am. Assoc. Petrol. Geol.* (1953) **37** (8), 1954–2026 (reprinted *in* "The Theory of Ground-Water Motion and Related Papers," Hafner Publishing Co., New York, 1969).

(20) Rossini, F. D., Wagman, D. D., Evans, W. H., Levine, Samuel, Jaffe, Irving, "Selected Values of Chemical Thermodynamic Properties," *Natl. Bur. Std. Circ. 500*, 1952.

(21) Stout, J. W., Robie, R. A., "Heat Capacity from 11° to 300°K. Entropy and Heat of Formation of Dolomite," *J. Phys. Chem.* (1963) **67**, 2248–52.

RECEIVED October 8, 1970. Publication authorized by the Director, U. S. Geological Survey.

4

Silica Variation in Stream Water with Time and Discharge

VANCE C. KENNEDY

U. S. Geological Survey, Menlo Park, Calif. 94025

Silica concentration in the Mattole River of northern California varies in a consistent manner during storm runoff as water from various sources enters the stream. Silica is more concentrated in water which seeps through the soil (subsurface flow) than in overland flow or in groundwater. Thus, during a stream rise, silica decreases initially while overland flow comprises much of the streamflow and then increases as subsurface flow becomes the major component of streamflow. With decreasing discharge, groundwater becomes an increasing proportion of streamflow, and silica concentration slowly decreases. Correlation between silica and stream discharge or specific conductance is poor during storm runoff. Data from other streams suggest that the pattern of silica variation observed in the Mattole River is present elsewhere.

Silica comprises a significant fraction of the dissolved solids in stream waters; however, there is relatively little detailed information on the variations in silica concentration with time and stream discharge, and few attempts have been made to explain such time-dependent variations as have been observed. The concentration of most dissolved constituents in stream water decreases with increasing discharge but, as Davis (*1*) has pointed out, the silica content is less variable than that of any other of the major dissolved constituents. This means that the proportion of silica in the dissolved solids becomes greater with increasing discharge and implies that the rate of silica release from soils increases more rapidly than that of the other dissolved solids during storm runoff. The fact that silica shows little or no correlation with discharge or specific conductance suggests that the controls of silica concentration in stream water are complex. Chemical-equilibrium models appear inadequate to explain the behavior of silica in natural waters during storm periods.

This report presents detailed information on the variation of silica with time, changes in stream discharge, and specific conductance in the water of the Mattole River of northern California and gives results of soil leaching studies that help in understanding the silica variations in the Mattole River. These results and data from other streams indicate that silica variations during the wet season may be related to varying rates of chemical reactions in the soil zone. Silica concentration in subsurface runoff waters is related to the length of time of contact with the soil, the soil:water ratio, and the rainfall history prior to the time of sampling.

Previous Work. Information pertinent to this investigation has been drawn from three types of data: the large body of published stream-water analyses, the basic work on solubilities of crystalline and amorphous silica and of various silicates, and a group of recent reports on dissolved silica in soil waters.

Palmer (2) thought that the more alkaline waters favored the retention of silica in solution in streams of the Piedmont Plateau and Gulf Coast. He also noted that the proportion of silica in the dissolved solids was helpful in comparing the chemistry of various stream waters. Hendrickson and Krieger (3) concluded that silica and specific conductance were poorly related for several streams in Kentucky. Davis (1) made a comprehensive study of silica in ground and surface waters, using many hundreds of published analyses. From these he concluded that silica in ground water—and, hence, that in stream water at low stages—is primarily related to the rocks and minerals contacting the water. He found no marked influence of pH, salinity, climatic regions, vegetation, or temperature on silica concentration. Storm runoff appeared to acquire most of its silica within a few days. Evidence for this was the fact that silica concentrations in stream water remained relatively constant during periods of high discharge despite the decrease in dissolved solids. He mentions several possible explanations for this, one of which was that water travelling through the upper part of the soil profile might contain appreciable silica leached from the soil and comprise much of the runoff during and shortly after storms. However, he felt that this method of obtaining silica required that silica be leached from the soil while other constituents were dissolved less rapidly. An explanation of this process posed difficulties. In a later study, Davis (4) found that surface runoff from rain can acquire 1–3 mg/liter silica during the first few minutes after rain contacts the soil. Feth and others (5) also showed that silica is released rapidly (that is, in a few days or so) to percolating, slightly acidic, snowmelt waters which supply ephemeral springs in the high Sierra Nevada.

One approach to interpreting the silica content of natural waters is to determine the rate of solution and solubility of various minerals which might be the source of the dissolved silica. Krauskopf (6) summarized previous work and pointed out that silicic acid (silica) is virtually undissociated below pH 9 and that silica in natural waters is predominantly in true solution as H_4SiO_4, monomeric silicic acid. In addition, he performed new experiments which indicated that the solubility of amorphous silica in distilled water and sea water is on the order of 120 mg/liter at 25°C. Time for equilibration was approximately 40 days for silica gel or colloidal silica but was greater than two months for opal. Siever (7) confirmed Krauskopf's work on solubility of amorphous silica as being 120–140 mg/liter and estimated from higher-temperature data that quartz solubility was about 10.8 mg/liter at 25°C, although, experimentally, quartz samples showed no measurable dissolution at 25°C after three years. Morey et al. (8) reported quartz solubility at 25°C as 6 mg/liter. Stöber (9) studied the solubility of various forms of silica and concluded that after initial release of silica from the solid, a layer of adsorbed silica forms and controls the final concentration of silica in solution.

Work has also been done on the rate of release of silica from silicate minerals and on equilibrium concentrations of silica in solutions in contact with naturally-occurring silicates. Garrels and Christ (10) have indicated that, considering the pH and aluminum concentrations present, the silica concentrations in most ground and stream waters (6–60 mg/liter) are in the range that might be expected from equilibrium with kaolinite. Polzer and Hem (11) found that silica concentrations were still increasing after two years in a dilute suspension of impure kaolinite at pH 3.3–3.7. At the end of the experiments, the silica concentration was 8–10 mg/liter. Part of the dissolved silica was attributed to solution of a free silica impurity. Mackenzie and Garrels (12) and Mackenzie et al. (13) showed that appreciable amounts of silica were released to sea water from various clay minerals within a 10-day period and concluded that the silica release was governed by an aluminous residue on the minerals. Correns and von Engelhardt (14), Nash and Marshall (15), and Garrels and Howard (16) in studying release of K from K-feldspars all visualize the development of a reaction film on the grain surface which is depleted in K (hence, relatively high in silica) and through which K ions must diffuse as they go from the mineral into solution. Wollast (17) studied the kinetics of silica release from K-feldspar and stated that "the weathering of feldspar under natural conditions can be described as a diffusion mechanism of H_4SiO_4 through a residual layer, constituted by slightly soluble $Al(OH)_3$ and subsequent reaction of these two substances to form a hydrated alumino–silicate." Luce (18) found solid-state diffusion to be the rate-controlling step for both silica and magnesium during early stages of leaching of magnesium silicates. In summarizing controls on silica concentration in soil waters, Kittrick (19) listed the following factors: rate of dissolution of unstable silicates, rate of precipitation of stable silicates, rate of movement of silica-bearing solutions out of the system, and rate of plant uptake.

Soil-leaching studies indicate that some silica is released from soil rather rapidly. McKeague and Cline (20) have shown that in soil–water mixtures at 100% water saturation, the silica in solution after 5 minutes was approximately half as great as that after 10 days. After the first day or two the silica concentration increased very slowly. They also demonstrated that pH has a marked effect on silica concentrations in soil solutions (21). These authors attributed the control of silica concentration to pH-dependent adsorption and indicated that, of the common soil minerals, iron and aluminum oxides have appreciable adsorption capacity. Jones and Handreck (22, 23) studied the effects of iron and aluminum oxides on silica concentrations in soil solutions and concluded that both caused a significant reduction in dissolved silica, with aluminum oxides being most effective. Minimum silica concentrations occurred at pH 9–10 in solutions in contact with iron and aluminum oxides. Harder and Flehmig (24) reported that the hydroxides of iron, aluminum, and other elements could remove silica from solutions containing as little as 0.5 mg/liter SiO_2.

Bricker and Godfrey (25) found that only a few hours to a few days were needed for the silica concentration to achieve a constant value in water recycled through a soil column. The same silica concentration was attained starting with water containing either more or less silica than that at "equilibrium." Similar conclusions regarding the stabilizing effect of soil on dissolved silica were reached by Miller (26).

The work described above shows that silica can be released or taken up rapidly—that is, within a few minutes or hours—and that the mechanism is not simply one of solubility.

Methods of Sample Treatment and Analysis

Water samples obtained in this study were collected near midstream in polyethylene bottles and filtered through 0.45-micron membrane filters as soon as possible after collection using compressed air. Normally, samples of 4–8 liters were passed through a 4-inch (10.16-cm) diameter filter, and the first 1–1.5 liters were discarded. Except under conditions of unusually clear water, the filter was partially clogged by sediment before any filtrate was retained, so the effective pore size of the filter was probably less than 0.2 micron for most samples and less than 0.1 micron for samples collected during larger flows when sediment concentrations were high. When samples containing more than 5000 mg/liter of suspended sediment were filtered, the time for filtration occasionally exceeded 6 hours but commonly 2–4 hours were required. After filtration, 4 ml of reagent grade concentrated HNO_3 was added to each gallon of filtrate, which was stored in a polyethylene bottle until it was analyzed for major constituents.

Suspended-sediment samples were obtained at low flow by compositing water samples collected at three points in the cross section. Under higher flow conditions, sampling was done at five points equally spaced in the cross section. The sampling device, a DH-59 hand sampler (27), was passed through the water column at the same rate at each sampling point so that the water collected represented the integrated effect of stream velocity and depth. Thus, all five samples could be composited to represent the average sediment concentration in the stream as a whole. Under very high flow conditions, when water velocities near the surface approached 15 feet (4.5 meters) per second (28), only samples of the upper 2 feet (0.5 meter) of flow could be obtained. However, the extreme turbulence under such conditions should have caused sufficient vertical mixing that the samples truly represented the average concentration of transported sediment of medium sand size and finer. The sizeable load of coarse sand and gravel moving near the streambed was not sampled. Suspended-sediment samples were always taken within 15–20 minutes of the collection of water samples.

Silica analyses were made on an automatic analyzer using a modification of the method of Mullin and Riley (29). Chloride analyses were also made on the automatic analyzer, using the method of O'Brien (30). Conductivity measurements usually were made on the pressure-filtered

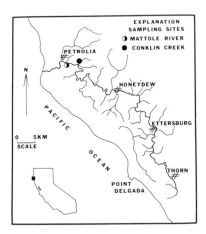

*Figure 1.   Index map of Mattole
River Basin*

samples. Enough determinations were made to establish that no change in conductivity could be detected between filtered and unfiltered samples. The specific conductance was used as an index of the concentration of dissolved electrolytes in the water and was determined for every sample collected. Ca, Mg, Na, and K were determined using an atomic absorption spectrophotometer.

Mattole Drainage Basin

The Mattole basin is in northern California (Figure 1), and an area of about 240 square miles lies upstream from the sampling point used in this study. Ranching and timber sales have been the major source of income but a little farming is done on suitable areas along the river. The towns generally consist of just a few rather old buildings with some other homes in the vicinity. Average population is probably less than two persons per square mile.

Topography and Geology. Topography is mature with a maximum relief of approximately 3500 feet and an average relief of perhaps 1500 feet. The rocks in the basin are mainly folded graywacke, shale, and conglomerate (*31*). A northwest strike of the folded and faulted sedimentary rocks causes the northwest trend of the stream basin. The downstream half of the basin is characterized in part by steep grassy slopes with shrubs and second growth trees located along some drainage lines and covering some whole tributary basins (Figure 2). The upstream half of the basin contains large areas of second-growth timber with scattered grasslands in an area of somewhat gentler topography.

Soils and Sediments. Soils in the Mattole basin range from gravelly to clay loam (*32*) and are acid with a pH generally in the range 4.6–6.0. In one 9-foot profile of upland soil sampled in this study, the pH ranged from 4.8–5.2. Where relatively little recent erosion has occurred, the soils are 10 feet or more in depth. However, in the last 20 years there has been extensive logging and road-building activity, and this has been a factor in the disturbance and resulting erosion of the soil. At present, erosion in the basin is rapid, and along drainage lines on some of the slopes the soil has been completely removed to bedrock. This erosion has caused aggradation of the river bed, and suspended-sediment concentrations in the river now exceed 10,000 mg/liter during high flows.

X-ray diffraction analyses of the < 2-micron size fraction of suspended sediment and soils from the Mattole basin indicate that kaolinite, vermiculite, and aluminum-interlayered vermiculite comprise the main clay minerals. Much smaller amounts of illite commonly are present. Chlorite is detectable in a few suspended-sediment samples.

Figure 2. View of lower Mattole River basin showing bridge from which sampling was done

Vermiculite identification is based upon the presence of a mineral having a 14-Å d-spacing with Mg saturation and glycol solvation which collapses to 10.5 Å with K saturation and air drying (33). There is little expansion above 14 Å when magnesium-saturated clay is solvated with ethylene glycol, which indicates that little, if any, montmorillonite is present. When heated to 550°C, only clay with about a 10-Å d-spacing remains in most samples, but a small amount of chlorite in a few sediment samples is indicated by the presence of 14-Å spacing (34). Evidence for aluminum-interlayered vermiculite (or possibly aluminum-interlayered montmorillonite) is the presence of a 14-Å mineral in Mg-saturated samples which fails to collapse to about 10 Å on K saturation (35) but which is neither montmorillonite nor chlorite, based on glycollation or heating to 550°C.

Precipitation. Rainfall averages 92 inches per year for the basin as a whole but within the basin ranges from about 50 to 110 inches. Of this, 76% appears as runoff (36). The bulk of the annual rainfall occurs in the months of November through March, with intense downpours associated with some storms. Normally, little or no rain falls from late May until late October.

The average composition of precipitation falling in the Mattole River basin is shown in Table I. Samples were collected near Petrolia, Honeydew, Ettersburg, and Thorn during part of the 1966–67 rainy

Table I.   Composition of Atmospheric Precipitation
in the Mattole River Basin

	Weighted Average Values (mg/liter)	
Constituent	*January–May* *1967*	*October–April* *1968–69*
Ca^{2+}	5.5	2.2
Mg^{2+}	0.4	0.3
Na^+	1.1	0.9
K^+	0.1	0.1
SO_4^{2-}	6.4	4.7
Cl^-	1.6	1.7
SiO_2	0.1	<0.1

season and for all of the 1968–69 rainy season. In calculating an average composition, the analyses were weighted according to the amount of rain falling during a sampling period and also according to the areal variation of rainfall within the basin. A total of 40 samples were collected.

Calcium and sulfate showed a definite tendency to be higher in concentration during periods of low rainfall, presumably because of dry fallout. Calcium concentrations were commonly less than 1 mg/liter and sulfate less than 2 mg/liter during periods of high rainfall.

Chemical Composition of Stream Water. Streams in the Mattole River basin contain relatively low concentrations of dissolved salts, as might be expected because of the high rainfall. Representative analyses of the river water are presented in Table II. Calcium, magnesium, and sodium are the major cations, and bicarbonate and sulfate are the major anions. Mean discharge for a 20-year period (1912–13, 1951–68) was 1331 $ft^3\ sec^{-1}$ (37.69 $m^3\ sec^{-1}$) and mean annual discharge was $1.189 \times 10^9\ m^3$ (37).

Silica vs. Time and Discharge in the Mattole

Diurnal Variation During Low Flow. The silica concentration in the Mattole River shows small but consistent diurnal changes, as do pH and specific conductance (Figure 3). Silica concentration reaches a maximum in the late afternoon along with pH, whereas specific conductance is at a minimum then. A minimum in silica occurs between 8 and 11 a.m. Both silica and specific conductance show about a 4% diurnal variation.

Samples taken the evening of October 5 at 1730 and at 0800 on October 6, 1968 showed a variation in silica, pH, and specific conductance similar to that observed on October 1–2 and shown in Figure 3. However, samples collected at 1730 on September 23 and at 0730 on September 24, 1969 showed the following results, respectively: pH, 8.8, 7.9; specific con-

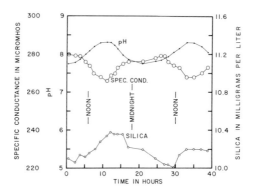

Figure 3.   Diurnal variation of pH, silica, and specific conductance on October 1–2, 1968

ductance, 241, 264; silica, 9.7, 9.2. Thus, a diurnal variation of about 5% in silica can occur in the Mattole River.

Such a variation is not unexpected; Barnes (38) has reported diurnal silica fluctuations of 15–20% in a stream, and Enright (39) has shown that the diatom population is lower in the evening than the morning in a coastal marine environment—a change which was attributed to diatom reduction by grazing of other organisms. Thus, the observed silica varia-

Table II.   Composition of Mattole River Water Collected under Various Flow Conditions

	Low Flow	*Medium Flow*	*High Flow*
Date	Nov. 11, 1966	May 1, 1967	Dec. 4, 1966
Discharge (cfs)	66	3500	41,150
Specific conductance (field meas., micromhos at 25°C)	286	109	80
Dissolved solids (mg/liter)[a]	172	70	53
pH (field meas.)	8.3	7.4	7.7
SiO_2 (mg/liter)	7.3	11.6	8.2
Electrolyte concentration (mg/liter)[b]			
Ca^{2+}	40	14	8.5
Mg^{2+}	6.7	2.4	1.4
Na^+	8.7	5.1	4.0
K^+	1.1	0.7	1.0
HCO_3^-	127	49	36
SO_4^{2-}	35	9.0	6.0
Cl^-	4.9	2.7	2.6

[a] Calculated from relation between specific conductance and dissolved solids for the Mattole River reported in U.S. Geological Survey publications.
[b] Analyses under supervision of J. W. Helms, U.S. Geological Survey.

tions in the Mattole River may result from diatom reproduction and growth during the night and early morning hours (*40*) and loss of diatoms during the day through consumption by other organisms.

The possibility exists that evaporation from the stream or transpiration through plants along the river bank may account for part or all of the increase in silica concentration during the daylight hours. However, replicate analyses of morning and evening samples for sodium and chloride indicate less than 1% difference, and a stream gage on the Mattole capable of detecting a 3% change in discharge showed no evidence of diurnal variation. Hence, the silica increase is believed owing to processes other than evaporation.

Variations for a Single Storm Event. An intensive study of one small stream rise was made during the period November 7–15, 1969 for the Mattole River and Conklin Creek (Figure 4). Conklin Creek is a tributary to the Mattole River which enters from the north about 2 miles upstream from the main sampling site on the Mattole. Its drainage area is 5–6 square miles.

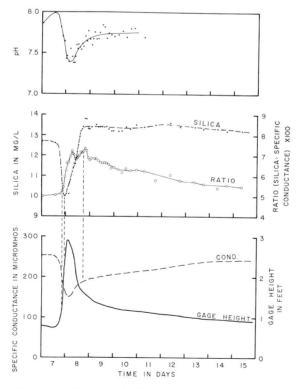

Figure 4. Variation in discharge and chemical composition of Conklin Creek, November 7–15, 1969

Table III.   Rainfall at Honeydew 2 WSW Station,
California, on November 7–8, 1969

Time Period	Rainfall,[a] Inches	Accumulated Rainfall on Nov. 7,[a] Inches
0000–1300	0.3	0.3
13–1400	0.2	0.5
14–1500	0.3	0.8
15–1600	0.1	0.9
16–1700	0	0.9
17–1800	0.3	1.2
18–1900	0.4	1.6
19–2000	0.3	1.9
20–2100	0.2	2.1
21–2200	0.6	2.7
2200–0600 (Nov. 8)	1.9	4.5

[a] Information from recording rain gage provided by K. E. Morgan, U.S. Weather Bureau.

After the summer dry period, two small storms had occurred, causing significant runoff. One storm causing about a 2-foot rise in stage on the Mattole River· was on October 15 and 16. A second storm resulted in 3.7 inches of rain at the Honeydew weather station about 12 miles upriver from the sampling site during the evening of November 4th and caused a river rise of about three feet. The river flow was still decreasing from this last storm when the intensive study began.

A little more than 0.1 inch of rain fell on the 5th and 6th of November. Light rain continued falling on the 7th, and 0.3 inch had accumulated by 1300. Table III presents data from the Honeydew recording rain gage for November 7 up until 2200 when the recording gage became inoperable. By combining information from a nonrecording rain gage at Honeydew and the available data from the recording gage, it appears that a total of about 4.5 inches of rain fell on November 7 and the first six hours on the 8th. Of the 1.8 inches that fell between 2200 on the 7th and 0600 on the 8th, almost all fell before 0200 on the 8th. This is on the basis of visual observation and the timing of the stream rise on Conklin Creek, which crested at 0300 on the 8th. Therefore, about 2.4 inches of rain fell in a 5-hour period at Honeydew between 2100 on the 7th and 0200 on the 8th. Normally, more rain falls at Honeydew than at the sampling point on the Mattole near Petrolia. An unofficial rain gage near our sampling point indicated 3.6 inches of rain for the storm, and this is probably a better value for the precipitation in the Conklin Creek drainage.

Although light rain fell the morning and early afternoon of November 7, the discharge of Conklin Creek continued to decrease following the previous storm, and specific conductance and pH increased. As rainfall increased during the late afternoon and evening and runoff from the acid soils began, the pH of the water dropped rapidly from 8.0 and reached a minimum of 7.4 at about the time of maximum discharge. A similar pattern of pH variation occurred on the Mattole (Figure 5). For both streams, the increase in pH after peak discharge was much slower than the decrease during rising discharge.

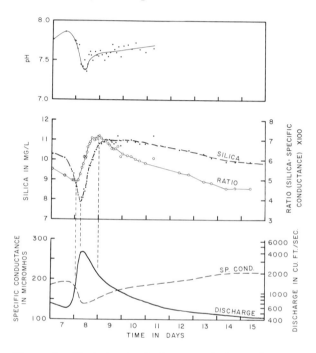

Figure 5. Variation in discharge and chemical composition of the Mattole River, November 7–15, 1969

The ratio silica to specific conductance (SiO_2:Spec. Cond.) is quite useful in showing changes in the relative rate of release of silica as compared to the major electrolytes, Ca^{2+}, Mg^{2+}, Na^+, HCO_3^-, SO_4^{2-}, and Cl^-. In Conklin Creek, as discharge began to rise, both the silica concentration and specific conductance (in this stream, a measure mainly of the amount of calcium and bicarbonate in solution) decreased at a rate such that the ratio SiO_2:Spec. Cond. remained relatively constant (Figure 4). When gage height was less than halfway to the maximum, the rate of silica decrease diminished compared with that of specific conductance. The result was a sharp increase in the SiO_2:Spec. Cond. ratio. This continued while silica concentration and then specific conductance reached a minimum. As specific conductance increased along with silica, the ratio tended to level off. After silica reached a maximum about 18 hours after peak discharge, specific conductance continued to increase, and the SiO_2:Spec. Cond. ratio decreased gradually. A similar trend in silica, specific conductance, and the ratio SiO_2:Spec. Cond. is apparent in Figure 5 for the Mattole River.

The decrease in silica concentration during rising discharge, rapid rise in silica near the discharge peak, and leveling off in silica concentration 18–24 hours after peak discharge are easily seen in the case of an isolated runoff event. However, when one storm follows another in rapid succession, the pattern of variation in silica and the ratio SiO_2:Spec. Cond. with time becomes complex. Nevertheless, if one keeps in mind the basic pattern outlined above, the reasons for the observed silica concentrations become evident.

Variations During a Complex Series of Storms. Changes of water chemistry and sediment concentration were monitored in detail for the Mattole River during the 1966–67 rainy season. Periodic samples were taken also during the following dry season. In Figure 6, water-quality variations are shown for the period November 24 to December 19, 1966. The first storm sequence of the rainy season ended on November 22, so the period November 24–27 represents a time of decreasing discharge and silica concentration but of increasing specific conductance. If one examines the changes in silica concentration and specific conductance with discharge during the rainy period November 28–December 13, the same relation between these parameters is seen as for the storm of November 7–8, 1969. Rainfall shown for each day on Figure 6 is the total reported for a 24-hour period ending at 0800. However, before discharge had

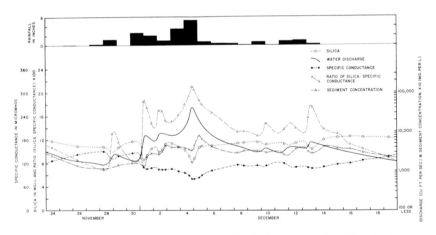

Figure 6. Water quality variation with discharge of the Mattole River, November 24 to December 19, 1966

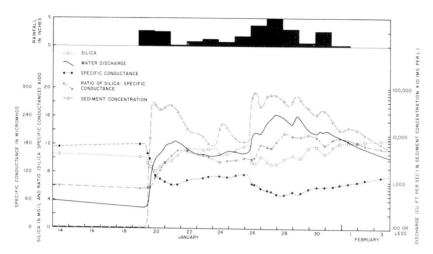

Figure 7. Water quality variation with discharge of the Mattole River, January 14 to February 3, 1967

declined very much after one rise, another storm arrived, the normal pattern of chemical changes was interrupted, and a new cycle begun. For example, on December 1 a sharp rise in discharge occurred, and silica concentration decreased by about 25%. As peak discharge was reached and flow began to decline, silica concentration showed a greater percentage rebound than did specific conductance, and this is reflected in the increase of the SiO_2:Spec. Cond. ratio. However, before discharge had declined very much, a new storm arrived early on December 2, and another peak in discharge resulted. Although little change in specific conductance was observed during and after the rise, silica followed its usual pattern of decreasing on the rise and rebounding rapidly soon after peak discharge.

Moderate rainfall continued on December 2 and gradually increased in intensity through December 3. During the day and early evening of December 4, about 5 inches of rain fell on soil that was already saturated. The result was a gradual but large rise in discharge. While flow increased on December 3 and 4, both silica and specific conductance declined at about the same rate, as shown by the relatively constant SiO_2:Spec. Cond. ratio, but, after the discharge peak, silica once again increased faster than specific conductance and then leveled off.

Sediment concentration and silica concentration show a consistent inverse relation during periods of storm runoff, and silica minima commonly tend to precede peak discharge by 2–4 hours, depending on how fast a rise occurs. Sediment concentration varies much more than does any of the dissolved constituents.

After more than a month of dry weather, heavy rains fell again beginning on January 19, 1967, and rain fell intermittently until February 1. The same patterns of sediment concentration, silica, specific conductance, and SiO_2:Spec. Cond. ratio are seen as for the other rainy periods described above (Figure 7).

Relation of Silica and the Ratio SiO_2:Spec. Cond. to Discharge. Silica concentration shows no correlation with discharge of the Mattole River (Figure 8). In Figure 9, the ratio of SiO_2:Spec. Cond. is plotted against log discharge. The ratio for samples collected more than four days after a peak in water discharge shows a rather well-defined positive correlation with log discharge. Samples collected during a rising stage tend to have a ratio which plots lower than those collected on a falling stage. This would be expected from the information presented in Figures 4–7. Points representing the first four days after peak discharge plot on, or somewhat lower than, the line of samples taken more than four days after peak discharge.

Seasonal Trends in Silica Concentration. The mean silica concentration (discharge weighted) for each month of the 1967 water year (Oct. 1966–Sept. 1967) is shown in Figure 10, along with the mean discharge.

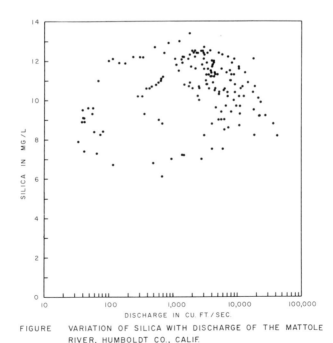

FIGURE   VARIATION OF SILICA WITH DISCHARGE OF THE MATTOLE
RIVER, HUMBOLDT CO., CALIF.

*Figure 8.   Variation of silica with discharge of the Mattole
River, Humboldt County, California*

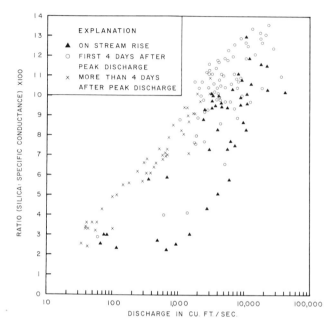

Figure 9.   Variation of the ratio silica:specific conductance
with discharge in the Mattole River, Humboldt County,
California

At the end of the 1966 dry season, discharge was low and so was silica concentration. When the rainy season began in mid-November, the surface runoff contained more silica than the ground water, so the average silica concentration in the river water increased. As the rains continued, the concentration of silica in runoff waters gradually increased. When average water discharge was decreasing during the period March to early July, silica concentrations remained relatively high. However, between early July and late August, the silica concentration decreased 25% from 12 to about 9 mg/liter.

Rates of Silica Release by Soils and Sediment

Silica Release in Suspensions with a High Solid:Liquid Ratio. Because silica variations with discharge suggested that silica was released by the soil at a rapid rate, some leaching studies were made of soils which had been collected from the Mattole River basin near the end of the summer dry period. Particles larger than 0.18 mm in diameter were separated by sieving and rejected. One composite sample representing several forest soils (0–2-inch depth) was mixed in a 1:1 ratio by weight with deionized water, stirred, let stand, sampled, and then occasionally stirred again with periodic sampling during a total period of four days. Aliquots were centrifuged, the supernatant tested for conductivity, and then filtered through a 0.1-micron membrane filter. A composite of grassland soils were subsequently treated the same way except that the mixture was stirred continuously at a slow rate using a magnetic stirring bar. The mixtures of water and soil were covered with a thin plastic sheet while standing to reduce evaporation.

The chemistry of the water in contact with the soil composites is shown in Figures 11 and 12. In the case of the forest-soil composite, there was little evidence of evaporation, either as condensation on the plastic container cover or as increased chloride content with time. In the case of the grassland-soil composite, the stirring equipment increased the temperature of the slurry 2–3°C above room temperature, and considerable condensation was noted on the loose-fitting plastic cover. Also, the chloride concentration gradually increased with time, suggesting evaporative losses. Because other leaching studies of chloride in Mattole basin soils had shown that almost all the chloride in a sample was soluble within a few hours' time, the increase in chloride was taken as a measure of evaporative losses. If specific conductance and silica concentration had been corrected accordingly in Figure 12, the silica concentrations and specific conductance would show a slight decrease on the fourth day.

The compositional patterns of the solutes were quite different for the forest- and the grassland-soil composites. Although both showed a rapid release of silica and other solutes from the soil during the first day,

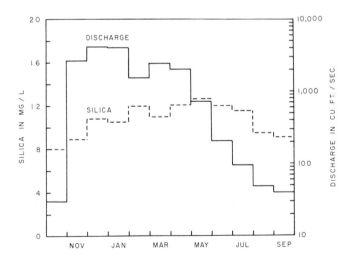

Figure 10.   Monthly mean discharge and mean silica concentration for the 1967 water year, Mattole River near Petrolia, California

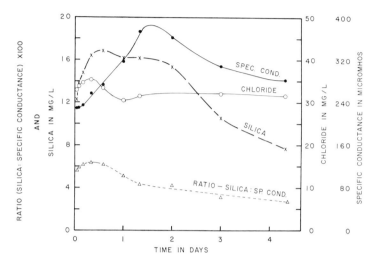

Figure 11.   Composition of the solution vs. time for a 1:1 mixture of water and forest soil

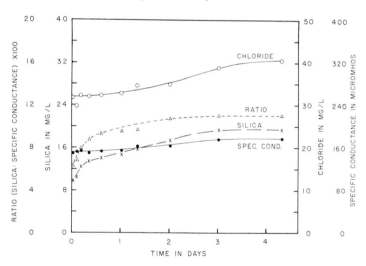

Figure 12.   Composition of the solution vs. time for a 1:1 mixture of water and grassland soil

the silica concentration in the forest-soil solution reached a maximum 16 hours after the experiment began and then decreased, slowly at first, and then more rapidly. In the grassland-soil solution, the silica concentration continued to rise slowly for about three days and may have decreased slightly on the fourth day. A surprising pattern was found in the specific conductance variation of the forest-soil solution with time. It increased in almost a linear relation with time for about a day and a half and then decreased again. The grassland-soil solution, on the other hand, showed very little indication of such a drop in conductance. The SiO_2:Spec. Cond. ratio showed an initial rise during the first 8 hours and then a gradual decrease for the forest soil. The grassland soil also showed a sharp increase in the ratio in the first 8 hours, and then this leveled off after almost three days. It is apparent, therefore, that these two types of soil displayed quite different responses to prolonged contact with water.

In another experiment, three soil composites, each representing about 10 surface soils from the southwestern part of the Mattole basin, were sieved to remove particles larger than 0.18 mm and then leached with

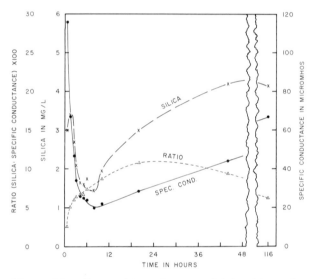

Figure 13. Composition of sequential leaching solutions in a 2:1 mixture of water and composite soil A

successive volumes of deionized water. Soil composite A was composed primarily of forest soils, with an organic carbon content of 9.6%. Soil composite B was in part from grassland and in part from brushland and contained 2.6% organic carbon. Soil composite C represented soils from brushland and areas of second-growth timber. Its content of organic carbon was 6.4%. In each case, duplicate 75-gram quantities of the air-dried samples were mixed with 150 ml of deionized water, stirred thoroughly, centrifuged, and the supernatant filtered through 0.2-micron membrane filters. Immediately after the supernatant was decanted, more water (110–115 ml) was added to bring the water–sediment mixture back to its original weight, the mixture was stirred, and centrifuged again. This procedure was repeated seven times, taking 40–65 minutes per leach. Beginning with the eighth leach, the water–sediment mixture was per-

mitted to stand after stirring for longer and longer periods before centrifugation was started.

The average composition of the duplicate leachates from composite
soil A is shown in Figure 13. Because interstitial water left behind from
one leach is mixed with the next leach in a ratio of about 1:3, very sharp
changes from leach to leach cannot be expected. However, the trends in
the various constituents are instructive. Highly-soluble electrolytes are
removed rapidly, as is some readily-leached silica, but with continued
leaching the ratio of SiO_2:Spec. Cond. climbs, indicating that the rate of
silica release increases more rapidly than that of the electrolytes. As the
leaching period increases, there is more opportunity for silica to reach a
"plateau" concentration and for the less soluble materials to release
electrolytes. Thus, when the time interval between leaches was 24–30
hours, the SiO_2:Spec. Cond. ratio reached a maximum, indicating that
silica and electrolytes were being leached temporarily in a constant ratio.
After a 3-day contact period, the silica concentration was increasing more
slowly than that of the electrolytes. The pH of the supernatant was 4.5,
5.0, and 5.8 after the first, seventh, and last leach, respectively.

A very similar pattern of release of silica and electrolytes was observed when the two other soil composites from the Mattole basin were
leached, so it appears that the type of variation seen in Figure 13 is
representative of at least three areas in the Mattole basin.

The data presented regarding the silica content of water contacting
Mattole basin soils are compatible with information found in the literature. For example, McKeague and Cline (20) showed (*see* Figure 14)

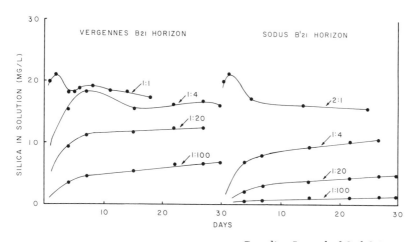

Canadian Journal of Soil Science

*Figure 14.   Concentration of silica vs. time for some soil–water mixtures
at different ratios; from McKeague and Cline (20)*

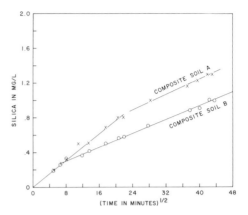

*Figure 15. Silica concentration vs. (time)$^{1/2}$
in 10,000 mg/liter suspensions of soil in
deionized water*

that in a 1:1 soil-to-water slurry, two different soils displayed a rapid
increase in dissolved silica during the first two days and then silica decreased, rapidly at first, then slowly with time. Slurries with soil-to-water
ratios lower than 1:4 did not show this pattern.

Silica Release in Suspension with a Low Solid:Liquid Ratio. The
possibility exists that a significant fraction of the silica in stream water
may have been derived from solution of suspended sediment during
stream transport. To establish an order-of-magnitude value for this leaching during transport, samples of soil and suspended sediment of less than
0.25-mm particle size were washed at least five times with deionized
water in rapid succession to remove easily-soluble material and then
given a final wash with ethyl alcohol to help speed the drying process. A
10,000-mg/liter suspension of three air-dried samples was prepared in
deionized water and stirred vigorously and continuously for about 31
hours and a 6000-mg/liter suspension of a fourth sample was stirred for
21 hours. Aliquots of the suspensions were removed periodically and
filtered immediately; that is, solids were separated from liquids within
4–5 minutes. The filtrates were then analyzed for silica content.

Silica concentration was plotted *vs.* the square root of time to test
for possible diffusion control (*17, 18*) in the release of silica. An allow-
ance was made for filtering time. Soil composite A displayed a release
pattern which suggested diffusion control at one rate for the first 8 hours
(Figure 15), and then the slope changed to a rate similar to that seen in
soil composite B. A break in slope in the line for soil B apparently
occurred at the third aliquot collected.

The 1966 sediment sample (Figure 16) was representative of the
suspended sediment in transport during a high-flow period. It probably

contained appreciably more clay-size material than the 1964 sample, and this may account for the greater amount of silica released per unit weight. The silica release shows a linear relation to the square root of time. The 1964 sediment was a sample of the fine material that settled out as river water passed through a flooded building during a period of very high flow in December 1964. It consisted mainly of silt with some clay and very fine sand. Presumably, most of the clay was carried away with the water. Just as in the case of the soil A (Figure 15), there appears to be a break in the slope of the line of silica concentration after 8–9 hours, but the linear relation between silica and the square root of time is still evident for the first 8 hours. This relationship suggests that the rate of solution of silica from both sediments and soil particles is controlled by a diffusion process, even though both soils and sediments are composed of a complex mixture of siliceous material.

Separate splits of the 1966 sample of suspended sediment were stirred with deionized water at concentrations of 2000 and 14,000 mg/liter for 31 hours, and the water was then removed and analyzed. After 31 hours, the silica in solution corresponded to 0.63, 1.80 (from extrapolation, Figure 16), and 2.84 mg/liter for the 2000, 6000, and 14,000 mg/liter suspensions, respectively. Silica in solution per 1000 mg/liter suspended sediment was 0.31, 0.30, and 0.20 for the 2000, 6000, and 14,000 mg/liter suspensions, respectively. Above about 6000 mg/liter, therefore, a linear relation between sediment concentration and silica released may not exist. This is not surprising if there is a back reaction owing to uptake of silica on mineral surfaces, as suggested by Wollast (17) and others.

In considering the probable effect of suspended sediment on silica in solution, one must take into account not only the concentration of the

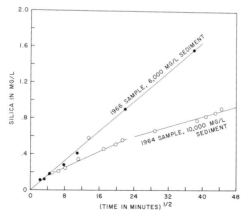

Figure 16.   Silica concentration vs. (time)$^{1/2}$ in suspensions of stream sediment in deionized water

sediment but also the contact time and concentration of silica already in solution. During high-flow periods when sediment concentrations may exceed 10,000 mg/liter, the water velocity in the stream is high, and the average elapsed time from actual rainfall until surface runoff arrives at the sampling point is about 5–6 hours. When rainfall results in only moderate runoff, sediment concentrations may be in the range 1000–3000 mg/liter, and water striking the land surface can take 10–12 hours on the average to reach the sampling point because of lower stream velocities. If 2–4 hours is allowed after sample collection for filtration, then total contact time of the sediment with the water would range from 7–16 hours.

The concentration of dissolved silica in the 6000 mg/liter suspension at the end of 10 hours was about 1 mg/liter. A reasonable estimate of silica concentration in the stream is 10 mg/liter when suspended sediment is 6000 mg/liter. Thus, a maximum of 10% of the dissolved silica in stream water might possibly have been derived from suspended sediment during transport. However, this would mean that 9 mg/liter of silica was already in solution when the sediment-bearing water entered a rivulet on the hillside. Such a silica concentration would be expected to slow the release of more silica from the sediment. Therefore, 6000 mg/liter suspended sediment probably contributes much less than 10% of the dissolved silica found in the Mattole River under moderately high flow conditions. Under more intense rainfall and higher flow conditions, greater concentrations of suspended sediment occur, and more rapid runoff offers less time for contact between soil and storm runoff. The net result of the higher sediment concentrations but shorter contact time would probably be lower total silica in solution in the stream water, but a greater proportion of the total dissolved silica would be attributable to release from suspended sediment. Even then, the silica derived from suspended sediment probably would not exceed 10% of the total in solution.

If wet soil at the land surface is thought of as simply a very concentrated suspension of sediment in water, the sediment concentration would be on the order of 1,600,000 mg/liter. If the rate of silica release into solution were related to total solid surface exposed per unit volume of water and thus proportional to the sediment concentration, it would take 3–4 minutes contact time to increase the silica concentration to 12 mg/liter based upon the data for soil A (Figure 15).

In fact, it is evident that back reactions would become increasingly significant as the silica concentration increased in the pore water. The soil suspensions shown in Figures 11 and 12 had effective sediment concentrations of about 730,000 mg/liter, and dissolved-silica concentration reached 12 mg/liter in 4 hours for the grassland-soil composite and in less than an hour for the forest-soil composite. Readily-soluble silica in both the grassland and forest soils undoubtedly contributed greatly to the rapid rise in silica concentration, but the fact remains that the very high solid-to-liquid ratio in wetted surface soil, even where leached by prior rain, would result in relatively rapid increase in the silica content of soil water. Thus, an hour or two of contact time with the soil can account for almost all of the dissolved silica found in Mattole river water during high flow. There is little need of, or evidence for, assuming that more than 10% of the dissolved silica is derived from suspended sediment.

Silicon–Cation Ratios in Solids and Stream Water. During low-flow periods of the Mattole river, the ratio of silicon (equivalents/liter) to alkalis plus alkaline earths (equivalents/liter) in solution is approximately 0.15–0.2, and following high flows it is in the range 0.8–0.9. These ratios can be compared with those in various common rock-forming minerals, as listed in Table IV.

Table IV. Ratio of Silicon to Alkalis Plus Alkaline Earths in Common Rock-Forming Minerals and Clay Minerals

Mineral	Range of the Ratio of Silicon: Alkalis + Alkaline Earths[a] (Equivalents: Equivalents)
Biotite	2–10
Chlorite	1–1.4
Montmorillonite	32–48
Muscovite	10–12
Vermiculite	2–14
K-feldspars	12
Plagioclases	4–12

[a] Calculated from data in Dana, (*41*) and Grim (*34*).

It is apparent even during high-flow periods, when most silica is carried away in solution, that the ratio of silicon:alkalis plus alkaline earths in stream water is much lower than that in the primary minerals. Hence, the fraction of the silicon contained in clay minerals, micas, and feldspars which is removed in solution in stream water must be small.

Factors Controlling Silica Concentration in the Mattole River

Silica concentration in the Mattole River displays a consistent pattern with respect to stream flow, specific conductance, and sediment concentration. However, the silica variation is "out of phase" with respect to both discharge and dissolved electrolytes, and silica varies inversely with sediment concentration. Any explanation of the silica variation must account for these facts.

When rain begins after a dry period, the water enters the pore spaces in the soil easily. This water dissolves readily-soluble materials and, as rain continues, the initial rainfall will be carried downward into the soil along with the readily-soluble salts. If rainfall increases in intensity or the soils have low permeability, the pores of the surface soil become saturated, and water flows over the land surface to produce "overland flow" (42, 43). A part of this overland flow infiltrates the soil farther downslope, and a part continues on to join rivulets and become part of a larger stream. Some of the water which entered the soil moves laterally down gradient through the upper soil horizons and emerges to join the overland flow. This "subsurface runoff" or "subsurface flow" can be expected to transport some of the readily-soluble soil materials initially and, as time of contact increases, some of the less readily-soluble materials will also go into solution and be carried along. With continuing rainfall, an increasing amount of subsurface runoff will join the overland flow to form the storm runoff. Subsurface runoff which has moved laterally through only a few inches of soil may rejoin surface flow within an hour or less (44), and it is probably this water which causes the initial increase in silica concentration observed during a stream rise. In many streams, the proportion of ground water in stream flow is very small during periods of storm runoff, so the stream chemistry is largely controlled at such times by chemical reactions at the land surface. This is probably the case for the Mattole River.

As rainfall decreases, overland flow also decreases, but subsurface runoff continues for several days after the storm as water gradually drains from the soil. This explanation assumes that three water sources of differing chemical characteristics contribute to stream flow, in contrast to the assumption of two water sources made by other authors, for example, Pinder and Jones (45).

The silica variation with time and discharge of the Mattole River and Conklin Creek is compatible with the sequence of events described above. This can be observed from the information shown in Figures 4 and 5. Although almost an inch of rain fell before 1700 hours on November 7, 1969, there was no evidence of increased flow in the Mattole River resulting from that rain. In fact, stream discharge was decreasing. The rain apparently infiltrated the soil as it fell. As rainfall became more intense, storm runoff entered the stream and discharge increased. Although the concentration of both silica and dissolved electrolytes decreased with increasing discharge, the decrease in concentration was much less than would result from dilution of groundwater inflow to the streams by rainfall alone. Thus, an appreciable amount of both silica and electrolytes must have been derived from surface soils as water passed over or through them.

The SiO$_2$:Spec. Cond. ratio of the stream water decreased slightly in the Mattole River and increased slightly in Conklin Creek during the initial stream rise, indicating that, on the average, the ratio of these constituents in the dissolved solids removed from surface soils was not greatly different from that of the dissolved solids present in the stream before the storm began. However, before stream discharge had increased halfway to its maximum and before silica reached a minimum, the SiO$_2$: Spec. Cond. ratio began to rise rapidly. This increase in the ratio is attributed to the influx of water to the stream from surface soils which had been leached of highly-soluble salts. Such water is characterized by a relatively high ratio of silica to electrolytes (*see* Figure 13).

Following the upturn in the SiO$_2$:Spec. Cond. ratio, the silica concentration continued to decrease for about 2.5 hours in the case of Conklin Creek and 5 hours in the case of the Mattole River before reaching a minimum. This minimum silica concentration, in turn, preceded maximum stream discharge by 3–4 hours in both Conklin Creek and the Mattole River. The increase in silica concentration while discharge was still rising is attributed to an increasing proportion of subsurface runoff in the stream flow and to an increasing concentration of silica in that subsurface runoff because of greater contact time with the soil.

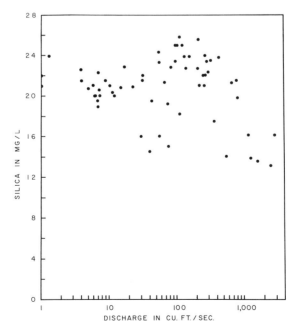

Figure 17.   Variation of silica with discharge of Pescadero Creek, San Mateo County, California; after Steele (46)

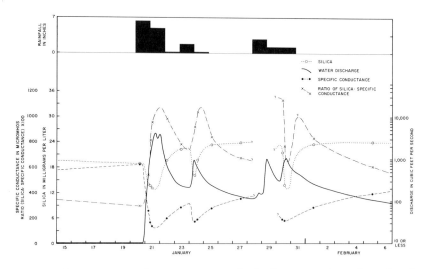

Figure 18. Water quality variation with discharge of Pescadero Creek, January 15 to February 6, 1967; after Steele (46)

Both silica concentration and the ratio SiO_2:Spec. Cond. continued to increase after peak stream discharge until the silica concentration began to level off while dissolved electrolytes continued to increase. At that point, the ratio reached a maximum and began to decline. The leveling off in silica concentration coincides with a change in slope of the discharge curve, suggesting that this marks the end of overland flow. The plateau silica concentration, then, indicates the "steady state" value for dissolved silica in soil water shortly after the storm.

The fact that the pattern of silica variation in Conklin Creek is very similar to that in the Mattole River, even though the stream basins are different by a factor of about 40 in size, indicates that the observed variation in the Mattole basin is not owing to travel time effects.

Silica Variation in Other Streams

The only data on other streams which are sufficiently detailed for comparison purposes are those obtained by Steele (*46*) on Pescadero Creek, San Mateo County, California. This stream has a drainage area of 45 square miles and the average annual rainfall in the basin is approximately 40 inches. As in the case of the Mattole basin, much of the rainfall occurs between November and March.

There is little correlation between silica concentration and discharge of Pescadero Creek (Figure 17). When the silica concentration, the ratio of SiO_2:Spec. Cond., and water discharge are plotted against time for a storm period (Figure 18), one can see that the silica variation and the ratio SiO$_2$:Spec. Cond. show the same relation to discharge as was observed in the Mattole River. This is true despite the fact that specific conductance and silica concentrations are appreciably higher in Pescadero Creek than in the Mattole River. As in the case of the Mattole River, samples taken on a rising stage tend to have a somewhat lower ratio of SiO$_2$:Spec. Cond. than average (Figure 19). Three of the four samples which plot low in Figure 19 and are underlined were taken within 2 hours after peak discharge and the fourth within 5 hours, so they were obtained just after a rise was completed.

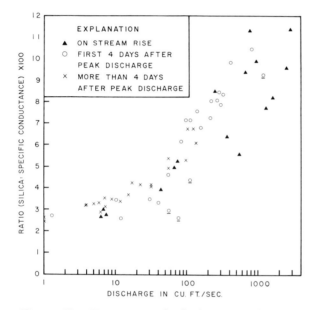

Figure 19. Variation with discharge of the ratio silica:specific conductance in Pescadero Creek, San Mateo County, California; after Steele (46)

If enough instantaneous samples are collected to be representative of all flow conditions, a plot of the ratio SiO$_2$:Spec. Cond. against discharge for a stream can be used to determine whether silica concentrations are relatively low during a rising stage and relatively high on a falling stage. Commonly, the period of sharply rising stage is short compared with the time of falling stage, so most samples are obtained on a falling stage. This is especially true for small streams.

A search of Water-Supply Papers of the U.S. Geological Survey was made to find streams for which there existed several years' record of instantaneous samples and for which determinations of silica, specific conductance, and instantaneous discharge had been made. The silica concentration and ratio SiO_2:Spec. Cond. were plotted against log discharge to determine trends. Records of mean daily discharge were examined to learn which samples were collected on a day on which a stream rise occurred and which were collected at a time of falling discharge. Results are shown in Table V. It appears that the ratio SiO_2: Spec. Cond. consistently increases with rising discharge although silica concentration itself may rise, fall, or show no trend. In most instances, the SiO_2: Spec. Cond. ratio is lower than normal for samples collected on days on which a rise occurred but, in the case of two streams, there were insufficient samples taken on a rise to establish a trend. These data provide good evidence that the rate at which silica is released from soils increases more rapidly than that of the electrolytes during storm runoff, just as in the Mattole River and Pescadero Creek.

Table V. Change in Silica Concentration and SiO_2:Spec. Cond. Ratio with Increasing Discharge of Various Streams

Stream Name	Change in Silica Concentration	Change in Ratio SiO_2: Spec. Cond.	Ratio SiO_2: Spec. Cond. Relatively Low on Rising Discharge?
Delaware R. at Montague, N.J.	Increases	Increases	Yes
Cape Fear R. at Lillington, N.C.	Increases	Increases	Yes
Pigeon R. near Hepco, N.C.	Decreases	Increases	Yes
French Broad R. at Blantyre, N.C.	Decreases	Increases	Yes
Green R. near Greensburg, Ky.	Trend uncertain	Increases	Yes
Licking R. at McKinneysburg, Ky.	Trend uncertain	Increases	Uncertain
Little Blue R. near Deweese, Neb.	Decreases	Increases	Yes
Goose Cr. below Sheridan, Wyo.	Trend uncertain	Increases	Uncertain

A plot of the data for the Cape Fear River at Lillington, N. C. is shown in Figure 20 as an example of the general relations described above. Some of the samples collected on the day of a stream rise show a normal SiO_2:Spec. Cond. ratio. This may result from collection of the sample on that day before the stream rise began. Nevertheless, the relatively low SiO_2:Spec. Cond. ratios commonly observed on rises suggest that the silica release pattern seen on the Mattole River is present also in the Cape Fear River, despite its much larger drainage area (3440 square miles).

Discussion

Mattole soils are acid, and water removed from a 1:1 soil–water mixture commonly has a pH in the range 4.5–5.5. This means that subsurface runoff from such soils should have a relatively low pH and suggests that the pH of the stream water should drop considerably when runoff is high. Figures 4 and 5 show that some decrease in pH occurs with increasing stream discharge, but much less than one might expect from the relatively large volume of storm runoff in the stream compared with ground water at the time of peak discharge. The excess H^+ must be lost rapidly, perhaps by reactions with soil minerals.

In attempting to understand the chemistry of Mattole stream water, the relative constancy of the SiO_2:Spec. Cond. ratio during the early part of a rise, regardless of the absolute level of the ratio before the rise began, needs to be considered further. If the rapid increase in the ratio observed in Figure 13 were indicative of what happened when storm runoff

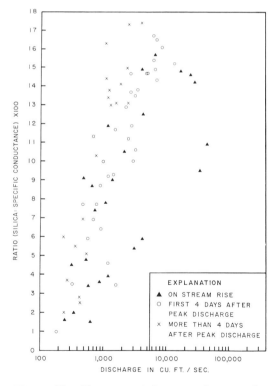

Figure 20.   Variation of the ratio silica:specific conductance with discharge of the Cape Fear River at Lillington, North Carolina; data from U. S. Geological Survey Water-Supply Papers

began, the ratio might be expected to increase in the Mattole River as soon as discharge increased. Examination of Figures 4–5 shows that this is not the case. One might call upon straight dilution of the stream water by rain water to cause a decrease in the concentrations of silica and electrolytes in a constant ratio, but such a mechanism would require that the dissolved solids be diluted in proportion to the increase in stream discharge. Such a great dilution does not occur, so an appreciable amount of dissolved solids must be acquired by storm runoff before it enters the stream. The explanation for the rather constant ratio, therefore, must be based upon the composition of the dissolved solids leached first from the surface soils.

Examination of the SiO_2:Spec. Cond. ratios in Figures 11–13 shows that, in each case, the ratio was initially low. This must mean that the electrolytes are released more quickly than the silica on the first contact of dry soil with water. If this happened in soil solutions at the beginning of a storm, the subsurface runoff contribution to stream flow might have a SiO_2:Spec. Cond. ratio which was as low as, or lower than, that present in the stream. After a few hours' contact time and further leaching by rainfall, this ratio would be expected to rise in the soil water as a result of removal of soluble salts, and this change would then be reflected in the stream chemistry. If the soil had been wet by a previous rain and stream flow consisted mainly of subsurface runoff when a new storm arrived, the first water flushed from the soil might well have a SiO_2:Spec. Cond. ratio similar to that in the stream. But with continued leaching, an increase in the ratio would be expected.

Figures 4 and 5 show that the increase in the SiO_2:Spec. Cond. ratio is delayed after the stream rise begins in both Conklin Creek and the Mattole River, but the upturn in the ratio in Conklin Creek preceeded that in the Mattole by about 5–6 hours. This delay is attributed to travel time effects in the river. Presumably, upstream tributaries show a pattern similar to that in Conklin Creek, but flow from them takes longer to arrive at the sampling site on the Mattole. The increasing ratio in Conklin Creek water is obscured initially at the Mattole sampling site by the much larger volume of flow coming from upstream.

Suspended sediment is an unlikely source for most of the silica found in the Mattole during high flow, and the only other possible source is the soil itself. Silica release from the soil must occur rapidly because the time for the rain to fall, pick up silica, and travel an average of perhaps 20 miles to the sampling site may be as little as 5–6 hours. A close look at the silica variation on and prior to December 4, 1966 (Figure 6) is enlightening. During the period November 1 to November 27, about 25 inches of rain fell near Honeydew (47), and in the storm period November 28 to 0800 December 4, an additional 11.0 inches fell. Then from

0800 on the 4th until 0800 on the 5th, 5.3 more inches fell, most of it during the afternoon of the 4th. Therefore, surface soils were thoroughly leached by the time of the big stream rise on the evening of the 4th. Yet silica concentration in the Mattole River decreased only to 8 mg/liter at peak runoff. The fact that silica concentrations reached a maximum of 12 mg/liter several days later, when there had been plenty of time for soil–water interactions, makes it probable that the maximum silica content in soil water during the height of the storm was less than 12 mg/liter. Because rainwater in the Mattole basin averages less than 0.1 mg/liter silica, and silica released from suspended sediment probably contributes less than 1 mg/liter to dissolved silica in the river, more than 7 mg/liter of the silica in the stream water must have been obtained during passage through the soil. To achieve this concentration, well over half of the water in the river at peak discharge must have been derived from subsurface flow. If that much subsurface flow occurs during periods of intense rainfall, then an even higher percentage of the stream flow must be attributable to subsurface runoff during periods of less intense rainfall.

Suspended-sediment concentrations increased sharply on rises and reached a maximum at or, commonly, shortly before peak discharge (Figures 6 and 7). Peak sediment concentrations consistently occurred at times of minima in silica concentration. This is to be expected if minimum silica and maximum sediment concentrations coincide with maximum contribution of overland runoff to stream flow, for at that time maximum dilution of silica-rich subsurface flow by direct runoff would occur and maximum erosive capability of direct runoff would exist. If this reasoning is correct, peak stream discharge would coincide with peak contribution from overland flow only when rainfall was intense or the surface soil was of very low permeability. Normally, peak direct runoff would precede peak stream discharge because of the significant contribution of subsurface flow at maximum stream discharge.

Because the chemistry of stream water reflects its source in the soil profile, detailed studies of runoff chemistry from small basins should be very useful in delineating the ways and rates at which various elements are released in weathering. In the Mattole basin, it is apparent that groundwater inflow to the stream at the end of the dry season has a relatively high pH (≈ 8), relatively high dissolved salts (Spec. Cond. $\geqslant 300$), and rather low silica (7–8 mg/liter). Thus, the alkalis and alkaline earths are preferentially removed as compared with silica from rock minerals at or near the water table. In surface soils, however, during storm runoff, quite different conditions prevail. The pH of the water may at first be low ($5\pm$), silica release is relatively rapid (8–12 mg/liter), and alkalis and alkaline earths are removed rather slowly (Spec. Cond. 70–150). The result is that the rate of silica removal compared with that of the major cations is about four times as great during storm runoff as at low flow. Even during storm runoff, however, the ratio of silica to alkalis and alkaline earths in solution is much less than the ratio in primary minerals.

From the standpoint of the annual load of dissolved silica, the lack of correlation between silica concentration and discharge, when considered on an annual basis (*see* Figure 8), means that silica load can be calculated as though it varied directly with discharge. This allows use of flow-duration curves to compute the time distribution of silica load carried by the Mattole River. Such information was obtained from the report of Rantz and Thompson (*36*), and the necessary computations were made. These show that about 25, 60, and 90% of the annual dissolved silica load is carried by the Mattole in 2, 10, and 35% of the time, respectively. Because storm runoff makes up most of the stream flow in the Mattole during at least 20% of the year, 75% or more of the dissolved silica transported by the Mattole River annually is derived from near-surface soils rather than from weathering of soils and rocks at depth.

Some tentative conclusions can now be drawn regarding the way in which silica is released from Mattole soils. The linear relation between silica concentration and the square root of time in soil and sediment suspensions, where dissolved silica is less than about 1 mg/liter, suggests that a diffusion mechanism controls the release of silica from mineral particles. Such a mechanism would be in agreement with studies by others (*14, 17, 18*). Those studies suggest that in the initial release of silica from feldspar only a diffusion mechanism would be apparent, but as the silica concentration increased a subsequent sorption (precipitation?) reaction on the altered solid surface would slow the net release of silica until a relatively steady condition existed. This appears to be a pattern that would explain the silica released from both low and high concentrations of prewashed Mattole soil and sediment in water.

The following sequence for the release of silica from soils and rocks of the Mattole basin is indicated by available data. At the end of the summer, surface soils contain weathering products that include readily-soluble silica. Early rains, which are too light to cause runoff, may carry much of this soluble material into the lower A and upper B soil horizons. But hydration of mineral surfaces encourages further weathering and the formation of more soluble silica which is available for removal with the first runoff. When rainfall is heavier than that required to saturate the surface soil, overland flow and subsurface flow, the latter containing some of the easily-soluble silica, join together to cause a stream rise. As rainfall continues, the surface soil is leached, and silica diffuses into solution in the soil water until it reaches a limiting concentration which represents a balance between the flushing rate and the rate of release from soil materials. When rain stops, the silica concentration in soil water increases to a new plateau level within 24 hours or less. If more rain falls on the moist soil, the dissolved silica and some sorbed or freshly reprecipitated silica form a reservoir of readily-available silica which can be flushed out first. With more rain and leaching, diffusion again becomes an important process, supplying the silica to the subsurface runoff. As the rainy season continues, the concentration of silica in soil water slowly increases. After the rainy season ends, water continues to drain from superficial deposits for several months, and silica concentrations in the river remain well above those observed at the end of the summer dry period. As ground water containing lower concentrations of silica becomes an increasing part of the stream flow, silica decreases in concentration. In late October or early November, a new cycle begins.

Summary and Conclusions

Silica concentrations in the Mattole River of northern California vary in a consistent pattern with relation to discharge and total concentration of electrolytes, as measured by specific conductance. During the initial part of a stream rise, both silica and electrolytes decrease in a nearly constant ratio, but as the rise continues, the rate of silica decrease slows relative to that of the electrolytes, causing the SiO_2:Spec. Cond. ratio to turn upward. Commonly, 2–4 hours before peak discharge, minimum silica concentration and maximum suspended sediment concentration occur. At peak discharge, the specific conductance is at a minimum but both silica and the ratio SiO_2:Spec. Cond. are rising and continue to do so for another 12–18 hours. Then silica concentration becomes almost constant while specific conductance continues to increase, causing the ratio between them to decrease. After several days, silica begins to decrease slowly. This cycle is repeated with each stream rise. Enough data from other streams are available to suggest that the pattern occurs elsewhere.

The silica-concentration pattern observed in the Mattole explains both the lack of high correlation between silica and discharge or specific conductance and the relatively small changes in silica concentration with discharge noted by investigators for other streams.

Because the minimum in silica concentration and maximum in sediment concentration normally precede peak discharge, peak overland flow probably also precedes peak discharge. Therefore, runoff that has spent an appreciable period of time in soil pores contributes a major part of the stream flow at peak discharge. The rather small decrease in silica concentration during a stream rise supports this interpretation. Because little silica can be obtained by interaction between water and stream sediment during transport, the relatively high concentration of soluble silica must result from percolation of water through the soil immediately after falling as rain. This apparently occurs on slopes of 15°–25° that are common in the Mattole basin.

If it can be shown that silica minima and sediment maxima mark peak overland flow elsewhere and that the end of overland flow is marked by a leveling off in silica concentration after peak discharge in stream flow, then monitoring of these parameters should be very helpful in separating storm runoff into the various components of flow.

The rapid achievement of a "constant" value of silica in soil water demonstrates that the opportunity for equilibration exists, but the electrolytes do not show any signs of reaching a plateau concentration. The electrolytes apparently do not equilibrate with the soil during or shortly after a storm period. Thus, during storm runoff, when more than 75% of the annual discharge occurs, there is not sufficient time for chemical equilibrium of the electrolytes to be reached in soil water, and it appears that reaction kinetics must be an important factor in determining the concentrations of the electrolytes.

Work on Mattole River water and studies of soil–water interactions indicate that detailed investigations of changes in water chemistry during storm runoff can be a powerful tool in obtaining a better understanding of weathering reactions. The use of ratios between constituents can be especially helpful because dilution is eliminated as a factor. Sampling intervals apparently should not exceed 2 hours during a stream rise, and 1-hour intervals are preferred. As basins increase in size, travel-time effects may obscure changes in chemistry owing to weathering reactions. It is desirable, therefore, to work with the smallest basin that can be considered representative of the area under study. Minor changes in concentration trends can be seen best if samples are run in sequence, using automatic equipment for analysis.

Acknowledgment

R. L. Malcolm worked with the author in collecting many of the water samples, often under adverse conditions. Water samples were also collected by Mr. and Mrs. T. E. Mathews and by John Schonrock. Many of the ideas expressed have been discussed and modified as a result of talks with E. A. Jenne, J. D. Hem, and K. V. Slack. E. A. Jenne, J. D. Hem, R. M. Garrels, S. N. Davis, P. B. Hostetler, and D. R. Schink read the manuscript and suggested improvements. Analyses were made by C. S. Barwis, E. A. Clarke, and D. K. MacDonald or were made in the Sacramento, California laboratory of the U. S. Geological Survey under the supervision of J. W. Helms. The study would have been impossible without the help of these individuals.

Literature Cited

(1) Davis, S. N., *Am. J. Sci.* (1964) **262**, 870.
(2) Palmer, Chase, *U. S. Geol. Surv. Bull.* (1911) **479**, 24.
(3) Hendrickson, G. E., Krieger, R. A., *U. S. Geol. Surv. Water-Supply Paper* (1964) **1700**, 93.
(4) Davis, S. N., *Univ. Hawaii Water Resources Res. Center Tech. Rept.* (1969) 29.
(5) Feth, J. H., Roberson, C. E., Polzer, W. L., *U. S. Geol. Surv. Water-Supply Paper* (1964) **1535–I**, 39.
(6) Krauskopf, K. B., *Geochim. Cosmochim. Acta* (1956) **10**, 1.
(7) Siever, R., *J. Geol.* (1962) **70**, 127.
(8) Morey, G. W., Fournier, R. O., Rowe, J. J., *Geochim. Cosmochim. Acta* (1962) **26**, 1029.
(9) Stöber, W., "Equilibrium Concepts in Natural Water Systems," ADVAN. CHEM. SER. (1967) **67**, 161–82.
(10) Garrels, R. M., Christ, C. E., "Solutions, Minerals, and Equilibria," p. 361, Harper and Row, New York, 1965.
(11) Polzer, W. L., Hem, J. D., *J. Geophys. Res.* (1965) **70**, 6223.
(12) Mackenzie, F. T., Garrels, R. M., *Science* (1967) **150**, 57.
(13) Mackenzie, F. T., Garrels, R. M., Bricker, O. P., Bickley, Frances, *Science* (1967) **155**, 1404.
(14) Correns, C. W., von Engelhardt, W., *Chem. Erde* (1938) **12**, 1.
(15) Nash, V. E., Marshall, C. E., *Missouri Univ. Res. Bull.* (1956) **613**, Pt. 1.
(16) Garrels, R. M., Howard, P. F., *Clays Clay Minerals* (1959) **6**, 68.
(17) Wollast, R., *Geochim. Cosmochim. Acta* (1967) **31**, 635.
(18) Luce, R. W., Ph.D. dissertation, Stanford University, 1969.
(19) Kittrick, J. A., *Clays Clay Minerals* (1969) **17**, 157.
(20) McKeague, J. A., Cline, M. G., *Can. J. Soil Sci.* (1963) **43**, 70.
(21) *Ibid.*, (1963) **43**, 83.
(22) Jones, L. H. P., Handreck, K. A., *Nature* (1963) **198**, 852.
(23) Jones, L. H. P., Handreck, K. A., *Advan. Agron.* (1967) **19**, 107.
(24) Harder, H., Flehmig, W., *Geochim. Cosmochim. Acta* (1970) **34**, 296.
(25) Bricker, O. P., Godfrey, A. E., "Trace Inorganics in Water," ADVAN. CHEM. SER. (1968) **73**, 128–42.
(26) Miller, R. W., *Soil Sci. Soc. Am. Proc.* (1967) **31**, 46.
(27) Inter-Agency Committee on Water Resources, Report No. 14, p. 60, Supt. of Documents, Washington, D. C., 1963.
(28) LaRue, G. W., oral communication, 1970.
(29) Mullin, J. B., Riley, J. P., *Anal. Chim. Acta* (1955) **12**, 1962.
(30) O'Brien, J. F., *Wastes Eng.* (1962) **33**, 670.
(31) Irwin, W. P., *Calif. Div. Mines Bull.* (1960) **179**, 42.
(32) McLaughlin, J., Harradine, F., "Soils of Western Humboldt Co., Calif.," p. 68, 71, 81–4, Univ. of Calif. at Davis, 1965.
(33) Walker, G. F., "X-Ray Identification and Structures of Clay Minerals," Ch. VII, pp. 199–223, Mineral Society of Great Britain Monograph, 1951.
(34) Grim, R. E., "Clay Mineralogy," 2nd ed., pp. 83, 105, McGraw-Hill, New York, 1968.
(35) Hathaway, J. C., *Clays Clay Minerals* (1955) **3**, 74.
(36) Rantz, S. E., Thompson, T. H., *U. S. Geol. Surv. Water-Supply Paper* (1967) **1851**, 38, 47.
(37) U. S. Geological Survey, "Water Resources Data for California, 1968 W. Y.," Pt. 1, 426 (1968).
(38) Barnes, Ivan, *Geochim Cosmochim. Acta* (1965) **29**, 85.
(39) Enright, J. T., *Ecology* (1969) **50**, 1070.

(40) Smith, G. M., "The Fresh Water Algae of the United States," 2nd ed., p. 451, McGraw-Hill, New York, 1950.
(41) Dana, E. D., Ford, W. E., "A Textbook of Mineralogy," 4th ed., pp. 536, 659, 663, Wiley, New York, 1947.
(42) Chow, V. T., "Handbook of Applied Hydrology," Sec. 14, p. 2, McGraw-Hill, New York, 1964.
(43) Storey, H. C., Hobba, R. L., Roas, J. M., "Handbook of Applied Hydrology," Sec. 22, p. 10, McGraw-Hill, New York, 1964.
(44) Jamieson, D. G., Amerman, C. R., *J. Hydrol.* (1969) **8**, 122.
(45) Pinder, G. F., Jones, J. F., *Water Resources Res.* (1969) **5**, 438.
(46) Steele, T. D., "Seasonal Variations in Chemical Quality of Surface Water in the Pescadero Creek Watershed," Ph.D. dissertation, Stanford University, 1968.
(47) U. S. Weather Bureau, "Climatological Data for California," Vols. 11 and 12, 1966.

RECEIVED June 29, 1970. Publication authorized by the Director, U. S. Geological Survey.

5

A Dynamic Model of the Phytoplankton Population in the Sacramento–San Joaquin Delta

DOMINIC M. DI TORO, DONALD J. O'CONNOR, and
ROBERT V. THOMANN

Environmental Engineering and Science Program, Manhattan College,
Bronx, N. Y. 10471

The quality of natural waters can be markedly influenced by the growth and distribution of phytoplankton, whose population development can be accelerated by the addition of nutrients resulting from man's activities or natural processes. The resulting fertilization provides more than ample inorganic nutrients for excessive phytoplankton growth. This is commonly referred to as eutrophication. This investigation presents a mathematical model of this phenomenon as a component in the solution of eutrophication problems. The model of the dynamics of phytoplankton populations is based on the principle of conservation of mass. The growth and death kinetic formulations of the phytoplankton and zooplankton are empirically determined by an analysis of existing experimental data. The resulting equations are compared with two years of data from the tidal portion of the San Joaquin River, California.

The quality of natural waters can be markedly influenced by the growth and distribution of phytoplankton. Utilizing radiant energy, these microscopic plants assimilate inorganic chemicals and convert them to cell material which, in turn, is consumed by the various animal species in the next tropic level. The phytoplankton, therefore, are the base of the food chain in natural waters, and their existence is essential to all aquatic life.

The quality of a body of water can be adversely affected if the population of phytoplankton becomes so large as to interfere with either water use or the higher forms of aquatic life. In particular, high concentrations of algal biomass cause large diurnal variations in dissolved oxygen which can be fatal to fish life. Also, the growths can be nuisances in themselves, especially when they decay and either settle to the bottom or accumulate in windrows on the shoreline. Phytoplankton can cause taste and odor problems in water supplies and, in addition, contribute to filter clogging in the water treatment plant.

The development of large populations of phytoplankton and, in some cases, larger aquatic plants can be accelerated by the addition of nutrients which result from man's activities or natural processes. The resulting fertilization provides more than ample inorganic nutrients, with the resulting development of excessive phytoplankton. This sequence of events is commonly referred to as eutrophication.

Generally, the management of water systems subjected to accelerated eutrophication because of waste discharges has been largely subjective. Extensive programs of nutrient removal have been called for, with little or no quantitative prediction of the effects of such treatment programs. A quantitative methodology is required to estimate the effect of proposed treatment programs that are planned to restore water quality or to predict the effects of expected future nutrient discharges. This methodology should include a model of the phytoplankton population which approximates the behavior of the phytoplankton in the water body of interest and, therefore, can be used to test the effects of the various control procedures available. In this way, rational planning and water quality management can be instituted with at least some degree of confidence that the planned results actually will be achieved.

This paper presents a phytoplankton population model in natural waters, constructed on the basis of the principle of conservation of mass. This is an elementary physical law which is satisfied by macroscopic natural systems. The use of this principle is dictated primarily by the lack of any more specific physical laws which can be applied to these biological systems. An alternate conservation law, that of conservation of energy, can also be used. However, the details of how mass is transferred from species to species are better understood than the corresponding energy transformations. The mass interactions are related, among other factors, to the kinetics of the populations, and it is this that the bulk of the paper is devoted to exploring.

Conservation of mass has been successfully applied to the modeling of the dissolved oxygen distribution in natural waters as well as the distribution of salinity and other dissolved substances. The resulting models have proved useful in guiding engineering and management decisions concerned with the efficient utilization of water resources and the protection of their quality. It is felt that the phytoplankton model presented herein can serve a similar purpose by providing a basis for predicting the effects of nutrient control programs on the eutrophication of natural waters.

Thus, the primary purpose of this paper is to introduce a quantitative model of phytoplankton population dynamics in natural waters. It is within this problem context that the simplifications, assumptions, and generally the structure of the model is formulated. An attempt is made to make the equations representative of the biological mechanisms while still retaining a sufficient simplicity so that the result is tractable and useful.

Review of Previous Models

The initial attempts to model the dynamics of a phytoplankton population were based on a version of the law of conservation of mass in which the hydrodynamic transport of mass is assumed to be insignificant. Let $P(t)$ be the concentration of phytoplankton mass at time t in a suitably chosen region of water. The principle of conservation of mass can be expressed as a differential equation

$$\frac{dP}{dt} = S$$

where S is the net source or sink of phytoplankton mass within the region. If hydrodynamic transport is not included, then the rate at which P increases or decreases depends only on the internal sources and sinks of phytoplankton in the region of interest.

The form of the internal sources and sinks of phytoplankton is dictated by the mechanisms which are assumed to govern the growth and death of phytoplankton. Fleming (1939), as described by Riley (1), postulated that spring diatom flowering in the English Channel is described by the equation

$$\frac{dP}{dt} = [a - (b + ct)]P$$

where P is the phytoplankton concentration, a is a constant growth rate, and $(b + ct)$ is a death rate resulting from the grazing of zooplankton. The zooplankton population, which is increasing owing to its grazing, results in an increasing death rate which is approximated by the linear increase of the death rate as a function of time.

A less empirical model has been proposed by Riley (1946) (1) based on the equation

$$\frac{dP}{dt} = [P_h - R - G]P$$

where P_h is the photosynthetic growth rate, R is the endogenous respiration rate of the phytoplankton, and G is the death rate owing to zooplankton grazing. A major improvement in Riley's equation is the attempt to relate the growth rate, the respiration rate, and the grazing to more fundamental environmental variables such as incident solar radiation, temperature, extinction coefficient, and observed nutrient and zooplankton concentration. As a consequence, the coefficients of the equations are time-variable since the environmental parameters vary throughout the year. This precludes an analytical solution to the equation, and numerical integration methods must be used. Three separate applications (2, 3, 4) of these equations to the near-shore ocean environment have been made, and the resulting agreement with observed data is quite encouraging.

A complex set of equations, proposed by Riley, Stommel, and Bumpus (1949) (5) first introduced the spatial variation of the phytoplankton with respect to depth into the conservation of mass equation. In addition, a conservation of mass equation for a nutrient (phosphate) was also introduced, as well as simplified equations for the herbivorous and carnivorous zooplankton concentrations. The phytoplankton and nutrient equations were applied to 20 volume elements which extended from the surface to well below the euphotic zone. In order to simplify the calculations, a temporal steady-state was assumed to exist in each volume element. Thus, the equations apply to those periods of the year during which the dependent variables are not changing significantly in time. Such conditions usually prevail during the summer months. The results of these calculations were compared with observed data, and again the results were encouraging.

Steele (1956) (6) found that the steady-state assumption did not apply to the seasonal variation of the phytoplankton population. Instead, he used two volume segments to represent the upper and lower water levels and kept the time derivatives in the equations. Thus, both temporal and spatial variations were considered. In addition, the differential equations for phytoplankton and zooplankton concentration were coupled so that the interactions of the populations could be studied, as well as the nutrient–phytoplankton dependence. The coefficients of the equations were not functions of time, however, so that the effects of time-varying solar radiation intensity and temperature were not included. The equations were numerically integrated and the results compared with the observed distribution. Steele applied similar equations to the vertical distribution of chlorophyll in the Gulf of Mexico (7).

The models proposed by Riley et al. and Steele are basically similar. Each consider the primary dependent variables to be the phytoplankton, zooplankton, and nutrient concentration. A conservation of mass equation is written for each species, and the spatial variation is incorporated by considering finite volume elements which interact because of vertical eddy diffusion and downward advective transport of the phytoplankton. Their equations differ in some details (for example, the growth coefficients that were used and the assumptions of steady state) but the principle is the same. In addition, these equations were applied by the authors to actual marine situations and their solutions compared with observed data. This is a crucial part of any investigation discussion wherein the assumptions that are made and the approximations that are used are difficult to justify *a priori*.

The models of both Riley and Steele have been reviewed in greater detail by Riley (*1*) in a discussion of their applicability and possible future development. The difficulties encountered in formulating simple

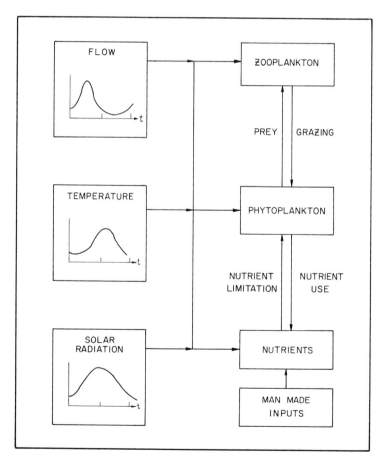

*Figure 1.   Interactions of environmental variables and the phyto-
plankton, zooplankton, and nutrient systems*

theoretical models of phytoplankton–zooplankton population models were discussed by Steele (8).

Other models have been proposed which follow the outlines of the equations already discussed. Equations with parameters that vary as a function of temperature, sunlight, and nutrient concentration have been presented by Davidson and Clymer (9) and simulated by Cole (10). A set of equations which model the population of phytoplankton, zooplankton, and a species of fish in a large lake have been presented by Parker (11). The application of the techniques of phytoplankton modeling to the problem of eutrophication in rivers and estuaries has been proposed by Chen (12). The interrelations between the nitrogen cycle and the phytoplankton population in the Potomac Estuary has been investigated using a feed-forward–feed-back model of the dependent variables, which interact linearly following first order kinetics (13).

The formulations and equations presented in the subsequent sections are modifications and extensions of previously presented equations which incorporate some additional physiological information on the behavior of phytoplankton and zooplankton populations. In contrast to the majority of the applications of phytoplankton models which have been made previously, the equations presented in the subsequent sections are applied to a relatively shallow reach of the San Joaquin River and the estuary further downstream. The motivation for this application is an investigation of the possibility of excessive phytoplankton growths as environmental conditions and nutrient loadings are changed in this area. Thus, the primary thrust of this investigation is to produce an engineering tool which can be used in the solution of engineering problems to protect the water quality of the region of interest.

Phytoplankton System Interactions

The major obstacle to a rigorous quantitative theory of phytoplankton population dynamics is the enormous complexity of the biological and physical phenomena which influence the population. It is necessary, therefore, to idealize and simplify the conceptual model so that the result is a manageable set of dependent systems or variables and their interrelations. The model considered in the following sections is formulated on the basis of three primary dependent systems: the phytoplankton population, whose behavior is the object of concern; the herbivorous zooplankton population, which are the predators of the phytoplankton, utilizing the available phytoplankton as a food supply; and the nutrient system, which represents the nutrients, primarily inorganic substances, that are required by the phytoplankton during growth. These three systems are affected not only by their interactions, but also by external environmental variables.

The three principal variables considered in this analysis are temperature, which influences all biological and chemical reactions, dispersion and advective flow, which are the primary mass transport mechanisms in a natural body of water, and solar radiation, the energy source for the photosynthetic growth of the phytoplankton.

In addition to these external variables, the effect of man's activities on the system is felt predominately in the nutrient system. Sources of the necessary nutrients may be the result of, for example, inputs of waste-water from municipal and industrial discharges or agricultural runoff. The man-made waste loads are in most cases the primary control variables which are available to affect changes in the phytoplankton and zooplankton systems. A schematic representation of these systems and their interrelations is presented in Figure 1.

In addition to the conceptual model which isolates the major interacting systems, a further idealization is required which sets the lower and upper limits of the temporal and spatial scales being considered. Within the context of the problem of eutrophication and its control, the seasonal distribution of the phytoplankton is of major importance, so that the lower limit of the temporal scale is on the order of days. The spatial scale is set by the hydrodynamics of the water body being considered. For example, in a tidal estuary, the spatial scale is on the order of miles whereas in a small lake it is likely a good deal smaller. The upper limits for the temporal and spatial extent of the model are dictated primarily by practical considerations such as the length of time for which adequate information is available and the size of the computer being used for the calculations.

These simplifying assumptions are made primarily on the basis of an intuitive assessment of the important features of the systems being considered and the experience gained by previous attempts to address these and related problems in natural bodies of water. The basic principle to be applied to this conceptual model, which can then be translated into mathematical terms, is that of conservation of mass.

Conservation of Mass

The principle of conservation of mass is the basis upon which the mathematical development is structured. Alternate formulations, such as those based on the conservation of energy, have been proposed. However, conservation of mass has proved a useful starting point for many models of the natural environment.

The principle of conservation of mass simply states that the mass of the substances being considered within an arbitrarily selected volume must be accounted for by either mass transport into and out of the volume or as mass produced or removed within the volume. The transport of mass in a natural water system arises primarily from two phenomena: dispersion, which is caused by tidal action, density differences, turbulent diffusion, wind action, etc.; and advection owing to a unidirectional flow —for example, the fresh water flow in a river or estuary or the prevailing currents in a bay or a near-shore environment. The distinction between the two phenomena is that, over the time scale of interest, dispersive mass transport mixes adjacent volumes of water so that a portion of the water in adjacent volume elements is interchanged, and the mass transport is proportional to the difference in concentrations of mass in adjacent volumes. Advective transport, however, is transport in the direction of the advective flow only. In addition to the mass transport phenomena, mass in the volume can increase resulting from sources within the volume. These sources represent the rate of addition or removal of mass per unit time per unit volume by chemical and biological processes.

A mathematical expression of conservation of mass which includes the terms to describe the mass transport phenomena and the source term is a partial differential equation of the following form

$$\frac{\partial P}{\partial t} = \nabla \cdot E \nabla P - \nabla \cdot QP + S_P \tag{1}$$

where P (x, y, z, t) is the concentration of the substance of interest—e.g., phytoplankton biomass—as a function of position and time; E is the diagonal matrix of dispersion coefficients; Q is the advective flow rate vector; S_P is the vector whose terms are the rate of mass addition by the sources and sinks; and ∇ is the gradient operator. This partial differential equation is too general to be solved analytically, and numerical techniques are used in its solution.

An effective approximation to Equation 1 is obtained by segmenting the water body of interest into n volume elements of volume V_j and representing the derivatives in Equation 1 by differences. Let V be the $n \times n$ diagonal matrix of volumes V_j; A, the $n \times n$ matrix of dispersive and advective transport terms; S_P, the n vector of source terms S_{Pj}, averaged over the volume V_j; and P, the n vector of concentrations P_j, which are the concentrations in the volumes. Then the finite difference equations can be expressed as a vector differential equation

$$V\mathring{P} = AP + VS_P \tag{2}$$

where the dot denotes a time derivative. The details of the application of this version of the dispersion advection equation to natural bodies of water has been presented by Thomann (14) and reviewed by O'Connor et al. (15).

The main interest in this report is centered on the source terms S_{Pj} for the particular application of these equations to the phytoplankton population in natural water bodies. It is convenient to express the source term of phytoplankton, S_{Pj}, as a difference between the growth rate, G_{Pj}, of phytoplankton and their death rate, D_{Pj}, in the volume V_j. That is

$$S_{Pj} = (G_{Pj} - D_{Pj})P_j \qquad (3)$$

where G_{Pj} and D_{Pj} have units [day^{-1}]. The subscript P identifies the quantities as referring to phytoplankton; the subscript j refers to the volume element being considered. The balance between the magnitude of the growth rate and death rate determines the rate at which phytoplankton mass is created or destroyed in the volume element V_j. Thus, the form of the growth and death rates as functions of environmental parameters and dependent variables is an important element in a successful phytoplankton population model.

Phytoplankton Growth Rate

The growth rate of a population of phytoplankton in a natural environment is a complicated function of the species of phytoplankton present and their differing reactions to solar radiation, temperature, and the balance between nutrient availability and phytoplankton requirements. The complex and often conflicting data pertinent to this problem have been reviewed recently by Hutchinson (1967) (16), Strickland (1965) (17), Lund (1965) (18), and Raymont (1963) (19). The available information is not sufficiently detailed to specify the growth kinetics for individual phytoplankton species in natural environments. Hence, in order to accomplish the task of constructing a growth rate function, a simplified approach is followed. The problem of different species and their associated nutrient and environmental requirements is not addressed. Instead, the population is characterized as a whole by a measurement of the biomass of phytoplankton present. Typical quantities used are the chlorophyll concentration of the population, the number of organisms per unit volume, or the dry weight of the phytoplankton per unit volume (20). With a choice of biomass units established, the growth rate expresses the rate of production of biomass as a function of the important environmental variables. The environmental variables to be considered below are light, temperature, and the various nutrients which are necessary for phytoplankton growth.

Light and Temperature. Consider a population of phytoplankton, either a natural association or a single species culture, and assume that the optimum or saturating light intensity for maximum growth rate of biomass is present and illuminates all the cells, and further that all the

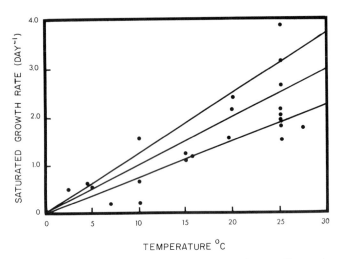

*Figure 2.   Phytoplankton saturated growth rate (base e)
as a function of temperature*

necessary nutrients are present in sufficient quantity so that no nutrient is in short supply. For this condition, the growth rate that is observed is called the maximum or saturated growth rate, K'. Measurements of K' (base e) as a function of temperature are shown in Figure 2 and listed in Table I. The experimental conditions under which these data were collected appear to meet the requirements of optimum light intensity and sufficient nutrient supply. The data presented are selected from larger groups of reported values, and they represent the maximum of these reported growth rates. The presumption is that these large values reflect the maximum growth rates achievable. From an ecological point of view, it is necessary to consider the species most able to compete, and, in terms of growth rate, it is the species with the largest growth rate which will predominate. A straight-line fit to this data appears to be a crude but reasonable approximation of the data relating saturated growth rate K' to temperature, T

$$K' = K_1 T \tag{4}$$

where K_1 has values in the range 0.10 ± 0.025 day^{-1} °C^{-1}. This coefficient indicates an approximate doubling of the saturated growth rate for a temperature change from 10° to 20°C, in accordance with the generally reported temperature-dependence of biological growth rates. The optimum temperature for algal growth appears to be in the range between 20° and 25°C, although thermophilic strains are known to exist (27). At higher temperatures, there is usually a suppression of the saturated growth rate, and the straight-line approximation is no longer valid. It should also be noted that the scatter in the data in Figure 2 is sufficiently large so that the linear dependence on temperature and also the magnitude of K' can vary considerably in particular situations.

In the natural environment, the light intensity to which the phytoplankton are exposed is not uniformly at the optimum value but it varies as a function of depth because of the natural turbidity present and as a function of time over the day. Thus, the phytoplankton in the lower layers are exposed to intensities below the optimum and those at the surface may be exposed to intensities above the optimum so that their growth rate would be inhibited. Figure 3b,c,d from Ryther (28) are plots of the photosynthesis rate normalized by the photosynthesis rate at the optimum or saturating light intensity *vs.* the light intensity, I, incident on the populations. Figure 3a is a plot of function

$$F(I) = \frac{I}{I_s} \exp\left[-\frac{I}{I_s} + 1 \right] \tag{5}$$

for $I_s = 2000$ ft-candles, proposed by Steele (8) to describe the light-dependence of the growth rate of phytoplankton.

The similarity between this function and data from Ryther is sufficient to warrant the use of this expression to express the influence of nonoptimum light intensity on the growth rate of phytoplankton. Other workers have suggested different forms for this relationship (29, 30).

Table I. Maximum Growth Rates as a Function of Temperature

Ref.	Organism	Temperature,	Saturated Growth Rate, $K'(Base_e, Day^{-1})$
21	Chlorella ellipsoidea	25	3.14
	(green alga)	15	1.2
22	Nannochloris atomus	20	2.16
	(marine flagellate)	10	1.54
23	Nitzschia closterium	27	1.75
	(marine diatom)	19	1.55
		15.5	1.19
		10	0.67
5	Natural association	4	0.63
		2.6	0.51
24	Chlorella pyrenoidosa	25	1.96
24	Scenedesmus quadricauda	25	2.02
25	Chlorella pyrenoidosa	25	2.15
25	Chlorella vulgaris	25	1.8
25	Scenedesmus obliquus	25	1.52
25	Chlamydomonas reinhardti	25	2.64
26	Chlorella pyrenoidosa	10	0.2
	(synchronized culture)	15	1.1
	(high-temperature strain)	20	2.4
		25	3.9

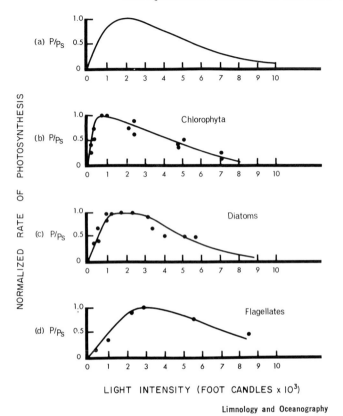

Figure 3. Normalized rate of photosynthesis vs. incident light intensity: (a) Theoretical curve after Steele (8); (b,c,d) Data after Ryther (28)

These variations approximately follow the shape of Equation 5 for low light intensities but differ for the region of high light intensities, usually by not decreasing after some optimum intensity is reached. In particular, Tamiya *et al.* (*21*) have investigated the growth rate of *Chlorella ellipsoidea* to various light and temperature regimes. The saturated growth rates as a function of temperature are included in Figure 2. The influence of varying light intensity fits the function

$$F(I) = \frac{I}{I + K'/\alpha} \tag{6}$$

where K' is the saturated growth rate and α is a constant ($\alpha = 0.45$ day^{-1} kilolux^{-1}). However, since K' is a function of temperature, the saturating light intensity for Equation 6 is also a function of temperature. Similar data obtained by Sorokin *et al.* (*26*) using a high-temperature strain of *Chlorella pyrenoidosa* support the temperature-dependence of the saturating light intensity for chlorella. Therefore, in using Equation 5, a temperature-dependent light saturation intensity may be warranted.

At this point in the analysis, the effect of the natural environment on the light available to the phytoplankton must be included. Equation 5 expresses the reduction in the growth rate caused by nonoptimum light intensity. This expression can therefore be used to calculate the reduction in growth rate to be expected at any intensity. However, this is too detailed a description for conservation of mass equations which deal with homogeneous volume elements, V_j, and the growth rate within these elements. What is required is averages of the growth rate over the volume elements.

In order to calculate the light intensity which is present in the volume V_j, the light penetration at the depth of water where V_j is located must be evaluated. The rate at which light is attenuated with respect to depth is given by the extinction coefficient, k_e. That is, at a depth z, the intensity at that depth, $I(z)$, is related to the surface intensity, I_0, by the formula

$$I(z) = I_0 \exp(-k_e z) \tag{7}$$

where $z = 0$ is the water surface and z is positive downward. Thus, the reduction of the saturated growth rate at any depth z resulting from the nonoptimum light intensity present is given by Equation 7, substituted into Equation 5.

$$F[I(z)] = \frac{I_0 e^{-k_e z}}{I_s} \exp\left[\frac{-I_0 e^{-k_e z}}{I_s} + 1\right] \tag{8}$$

To apply this equation to the finite volume elements, within which it is assumed that the phytoplankton concentration is uniform, it is necessary to average this reduction factor throughout the depth of the volume element V_j. Let H_{0j} and H_{1j} be the depths of the surface and bottom, respectively, of the volume element V_j. For example, if the volume element V_j extends from the water surface to the bottom of the water body, then $H_{0j} = 0$ and H_{1j} is the water depth at the location of V_j. For the sake of simplicity, it is assumed that this is the case. If $H_{0j} \neq 0$, a straightforward generalization of the following average is required.

In addition to an average over depth, it is also expedient to average the phytoplankton growth rate over a time interval. Since the time scale within which this analysis is addressed is the week-to-week change in the population over a year, a daily average is appropriate. For simplicity, it is assumed that the incident solar radiation as a function of time over a day is given by the function

$$I_o(t) = I_a \qquad 0 < t < f$$
$$I_o(t) = 0 \qquad f < t < 1 \tag{9}$$

where f is the daylight fraction of the day (*i.e.*, the photo period) and I_a is the average incident solar radiation intensity during the photo period.

Let r_j be the reduction in growth rate attributed to nonoptimum light conditions in volume V_j, averaged over depth and time. Then r_j is given by

$$r_j = \frac{1}{H_j} \int_0^{H_j} \frac{1}{T} \int_0^f \frac{I_a e^{-k_{ej}z}}{I_s} \exp\left[\frac{-I_a e^{-k_{ej}z}}{I_s} + 1\right] dt \, dz \qquad (10)$$

where $T = 1$ day, the time-averaging interval, $H_{1j} = H_j =$ the depth of segment V_j, and k_{ej} is the extinction coefficient in V_j. The result is

$$r_j = \frac{1}{k_{ej}H_j} e^{-\alpha_{1j}} - e^{-\alpha_{0j}} \qquad (11)$$

where

$$\alpha_{1j} = \frac{I_a}{I_s} e^{-k_{ej}H_j} \qquad (12)$$

$$\alpha_{0j} = \frac{I_a}{I_s}$$

The integral given by Equation 10 is a form of an integral used by Steeman Nielson (1952), Talling (1957), and Ryther and Yentsch (1957), as described by Vollenweider (1958) (*30*), and, in particular, Steele (*8*), for the purpose of relating an instantaneous rate (*e.g.*, growth, photosynthesis, etc.) to an average day rate and an average depth rate.

The reduction factor r_j is a function of the extinction coefficient k_{ej} of the volume V. However, the extinction coefficient is a function of the phytoplankton concentration present if their concentration is large. Thus, an important feedback mechanism exists which can have a marked effect on the growth rate of phytoplankton. As the concentration of phytoplankton in a volume element increases, the extinction coefficient, particularly at the green wavelengths, starts to increase. This mechanism is called self-shading. The most straightforward approach to including this effect into the growth rate expression is to specify the extinction coefficient as a function of the phytoplankton concentration

$$k_{ej} = k'_{ej} + h(P_j) \qquad (13)$$

where k'_{ej} is the extinction coefficient attributable to other causes and k_{ej} includes the phytoplankton's contribution. The function $h(P_j)$ has been investigated by Riley (*31*), who found that it can be approximated by

$$h(P_j) = 0.0088 \, P_j + 0.054 \, P_j^{2/3} \tag{14}$$

where P_j has the units μg/liter chlorophyll$_a$ concentration and h has units m^{-1}. A more recent investigation (32) shows that this relationship applies to coastal waters of Oregon for a range in chlorophyll$_a$ concentration of from 0 to 5.0 mg Chl$_a$/m^3.

A theoretical basis for this relationship is the Beer-Lambert law, which expresses the extinction coefficient in terms of the concentration of light-absorbing material. For dense algal cultures, this law has been experimentally verified (33). A similar relationship based on this law has been proposed by Chen (12) from the data of Azad and Borchardt (34)

$$h(P_j) = 0.17 \, P_j \tag{15}$$

For h in m^{-1} and P_j, the phytoplankton concentration is mg/liter of dry weight. This expression gives values comparable with Equation 14 for a reasonable conversion factor for the units involved.

To summarize the analysis to this point, the saturated growth rate K' has been estimated from available data and its temperature dependence established. The reduction to be expected from nonoptimum light intensities has been quantified and used to calculate the reduction in growth rate, r_j, to be expected in each volume element V_j as a function of the extinction coefficient and the depth of the segment. The mechanism of self-shading has been included by specifying the chlorophyll dependence of the extinction coefficient. It remains to evaluate the effect of nutrients on the growth rate.

Nutrients. The effects of various nutrient concentrations on the growth of phytoplankton has been investigated and the results are quite complex. As a first approximation to the effect of nutrient concentration on the growth rate, it is assumed that the phytoplankton population in question follow Monod growth kinetics with respect to the important nutrients. That is, at an adequate level of substrate concentration, the growth rate proceeds at the saturated rate for the temperature and light conditions present. However, at low substrate concentration, the growth rate becomes linearly proportional to substrate concentration. Thus, for a nutrient with concentration N_j in the jth segment, the factor by which the saturated growth rate is in the jth segment reduced is: $N_j/(K_m + N_j)$. The constant, K_m, which is called the Michaelis or half saturation constant, is the nutrient concentration at which the growth rate is half the saturated growth rate. There exists an increasing body of experimental evidence to support the use of this functional form for the dependence of the growth rate on the concentration of either phosphate (35), nitrate, or ammonia (36) if only one of these nutrients is in short supply. An

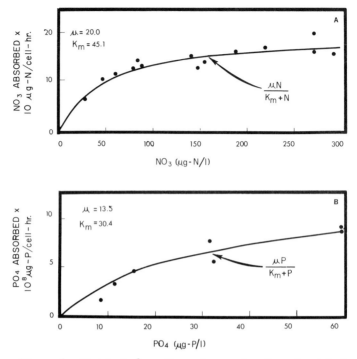

Figure 4. Nutrient absorption rate as a function of nutrient concentration: comparison of Michaelis Menton theoretical curve with data from Ketchum (37)

example of this behavior, using the data from Ketchum (37), is shown in Figure 4a for the nitrate uptake rate as a function of nitrate concentration and in Figure 4b for the phosphate uptake as a function of phosphate concentration. These results are from batch experiments. Similar results from chemostat experiments, which seem to be more suitable but more lengthy for this type of analysis, have also been obtained. Table II is a listing of measured and estimated Michaelis constants for ammonia, nitrate, and phosphate. The estimates are obtained by taking one-third the reported saturation concentration of the nutrients. These measurements and estimates indicate that the Michaelis constant for phosphorus is approximately 10 μg P/liter and for inorganic nitrogen forms in the range from 1.0 to 100 μg N/liter, depending on the species and its previous history.

The data on the effects of the concentration of other inorganic nutrients on the growth rate is less complete. Since algae use carbon dioxide as their carbon source during photosynthesis, this is clearly a nutrient which can reduce the growth rate at low concentrations (43). Reported saturation concentration for Chlorella is < 0.1% atm (24).

The silicate concentration is a factor in the growth rate of diatoms for which it is an essential requirement. The saturated growth rate concentration is in the range of 50–100 μg Si/liter (*17*).

There are a large number of trace inorganic elements which have been implicated in the growth processes of algae, among which are iron [for which a Michaelis constant of 5 μg/liter for reactive iron has been reported (*39*)], manganese, calcium, magnesium, and potassium (*18*). However, the significance of these elements in the growth of phytoplankton in natural waters is still unclear. Trace organic nutrients have also been shown to be necessary for most species of algae: 80% of the strains studied require vitamin B_{12}, 53% require thiamine, and 10% require biotin (*44*). Presumably, these nutrients are available in sufficient quantities in natural waters so that their concentration does not appreciably affect the growth rate.

In the preceeding discussion of nutrient influences on the growth rate, it is tacitly assumed that only one nutrient is in short supply and all the other nutrients are plentiful. This is sometimes the case in a natural body of water. However, it is also possible that more than one nutrient is in short supply. The most straightforward approach to formulating the growth rate reduction caused by a shortage of more than one nutrient is to multiply the saturated growth rate by the reduction factor for each nutrient. This approach has also been suggested by Chen (*12*).

Table II.   Michaelis Constants for Nitrogen and Phosphorus

Ref.	*Organism*	*Nutrient*	*Michaelis Constant, μg/Liter as N or P*
38	*Chaetoceros gracilis* (marine diatom)	PO_4	25
39	*Scenedesmus gracile*	total N	150
		total P	10
40	Natural association	PO_4	6[a]
	Microcystis aeruginosa (blue-green)	PO_4	10[a]
41	*Phaeodactylum tricornutum*	PO_4	10
36	Oceanic species	NO_3	1.4–7.0
	Oceanic species	NH_3	1.4–5.6
36	Neritic diatoms	NO_3	6.3–28
	Neritic diatoms	NH_3	7.0–120
36	Neritic or littoral	NO_3	8.4–130
	Flagellates	NH_3	7.0–77
42	Natural association Oligotrophic	NO_3	2.8
		NH_3	1.4–8.4
42	Natural association Eutrophic	NO_3	14
		NH_3	18

[a] Estimated.

As an example, the data from Ketchum (37) for the rate of phosphate absorption as a function of both phosphate and nitrate concentration can be satisfactorily fit with a product of two Michaelis–Menton expressions. The resulting fit, obtained by a multiple nonlinear regression analysis, is shown in Figure 5. The Michaelis constants that result are 28.4 μg NO_3–N/liter and 30.3 μg PO_4–P/liter, with a saturated absorption rate of 15.1×10^{-8} μg PO_4–P/cell-hr. This approximation to the growth rate behavior as a function of more than one nutrient must be regarded as only a first approximation, however, since the complex interaction reported between the nutrients is neglected.

The result of the above investigation is the following growth rate expression. For the case of one limiting nutrient, N, with Michaelis constant K_m, the growth expression for the rate in the jth segment is

$$G_{Pj} = K_1 T_j \left(\frac{2.718\,f}{k_{ej}H_j} \left(e^{-\alpha_{1j}} - e^{-\alpha_{0j}} \right) \right) \left(\frac{N_j}{K_m + N_j} \right) \tag{16}$$

in which Equations 4 and 11 have been combined. This is the functional

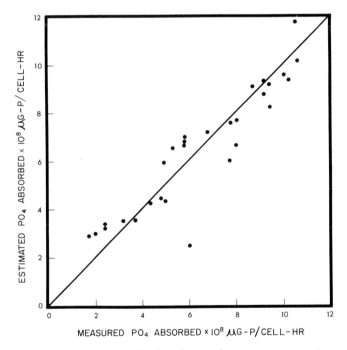

Figure 5. Measured phosphate absorption rate, after Ketchum (37), vs. phosphate absorption rate estimated using $\mu N_1 N_2 / (K_{m1} + N_1)(K_{m2} + N_2)$ where N_1 and N_2 are the nitrate and phosphate concentrations, respectively

form that is used subsequently in the applications of these equations to natural phytoplankton populations.

Comparison with Other Growth Rate Expressions

The most extensive investigation of the relationship between the growth rate of natural phytoplankton populations and the significant environmental variables, within the context of phytoplankton models, is that of Riley *et al.* (1949) (5). The expression which results from their work is

$$\log \left[\frac{G_P}{K'I_0 - G_P} \right] = 22.884 + \log \nu_p - \log I_0 - \frac{6573.8}{T'} \tag{17}$$

where G_P is the growth rate (day^{-1}), $K' = 7.6$, I_0 = average daily incident solar radiation (1y/min), T' = temperature in °K, and ν_p is the nutrient reduction factor for phosphate concentration, N_P, defined as

$$\nu_p = 1.0 \qquad\qquad N_P > 0.55 \text{ mg-at}/m^3 \tag{18}$$

$$\nu_p = (0.55)^{-1}N_P \qquad N_P < 0.55 \text{ mg-at}/m^3$$

In order to compare this expression with that in the previous section, let the nutrient reduction factor be replaced by a Michaelis–Menton expression.

$$\nu_p = \frac{N_p}{K_{mp} + N_p} \tag{19}$$

where K_{mp} is the Michaelis constant for phosphate. To be comparable with Equation 16, K_{mp} should equal approximately 0.20 mg-at/m^3 (6.2 mg P/m^3). Using Equation 19 for ν_p, the growth rate expression becomes

$$G_P = K'I_0 \left[\frac{\gamma_i(T)}{\gamma_i(T) + I_0} \right] \left[\frac{N_P}{K_m \left[\frac{I_0}{\gamma_i(T) + I_0} \right] + N_p} \right] \tag{20}$$

where

$$\log_{10} \gamma_i(T) = \frac{22.9\ (T) - 336.4}{T + 273} \tag{21}$$

and T is temperature in degrees centigrade. To compare this expression with that proposed in the previous section, consider first the nutrient saturated growth rate as a function of solar radiation intensity and temperature. The equations are compared in Figure 6a as a function of total daily solar radiation for three temperatures. The dotted line is Equation

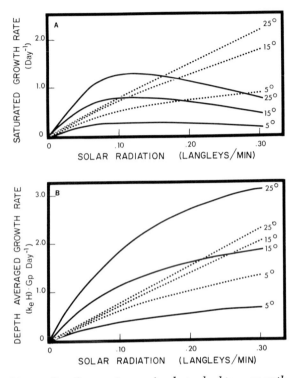

Figure 6. Comparison of phytoplankton growth rates as a function of incident solar radiation intensity and temperature

20, and the solid line is the product of Equations 4 and 5. The rate expressions are comparable, although two differences are apparent. In Riley's expression the effect of temperature is less pronounced in the 15° to 25°C range, and the effect of higher daily average solar radiation intensities is opposite (*i.e.*, tends to increase the rate) to that of Equation 5 based on Steele's expression. The growth rate equations averaged over depth are compared in Figure 6b. The depth average rate resulting from Riley's expression is

$$G_P = \frac{K' r_i(T)}{k_e H} \ln \left[1 + I_0 / r_i(T) \right] \tag{22}$$

which is compared with Equation 16. The differences are now more pronounced. In particular, the higher growth rates at lower light intensities given by Equation 16 are reflected in the increased depth average growth rate. This is not unexpected since the majority of the population is exposed to lower light levels at the lower depths. Also, the dependence on temperature is quite different, being linear in the case of Equation 16 but practically suppressed in Equation 22.

An interesting feature of Riley's Equation 20 is the multiplication of the Michaelis constant by an expression which depends on temperature and light intensity. The effect is to lower the Michaelis constant at high temperatures and at high light intensity levels, which seems to be a reasonable behavior for a phytoplankton population.

More elementary growth rate formulations have been proposed which do not span the range of conditions attempted in Equations 16 and 20. In particular, a common proposal is to make the growth rate linearly proportional to the various environmental variables. For example, Davidson and Clymer (1966) (9) assumed that the growth rate is proportional to phosphate concentration and photo period and a temperature factor given by $\exp\left[-(T - 18)^2/18\right]$. This temperature factor is quite different from all others proposed and greatly magnifies the effect of temperature on the growth rate. For example, at $T = 18°C$, the factor equals 1.0, whereas at $T = 9°C$, the factor drops to 0.01, a 100-fold decrease, compared with approximately a 2-fold decrease predicted by Equations 16 and 20. This exaggerated effect seems to be unrealistic.

A complete investigation of the environmental influences on the growth rate is still to be made. In particular, it has been emphasized that there is an interaction between nitrogen and phosphorus limitations as well as other effects which influence the phytoplankton growth rate. Also, these effects are different for differing species. The species-dependent effects are important in the problem of the seasonal succession of phytoplankton species.

For any particular application, it is advisable to investigate the growth rate of the already-existing population, as the resulting expression may differ significantly from the general over-all behavior as described by Equations 16 and 20. Also, in dealing with natural associations of species of phytoplankton, the various constants which result from such an investigation can be considered to be averages over the population, and so they represent in some average way the population behavior as a whole.

Phytoplankton Death Rate

Numerous mechanisms have been proposed which contribute to the death rate of phytoplankton: endogenous respiration rate, grazing by herbivorous zooplankton, a sinking rate, and parasitization (27). The first three mechanisms have been included in previous models for phytoplankton dynamics, and they have been shown to be of general importance.

respiration rate of the population as a whole is greater than the photosynthesis or growth rate, there is a net loss of phytoplankton carbon, and the population biomass is reduced in size. The respiration rate as a function of temperature has been investigated, and some measurements are presented in Figure 7 and Table III. A straight line seems to give an adequate fit of the data; that is, Respiration Rate $= K_2 T$. For the respiration rate in days^{-1} and T in °C, the value of K_2 is in the range 0.005 $\pm$ 0.001. The lack of any more precise data precludes exploring the respiration rate's dependence on other environmental variables. However, an important interaction has been suggested by Lund (18). During nutrient-depleted conditions, "many algae pass into morphological or physiological resting stages under such unfavorable conditions. Resting stages are absent in *Asterionella formosa*, and this is why a mass death occurs in the nutrient-depleted epilimnion after the vernal maximum." In terms of the respiration rate, the resting stage corresponds to a lowering of the respiration rate as the nutrient concentrations decrease. Thus, a Michaelis–Menton expression for the respiration rate nutrient dependence may also be required, and this dependence should be investigated experimentally. This mechanism is quite significant from a water quality point of view since the deaths of algae after a bloom is of primary concern in protecting

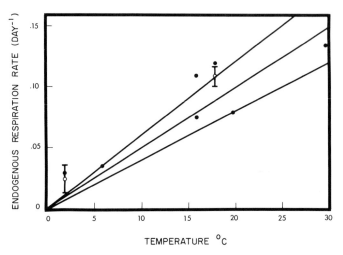

Figure 7.   Endogenous respiration rate of phytoplankton vs. temperature; after Riley (5)

Table III. Endogenous Respiration Rates of Phytoplankton (5)

Organism	Temperature, °C	Endogenous Respiration Rate, Day^{-1} (Base$_e$)
Nitzschia closterium	6	0.035
	35	0.170
Nitzschia closterium	20	0.08
Coscinodiscus excentricus	16	0.075
	16	0.11
Natural association	2	0.03
	18	0.12
	2.0	0.024 ± 0.012
	17.9	0.110 ± 0.007

the quality of natural bodies of water. The resulting mass of dead algal cells becomes a sink of dissolved oxygen which can dangerously lower the available oxygen for fish and other aquatic animals.

Grazing by Zooplankton. The interaction between the phytoplankton population and the next trophic level, the herbivorous zooplankton, is a complex process for which only a first approximation can be given. A basic mechanism by which zooplankters feed is by filtering the surrounding water and clearing it of whatever phytoplankton and detritus is present. Thus, the presence of zooplankton contribute to the death rate of phytoplankton since many species of zooplankton prey on phytoplankton as a food source. The filtering or grazing rate of some species of zooplankton have been measured and are presented in Table IV. The grazing rate is sometimes reported as a volume of water filtered per unit time per individual. In order to be applicable to a natural zooplankton population consisting of differing species, these grazing rates are converted to a filtering rate per unit biomass of zooplankton and denoted by C_g. A convenient biomass unit for zooplankton concentration is their dry weight. As can be seen from Table IV, the resulting values of C_g vary considerably. This variation is not unexpected since the measurement of grazing rates of zooplankters is a difficult procedure and subject to large variations in the estimates.

Variations of the filtering rate with temperature change have been reported (47). Examples of this variation are presented in Figure 8 for four species of genus *Daphnia*, a small crustacea (49); two species of *Acartia* (52); and two species of *Centropages* (47), both copepods. The copepods show a marked grazing rate temperature-dependence while the grazing rates of the Daphnia do not vary as markedly. The filtering rate also varies as a function of the size of the phytoplankton cell being ingested (53), the concentration of phytoplankton (51), and the amount of particulate matter present (54). Selective grazing of certain phytoplankton species has also been reported (49). The complexity of this aspect of phytoplankton mortality is such that the use of one grazing coefficient to represent the process must be viewed as a first approximation. However, since this mathematical expression does represent a physiological mechanism that has been investigated and for which reported values of C_g are available, this approximation is a realistic first step. Also, it is difficult to see, aside from refinements as to temperature and phytoplankton concentration dependence, what further improvements could be made in the formulation so long as the phytoplankton and zooplankton population are represented by a biomass measurement which ignores the species present and their individual characteristics. For simplicity in this investigation, the grazing rate is assumed to be a constant. The death rate of phytoplankton resulting from the grazing of zooplankton is given by the expression $C_g Z_j$, where Z_j is the concentration of herbivorous zooplankton biomass in the jth volume element.

For models of the phytoplankton populations in coastal oceanic waters and in lakes, the sinking rate of phytoplankton cells is an important contribution to the mortality of the population. The cells have a net downward velocity, and they eventually sink out of the euphotic zone to the bottom of the water body. This mechanism has been investigated and included in phytoplankton population models (5, 12). However, for the application of these equations to a relatively shallow vertically well mixed river or estuary, the degree of vertical turbulence is sufficient to eliminate the effect of sinking on the vertical distribution of phytoplankton.

Table IV. Grazing Rates of Zooplankton

Ref.	Organism	Reported Grazing Rate	Grazing Rate, Liter/Mg Dry Wt.-Day
	Rotifer		
16	Brachionus calyciflorus	0.05–0.12[a]	0.6–1.5
	Copepod		
5	Calanus sp.	67–208[b]	0.67–2.0
45	Calanus finmarchicus	27[a]	0.05
46	Rhincalamus nasutus	98–670[a]	0.3–2.2
47	Centropages hamatus		0.67–1.6
	Cladocera		
48	Daphnia sp.		1.06
49	Daphnia sp.		0.2–1.6
50	Daphnia magna	81[a]	0.74
51	Daphnia magna	57–82[a]	0.2–0.3
	National Association		
5	Georges Bank	80–110[b]	0.8–1.10

[a] Ml/ animal-day.
[b] Ml/mg wet wt-day.

Therefore, considering only the phytoplankton respiration and the predation by zooplankton, the death rate of phytoplankton is given by the equation

$$D_{Pj} = K_2 T + C_g Z_j \qquad (23)$$

and for a zooplankton biomass concentration Z_j, the mortality rate can be calculated from this equation.

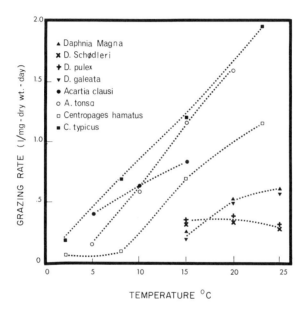

Figure 8.   Grazing rates of zooplankton vs. temperature

This completes the specification of the growth and death rates of the phytoplankton population in terms of the physical variables: light and temperature, the nutrient concentrations, and the zooplankton present. With these variables known as a function of time, it is possible to calculate the phytoplankton population resulting throughout the year. However, the zooplankton population and the nutrient concentrations are not known *a priori* since they depend on the phytoplankton population which develops. That is, these systems are interdependent and cannot be analyzed separately. It is therefore necessary to characterize both the zooplankton population and the nutrients in mathematical terms in order to predict the phytoplankton population which would develop in a given set of circumstances.

The Zooplankton System

As indicated in the previous section, the zooplankton population exerts a considerable influence on the phytoplankton death rate by its feeding on the phytoplankton. In some instances, it has been suggested that this grazing is the primary factor in the reduction of the concentration of phytoplankton after the spring bloom. In the earlier attempts to model the phytoplankton system, the measured concentration of zooplankton biomass was used to evaluate the phytoplankton death rate resulting from grazing. However, it is clear that the same arguments used to develop the equation for the conservation of phytoplankton biomass can be applied directly to the zooplankton system. In particular, the source of zooplankton biomass S_{Zj} within a volume element V_j can be given as the difference between a zooplankton growth rate G_{Zj} and a zooplankton death rate D_{Zj}. Thus, the equation for the source of zooplankton biomass, which is analogous to Equation 3, is

$$S_{Zj} = (G_{Zj} - D_{Zj})Z_j \qquad (24)$$

where G_{Zj} and D_{Zj} have units day^{-1} and Z_j is the concentration of zooplankton carbon in the volume element V_j. The magnitude of the growth rate in comparison with the death rate determines whether the net rate of zooplankton biomass production in V_j is positive, indicating net growth rate, or negative, indicating a net death rate.

As in the case of the phytoplankton population, the growth and death rates, and in fact the whole life cycle of individual zooplankters, are complicated affairs with many individual peculiarities. The surveys by Hutchinson (*16*) and Raymont (*19*) give detailed accounts of their complex biology. It is, however, beyond the scope of this paper to summarize all the differences and species-dependent attributes of the many zooplankton species. The point of view adopted is macroscopic, with the population characterized in units of biomass. The resulting growth and death rates can be thought of as averages over the many species present. These simplifications are made in the interest of providing a model which is simple enough to be manageable and yet representative of the over-all behavior of the populations.

Growth Rate. The grazing mechanism of the zooplankton provides the basis for the growth rate of the herbivorous zooplankton, G_{Zj}. For a filtering rate C_g, the quantity of phytoplankton biomass ingested is $C_g P_j$, where P_j is the phytoplankton biomass concentration in V_j. To convert this rate to a zooplankton growth rate, a parameter which relates the phytoplankton biomass ingested to zooplankton biomass produced, a utilization efficiency, a_{ZP}, is required. However, this utilization efficiency or yield coefficient is not a constant. At high phytoplankton concentrations, the zooplankton do not metabolize all the phytoplankton that they graze, but rather they excrete a portion of the phytoplankton in undigested or semidigested form (55). Thus, utilization efficiency is a function of the phytoplankton concentration. A convenient choice for this functional relationship is $a_{ZP}K_{mP}/(K_{mP} + P_j)$ so that the growth rate for the zooplankton population is

$$G_{Zj} = a_{ZP}C_gK_{mP}\left(\frac{P_j}{K_{mP} + P_j}\right) \tag{25}$$

The resulting growth rate has the same form as that postulated for the nutrient—phytoplankton relationship, namely, a Michaelis–Menton expression with respect to phytoplankton biomass. In fact, the argument which is used to justify its use in Equation 16 can be repeated in this context. The difference is that in this case the substrate or nutrient is phytoplankton biomass, and the microbes are the zooplankton. The Michaelis constant K_{mP} is the phytoplankton biomass concentration at which the growth rate G_{Zj} is one-half the maximum possible growth rate $a_{ZP}C_gK_{mP}$. The fact that at high phytoplankton concentrations the zooplankton growth rate saturates was incorporated by Riley (1947) (55) in the first model proposed for a zooplankton population.

The assimilation efficiency of the zooplankton at low phytoplankton concentrations, a_{ZP}, which is the ratio of phytoplankton organic carbon utilized to zooplankton organic carbon produced has been estimated by Conover (56) for a mixed zooplankton population. The results of 26 experiments gave an average of 63% and a standard deviation of 20%. Other reported values are within this range. Experimental values for K_{mP}, which in effect set the maximum growth rate of zooplankton, are not available and would probably be highly species-dependent. Perhaps a more effective way of estimating K_{mP} is first to estimate the maximum growth rate at saturating phytoplankton concentrations, $a_{ZP}C_gK_{mP}$, and then calculate K_{mP}. Growth rates for copepods through their life cycle average 0.18 day^{-1} (46). For the Georges Bank population, Riley used 0.08 day^{-1} (55) for the maximum zooplankton growth rate. For a value of the grazing coefficient C_g of 0.5 liter/mg-dry wt-day and an assimilation coefficient of 65%, the Michaelis constant for zooplankton assimilation, K_{mP}, ranges between 0.25 and 0.55 mg-dry wt/liter of phytoplankton biomass. However, these values should only be taken as an indication of the order of magnitude of K_{mP}. It is probable that its value can vary substantially in different situations.

The fact that the growth rate reaches a maximum or saturates is an important feature of the formulation of the zooplankton growth rate since in some cases the phytoplankton concentration during part of the year exceeds that which the zooplankton can effectively metabolize. If the zooplankton growth rate is not limited in some way and, instead, is assumed simply to be proportional to the phytoplankton concentration, as proposed in simpler models, the resulting zooplankton growth rate during phytoplankton blooms can be very much larger than is physiologically possible for zooplankton, an unrealistic result. The saturating growth rate also has implications in the mathematical properties of the resulting equations. In particular, the behavior differs significantly from the classical Volterra Preditor–Prey equations (57). This is discussed further in a subsequent section.

The growth of the zooplankton population as a whole, of which the herbivorous zooplankton are a part, is complicated by the fact that some zooplankters are carnivorous or omnivorous. Thus, the nutrient for the total population should include not only phytoplankton but also organic detritus as a food source since this is also available to the grazing zooplankton. However, for cases where the phytoplankton are abundant and the growth rate saturates for the significant growing periods, the simplification introduced by ignoring the detritus is probably acceptable.

Death Rate. The death rate of herbivorous zooplankton is thought to be caused primarily by the same mechanisms that cause the death of the phytoplankton, namely, endogeneous respiration and predation by higher trophic levels. The endogeneous respiration rate of zooplankton populations has been measured and the results of some of these experiments are presented in Figure 9 and Table V.

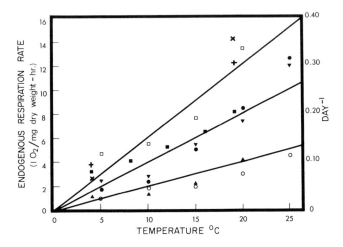

Figure 9.   Endogenous respiration rate of zooplankton vs.
temperature

Table V.   Endogenous Respiration Rate of Zooplankton

Ref.	Organism	Plotting Symbol	Temp., °C	Respiration Rate, Ml O_2/Mg Dry Wt-Day
58	Cladocerans	×	18	14.2
			4	2.7
58	Copepods	+	18	12.2
			4	3.8
58	Copepods	■	18	8.2
			16	6.5
			12	5.2
			8	4.1
			4	3.4
1	*Calanus finmarchicus*	▲	20	4.2
			15	2.3
			10	1.4
			4	1.3
59	*Diaptomus leptopus*	▼	25	12.1
			20	7.4
			15	5.3
			10	2.8
			5	2.5
59	*D. clavipes*	●	25	12.5
			20	8.5
			15	5.1
			10	2.4
			5	1.8
59	*D. siciloides*	□	25	21
			20	13.5
			15	7.8
			10	5.5
			5	4.8
59	*Diaptomus sp.*	○	25	4.3
			20	3.0
			15	2.1
			10	1.7
			5	1.1

It is clear from these measurements that the respiration rate of zooplankters is temperature-dependent. It is also dependent on the weight of the zooplankter in question and its life cycle stage (59). As a first approximation, a straight line dependence is adequate, and the endogeneous respiration rate is given by the equation: respiration rate = $K_3 T$ where $K_3 = 0.2 \pm 0.1$ (day °C)$^{-1}$. The conversion from the reported units to a death rate is made by assuming that 50% of the zooplankton dry weight represents the carbon weight and that carbohydrate (CH_2O) is being oxidized. The data are somewhat variable and serve only to establish a range of values within which the respiration rate of a natural zooplankton association might be expected.

The death rate attributed to predation by the higher trophic levels, specifically the carnivorous zooplankton, has been considered by previous models in a more or less empirical way. The complication resulting from another equation and the uncertainty as to the mechanisms involved are quite severe at this trophic level. In particular, it is probable that an equation for organic detritus is necessary to describe adequately the available food. Hence, it is expedient to break the causal chain at this point and assume that the herbivorous zooplankton death rate resulting from all other causes is given by a constant, the magnitude of which is to be determined empirically. The severity of this assumption can be tested by examining the sensitivity of the solutions of the phytoplankton and zooplankton equations to the magnitude of this constant. Hence, the resulting zooplankton death rate is given by

$$D_{Zj} = K_3 T + K_4 \qquad (26)$$

where K_4 is empirically determined.

With the growth and death rates given by Equations 25 and 26, respectively, the source term for herbivorous zooplankton biomass is given by Equation 24. The conservation of mass equation which describes the behavior of Z_j is given by Equation 2, with Z_j as the dependent variables replacing P_j and S_{Zj} replacing S_{Pj} as the source terms.

This completes the formulation of the equations which describe the zooplankton system. The equations for the nutrient system remain to be formulated.

The Nutrient System

The conservation of mass principle is applied to the nutrients being considered in the same way as it has been previously applied to the phytoplankton and zooplankton biomass within a volume segment. The number of mass conservation equations required is equal to the number of nutrients that are explicitly included in the growth rate formulation for the phytoplankton. For the sake of simplicity, the formulation for only one nutrient is discussed below.

The source term S_{Xj} in the conservation of mass equation for the concentration of the nutrient N_j in the jth volume segment V_j is the sum of all sources and sinks of the nutrient within V_j. The primary interaction between the nutrient system and the phytoplankton system is the reduction or sink of nutrient connected with phytoplankton growth. The rate of increase of phytoplankton biomass is $G_{Pj}P_j$. To convert this assimilation rate to the rate of utilization of the nutrient, the ratio of biomass production to net nutrient assimilated is required. Over a long time interval, this ratio approximates the nutrient-to-biomass ratio of the phytoplankton population. For example, if the nutrient being considered is total inorganic nitrogen and the phytoplankton biomass is characterized in terms of dry weight, then this ratio is the nitrogen-to-dry-weight ratio of the population. For both nitrogen and phosphorus, these ratios have been studied for a number of phytoplankton species and natural associations. An example of this information is presented in Table VI, condensed from Strickland (17). If a_{NP} is the nutrient-to-phytoplankton biomass ratio of the population, then the sink of the nutrient owing to phytoplankton growth is $a_{NP}G_{Pj}P_j$.

A secondary interaction between the biological systems and the nutrient systems is the excretion of nutrients by the zooplankton and the release of nutrients in an organic form by the death of phytoplankton and zooplankton. The excretion mechanism has been considered by Riley

Table VI. Dry Weight Percentage[a] of Carbon, Nitrogen, and Phosphorus in Phytoplankton[b]

Phytoplankter	% Carbon		% Nitrogen		% Phosphorus	
	Average	Range	Average	Range	Average	Range
Myxophyceae	36	(28–45)	4.9	(4.5–5.8)	1.1	(0.8–1.4)
Chlorophyceae	43	(35–48)	7.8	(6.6–9.1)	2.9	(2.4–3.3)
Dinophyceae	43	(37–47)	4.4	(3.3–5.0)	1.0	(0.6–1.1)
Chrysophyceae	40	(35–45)	8.4	(7.8–9.0)	2.1	(1.2–3.0)
Bacillariophyceae	33	(19–50)	4.9	(2.7–5.9)	1.1	(0.4–2.0)

[a] The units are (mg of carbon, nitrogen, or phosphorus)/(mg dry weight of phytoplankton) × 100%.
[b] Condensed from Strickland (17).

(40) in a generalization of the equations of Steele. The rate of phosphorus excretion has also been measured experimentally (60). Using the formulation for zooplankton growth rate proposed herein, the rate of nutrient excretion is the rate grazed, $a_{NP}C_gP_jZ_j$, minus the rate metabolized, $a_{NP}G_{Zj}Z_j$; that is, the excretion rate is

$$a_{NP}C_gZ_jP_j\left(1 - \frac{a_{ZP}K_{mP}}{K_{mP} + P_j}\right) \qquad (27)$$

At high phytoplankton concentrations, almost all the grazed phytoplankton is excreted since the bracketed term in Equation 27 approaches unity.

There is a difficulty, however, in using this term directly as a source of nutrient. To illustrate this difficulty, assume that the nutrient is inorganic nitrogen. A part of the excreted nitrogen, however, is in organic form, and a bacterial decomposition into the inorganic forms must precede utilization by the phytoplankton. The same is true for the nutrient released by the death of phytoplankton, $a_{NP}K_2TP_j$, and that released by the death of zooplankton, $a_{NZ}K_3TZ_j$, where a_{NZ} is the nutrient-to-zooplankton biomass ratio. Therefore, strictly speaking, a conservation of mass equation for the organic form of the nutrient is required. The organic form is then converted to the inorganic form. For the case of nitrogen, the kinetics of this conversion have been investigated and applied to stream and estuarine situations (13). If the conversion rate is large by comparison with the other rates in the phytoplankton and zooplankton equations, then the direct inclusion of these sources is approximatey correct.

The sources of nutrients arising from man-made inputs, such as wastewater discharges and agricultural runoff, are included explicitly into the source term since these sources are usually the major control variables available to influence the biological systems. An extensive review of the magnitude and relative importance of these sources of nutrients, primarily nitrogen and phosphorus, has recently been made (61). A useful distinction is made between diffuse sources such as agricultural runoff loads and ground water infiltration, which are difficult to measure and control, and point sources such as wastewater discharges from municipal and industrial sources, for which more information is available. The nitrogen and phosphorus loads from agricultural runoff are quite variable and depend on many variables such as soil type, fertilizer application, rainfall, and irrigation practice. The nutrient sources from point loads can be estimated more directly. For example, the nutrient load from biologically treated municipal wastewater is in the order of 10 g/capita-day total nitrogen and 2 g/capita-day total phosphorus. The ratio of per capita phosphorus to physiologically-required phosphorus is approximately 2 to 3, the excess being primarily the result of detergent use. Industrial loads can also be important, especially effluents from food processing industries. If the required loading rates are available, their loads should be included in the nutrient mass balance equations. In particular, if the investigation of the phytoplankton population is directed at the probable effects of increasing or decreasing the nutrient load, these loads must be explicitly identified and their magnitude assessed.

Let W_{Nj} be the rate of addition of the nutrient to the jth volume element. This source is then included as a component in the nutrient source term in the mass balance equation.

An important additional source of inorganic nutrients which may influence the availability of nutrients is the interaction of the overlying

water either with the underlying mineral strata if exposed or with whatever sediment is present. These interactions can complicate the source
term but they should be included if they add significantly to the available
nutrient.

The source term which results from the inclusion of the phytoplankton utilization sink, the zooplankton excretion and the mortality sources,
and the man-made additions is

$$S_{Nj} = \frac{W_{Nj}}{V_j} - a_{NP}G_{Pj}P_j + a_{NP}C_gZ_jP_j\left(1 - \frac{a_{ZP}K_{mP}}{K_{mP} + P_j}\right)$$

$$+ a_{NP}K_2TP_j + a_{NZ}K_3TZ_j \tag{28}$$

Any additional sources and sinks that contribute can be added to the
source term as needed. With the source term formulated, the conservation of nutrient mass equation is given by Equation 2 with N_j as the
dependent variable replacing P_j and S_{Nj} replacing S_{Pj}.

The Equations of the Model

In the previous sections, the equations for phytoplankton and zooplankton biomass and nutrient concentration within one volume element
have been formulated. The resulting equations are an attempt to describe
the kinetics of the growth and death of the phytoplankton and zooplankton populations and their interaction with the nutrients available. The
form of the equations for the volume V_j are as follows:

$$\overset{\circ}{P}_j = [G_{Pj}(P_j, N_j, t) - D_{Pj}(Z_j, t)]P_j + S_{Pj}(P_j, Z_j, N_j, t) \tag{29}$$

$$\overset{\circ}{Z}_j = [G_{Zj}(P_j, t) - D_{Zj}(t)]Z_j + S_{Zj}(P_j, Z_j, t) \tag{30}$$

$$\overset{\circ}{N}_j = S_{Nj}(P_j, Z_j, t) \tag{31}$$

where G_{Pj} and D_{Pj} are given by Equations 16 and 23, G_{Zj} and D_{Zj} are
given by Equations 25 and 26, and S_{Nj} by Equation 28. The dependence
of the growth and death rates on the concentration of the three dependent
variables and time is made explicit in this notation.

These equations describe only the kinetics of the populations in a
single volume element V_j. However, in a natural water body there exists
significant mass transport as well. The mass transport mechanisms can be
conveniently represented by the matrix A with elements a_{ij}. If for particular segments i and j the matrix element a_{ij} is nonzero, then the
volume segments V_i and V_j interact, and mass is transported between
the two segments. Letting P, Z, and N be the vectors of elements
P_j, Z_j, and N_j and letting S_P, S_N, S_Z be the vectors of elements S_{Pj}, S_{Zj},

and S_{Nj}, the conservation of mass equations for the three systems including the mass transport and kinetic interactions are

$$V\overset{\circ}{P} = AP + VS_P \tag{32}$$

$$V\overset{\circ}{Z} = AZ + VS_Z \tag{33}$$

$$V\overset{\circ}{N} = AN + VS_N \tag{34}$$

where V is the diagonal matrix of the volumes of the segments. These are the equations which form the basis for the phytoplankton population model. The detailed formulation and evaluation of the mass transport matrix has been discussed elsewhere (14, 15, 62).

The form of Equations 32–34 makes explicit the linear and nonlinear portions of the equations. In the equation for P, the phytoplankton biomass, the concentration P_j, in the volume element V_j, is linearly coupled to the other P_k's through the matrix multiplication by A. However, there is no nonlinear interaction between P_j and any other P_k, $k \neq j$. The reason is that the transport processes are described by linear equations. It is usually the case, however, that the A matrix is a function of time, since at least the advective terms usually vary in time. The nonlinear terms in the vector S_P involve P_j itself and the corresponding Z_j and N_j. Hence, the P equation is coupled to the Z and N equations through this term. Note, however, that P_j is not coupled to the Z_k, $k \neq j$, in any other segment, so that the coupling takes place only within each volume segment.

Therefore, the nonlinearities provide the coupling between the phytoplankton, zooplankton, and nutrient systems. This coupling is accomplished within each volume and does not extend beyond the volume boundary. The coupling among the volumes is accomplished by the linear transport interaction represented by the matrix A. This matrix may be time-varying but its elements are not functions of the phytoplankton, zooplankton, or nutrient concentrations. Hence, in many ways these equations behave linearly. In particular, their spatial behavior is unaffected by the nonlinear source terms. However, the temporal behavior and the relationships between each P_j, Z_j, and N_j are distinctly nonlinear.

Comparison With Lotka–Volterra Equations

The classical theory of predator–prey interaction as formulated by Volterra involves two equations which express the growth rate of the prey and the predator (57). Within the context of phytoplankton and zooplankton population, the prey is the phytoplankton and the predator the zooplankton. In the notation of the previous sections, for a one-volume system, the Lotka–Volterra equations are:

$$\frac{dP}{dt} = (G_P - D_P')P - C_gPZ \tag{35}$$

$$\frac{dZ}{dt} = -D_ZZ + a_{ZP}C_gPZ \tag{36}$$

where all the coefficients, G_P, D_P', C_g, D_Z, and a_{ZP} are assumed to be constants and $G_P > D_P'$. This is a highly simplified situation since, as indicated previously, the growth and death rates are functions of time and, in the case of the phytoplankton growth rate, of the phytoplankton and nutrient concentrations as well. However, for a situation with adequate nutrients and low initial phytoplankton concentration, the nonlinear interaction is small initially, and the time variation of G_P can be small during the summer months. In any case, the analysis of this simplified situation is quite instructive.

Although no analytical solution is available for these simplified equations, their properties are well understood (*63*). In particular, the equations have two sets of singular points corresponding to the solution of the righthand side of Equations 35 and 36 equated to zero: the trivial solutions $P^* = 0, Z^* = 0$, and

$$P^* = \frac{D_Z}{a_{ZP}C_g}, \qquad Z^* = \frac{G_P - D_P'}{C_g} \tag{37}$$

A perturbation analysis of Equations 35 and 36 about this singular point shows that the solutions whose initial conditions are close to P^*, Z^*, oscillate sinusoidally about this singular point. Hence, no constant solution is possible. The prey and predator populations continually oscillate and are out of phase with each other. When the predator predominates, the prey is reduced, which in turn causes the predator to die for lack of food, which allows the prey to proliferate for lack of predator, which then causes the predator to grow because of the prey available as a food supply, and so on. The interesting feature is that these oscillations continue indefinitely.

The classical Lotka–Volterra equations assume an isolated population with no mass transport into or out of the volume being considered. To simulate the effect of mass transport into the volume, assume that an additional source term of phytoplankton biomass exists and has constant magnitude P_0. For this situation, the equations become

$$\frac{dP}{dt} = (G_P - D_P')P - C_gPZ + P_0 \tag{38}$$

$$\frac{dZ}{dt} = D_ZZ + a_{ZP}C_gPZ \tag{39}$$

The nontrival singular point for these equations is

$$P^* = \frac{D_Z}{a_{ZP}C_g}, \quad Z^* = \frac{a_{ZP}P_o}{D_Z} + \frac{(G_P - D'_P)}{C_g} \tag{40}$$

A perturbation analysis about this singular point yields a second order linear ordinary differential equation whose characteristic equation has the roots λ_1 and λ_2 where

$$\lambda_{1,2} = -\frac{a_{ZP}P_0C_g}{2\,D_Z} \pm \sqrt{\left[\frac{a_{ZP}P_0C_g}{2\,D_Z}\right]^2 - a_{ZP}C_gP_0 - (G_P - D'_P)\,D_Z} \tag{41}$$

Since for $P_0 > 0$, these roots have negative real parts, this singular point is a stable focus, and the steady state values given by Equation 40 are approached either by a damped sinusoid or an exponential (63). Note that for $P_0 = 0$, the classical case, the roots are purely imaginary, and the oscillation persists indefinitely.

This analysis suggests that the effect of transport into the system stabilizes the behavior of the equations and in particular allows the solutions to achieve a constant solution. This is in marked contrast to the behavior of the classical Lotka–Volterra equations.

Another modification, which has been introduced into the zooplankton equations, changes the behavior of the proposed equations in contrast to the Lotka–Volterra equations. It has been argued that the zooplankton growth rate resulting from grazing must approach its maximum value when the phytoplankton population becomes large since the zooplankters cannot metabolize the continually increasing food that is available. Thus, the growth rate $a_{ZP}C_gPZ$ is replaced by $a_{ZP}C_gZP\,K_{mP}/(P + K_{mP})$ where K_{mP} is a Michaelis constant for the reaction. The equations then become

$$\frac{dP}{dt} = (G_P - D'_P)P - C_gPZ + P_0 \tag{42}$$

$$\frac{dZ}{dt} = -D_ZZ + \frac{a_{ZP}C_gZPK_{mP}}{P + K_{mP}} \tag{43}$$

The nonzero singular points are

$$P^* = \frac{D_ZK_{mP}}{a_{ZP}C_gK_{mP} - D_Z} \tag{44}$$

$$Z^* = \frac{P_0}{C_gP^*} + \frac{(G_P - D'_P)}{C_g} \tag{45}$$

This solution reduces to the previous situation, Equation 40, for large K_{mP}. This is expected since for K_{mP} large with respect to P, the expression $K_{mP}/(P + K_{mP})$ approaches one. However, an interesting modification from classical predator–prey behavior occurs if the following condition is met

$$a_{ZP} C_g K_{mP} = D_Z + \varepsilon \qquad (46)$$

where ε is a small positive number. For this condition, P^* is large and positive. What happens in this case is that the zooplankton population, although it continues to grow exponentially, cannot grow quickly enough to terminate the phytoplankton growth by grazing, and the phytoplankton continue to grow exponentially until P^* is reached. Of course, in the actual situation, for which G_P is not a constant, other phenomena such as nutrient depletion and self-shading exert their effect, and the growth may be stopped sooner. However, if the growth rate of zooplankton at a phytoplankton concentration equal to the Michaelis constant K_{mP} is only slightly larger than their death rate D_Z, then the zooplankton alone do not rapidly terminate the bloom.

This condition is an important dividing line for the possible behavior of the phytoplankton–zooplankton equations set forth in the previous sections. In particular, it indicates what must be true for a system wherein the zooplankton are a significant feature in the resulting phytoplankton solution. However, if Equation 46 is satisfied, then the zooplankton are not the dominant control of the phytoplankton population.

Application—San Joaquin River

As an example of the application of the equations proposed herein, consider the phytoplankton and zooplankton population observed at Mossdale Bridge on the San Joaquin River in California during the two years 1966–1967. Mossdale is located approximately 40 miles from the confluence of the San Joaquin and the Sacramento Rivers. The data presented below have been supplied to the authors by the Department of Water Resources, State of California (64), as part of an ongoing project to assess the effects of proposed nutrient loads and flow diversions on the water quality of the San Francisco Bay Delta. A more complete report of this investigation is forthcoming (62).

In order to simplify the spatial segmentation and the calculations, a one-volume segment is chosen for the region of the San Joaquin for which Mossdale is representative. The volume of this segment is, of course, somewhat arbitrary, and a more representative spatial segmentation would remove this uncertaintly. However, it is instructive to consider the behavior of the solution of this simplified model.

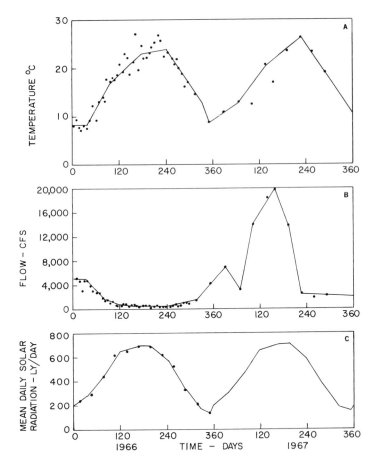

*Figure 10. Temperature, flow, and mean daily solar radiation;
San Joaquin River, Mossdale, 1966–1967*

The nutrient data available indicate that phosphate, bicarbonate, silicate, calcium, and magnesium are available at concentrations well above the levels for which it has been suggested that these nutrients limit growth. Only the ammonia and nitrate concentrations are low, and they both decrease markedly during the 1966 spring bloom. Hence, these nutrients must be considered explicitly. To simplify the computations, the ammonia and nitrate nitrogen are combined, and the nutrient considered is total inorganic nitrogen.

There is some uncertainty concerning the magnitude and the temporal variation of the inorganic nitrogen load being discharged to the system during the time interval of interest. For lack of a better assumption, the inorganic nitrogen load W_N being discharged into the volume is assumed to be a constant, the magnitude of which is determined by comparison with the observed inorganic nitrogen concentration data at Mossdale.

The variation of the environmental variables being considered—namely, temperature, solar radiation, and advective flow in the San Joaquin during the two-year period of interest—and the straight line approximations that are used directly in the numerical computation are shown in Figure 10. The influent advective flow, which is assumed to have constant concentrations of phytoplankton and zooplankton biomass and inorganic nitrogen, is routed through the volume. Since Mossdale is located above the saline portion of the San Joaquin, no significant dispersive mass transfer is assumed to exist by comparison with the advective mass transfer.

The equations which represent this one-segment model are

$$\overset{\circ}{P} = (G_P - D_P)P + \frac{Q}{V}(P_0 - P) \tag{47}$$

$$\overset{\circ}{Z} = (G_Z - D_Z)Z + \frac{Q}{V}(Z_0 - Z) \tag{48}$$

$$\overset{\circ}{N} = -a_{NP}G_PP + \frac{W_N}{V} + \frac{Q}{V}(N_0 - N) \tag{49}$$

where $Q = Q(t)$ is the advective flow entering and leaving the volume; V is the volume of the segment; P_0, Z_0, and N_0 are the phytoplankton, zooplankton, and inorganic nitrogen concentration of the flow entering the volume. The remaining terms have been defined previously by Equations 16, 23, 25, and 26. In the nutrient equation, only the direct source of inorganic nitrogen, W_N, has been included; the organic feedback terms representing excreted nitrogen, etc., Equation 28, have been dropped. Since the magnitude of W_N is uncertain and is assigned by comparison with observed data and computed model output, these feedback terms can be thought of as being incorporated in the value obtained for W_N.

The solution of Equations 47, 48, and 49 requires numerical techniques. For such nonlinear equations, it is usually wise to employ a simple numerical integration scheme which is easily understood and pay the price of increased computational time for execution rather than using a complex, efficient, numerical integration scheme where unstable behavior is a distinct possibility. A variety of simple methods are available for integrating a set of ordinary first order differential equations. In particular, the method of Huen, described in Ref. 65, is effective and stable. It is self-starting and consists of a predictor and a corrector step. Let $y = f(t,y)$ be the vector differential equation and let h be the step size.

The predictor is that of Euler: with y_0 the initial condition vector at t_0, the predictor value of y at $t_0 + h = t_1$ is

$$y_1^* = y_0 + hf(t_0, y_0) \tag{50}$$

The corrector value is simply

$$y_1 = y_0 + \frac{h}{2} [f(t_0, y_0) + f(t_1, y_1^*)] \tag{51}$$

That is, the corrector uses the predictor value at t_1 to estimate the slope at t_1 which is averaged with the slope at t_0 to provide the slope of the straight line approximation. A variation of this method is discussed at some length by Hamming (66).

Another simple two-step method is that of Runge, described in Ref. 67. The Euler predictor is used with a half-step integration.

$$y^* = y_0 + \frac{h}{2} f(t_0, y_0) \tag{52}$$

This value of y is used to estimate the slope at the midpoint of the interval, which is then used as the slope of the straight line approximation

$$y_1 = y_0 + hf(t_0 + \frac{h}{2}, y^*) \tag{53}$$

Both of these methods are second order methods, being accurate to terms of order Δt^2 in a comparison of Taylor series expansions of the exact and approximate values, and both methods require two derivative evaluations per step. The method of Runge has been used in the calculations presented below.

The equations themselves are programmed for solution using a continuous simulation language and a digital computer. The language, in this case CSMP/1130, is based on a block diagram, analog computer, representation of the differential equations. The flexibility of these languages which allow changes in the equation structure to be made easily is an asset in modeling complex systems.

The biomass variables used in the calculations are total cell counts for the phytoplankton and rotifer counts for the zooplankton. The rotifer population represented the large majority of the zooplankton present on a weight basis as well. In order to relate these variables to comparable units, a series of conversion factors have been used. The phytoplankton count–chlorophyll concentration ratio was measured. However, the carbon–chlorophyll or dry weight–chlorophyll conversions are unknown. Hence, the conversion to an organic carbon basis is made rather arbi-

Table VII. Parameter Values for the Mossdale Model

Notation	Description	Parameter Value
K_1	Saturated growth rate of phytoplankton	0.1 day^{-1} °C
I_s	Light saturation intensity for phytoplankton	300 ly/day
k'_e	Extinction coefficient	4.0 m^{-1}
H	Depth	1.2 m
K_m	Michaelis constant for total inorganic nitrogen	0.025 mg N/liter
f	Photoperiod	$0.5 + \sin [0.0172 (t - 165)]$ day
K_2	Endogenous respiration rate of phytoplankton	0.005 day^{-1} °C^{-1}
C_g	Zooplankton grazing rate	0.13 liter/mg - C - day
P_0	Influent phytoplankton chlorophyll concentration	5.0 μg Chl/liter
a_{ZP}	Zooplankton conversion efficiency	0.6 mg C/mg - C
K_{mP}	Phytoplankton Michaelis constant	60 μg Chl/liter
D_Z	Zooplankton death rate	0.075 day^{-1}
Z_0	Influent zooplankton carbon concentration	0.05 mg C/liter
a_{NP}	Phytoplankton nitrogen–carbon ratio	0.17 mg N/mg - C
C/Chl	Phytoplankton carbon to total chlorophyll ratio	50 mg C/mg - Chl
N_0	Influent total inorganic nitrogen concentration	0.1 mg N/liter
W_N	Direct discharge rate of nitrogen	12500 lbs/day
V	Segment volume	9.7×10^8 ft^3
	Phytoplankton total cell count/phytoplankton	100 cells/ml = 1.75 μg Chl/liter
	Zooplankton count/ zooplankton carbon ratio	10^4 No./liter = 1.30 mg C/liter

trarily. However, the carbon-to-chlorophyll ratio which results (*see* Table VII) is within the range reported in the literature. The same problem exists with the rotifer counts to rotifer carbon conversion; the value used is given in Table VII.

The comparison of the model output and the observed data for the two-year period for which data are available is shown in Figure 11. The parameter values used in the equations are listed in Table VII.

It is clear from both the data and the model results that a classical predator–prey situation is observed in 1966: the spring bloom of phytoplankton resulting from favorable temperature and light intensity provides the food for zooplankton, which then reduce the population during

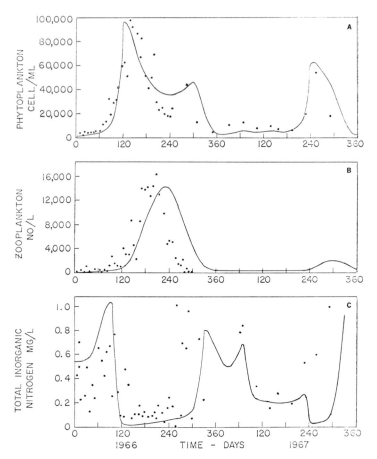

*Figure 11.   Phytoplankton, zooplankton, and total inorganic ni-
trogen; comparison of theoretical calculations and observed data;
San Joaquin River, Mossdale, 1966–1967*

the summer. The decrease of the zooplankton and the subsequent slight
secondary bloom of phytoplankton complete the cycle for the year. It is
not clear, however, from a casual inspection of the data, whether the
zooplankton population terminated the phytoplankton growth, as in
classical predator–prey situations, whether the nutrient concentration
dropped to a limiting value that reduced the growth rate, or a combination of the two. This point is elaborated in the next section.

The situation in 1967 is quite different. No significant phytoplankton
growth is observed until late in the year. The controlling variable in this
case is the large advective flow during the spring and summer of 1967
(*see* Figure 10) which effectively washes out the population in the region.
Only when the flow has sufficiently decreased so that a population can

develop do the phytoplankton show a slight increase. However, the dropping temperature and light intensity level terminate the growth for the year.

Growth Rate–Death Rate Interactions

The behavior of the equations which represent the phytoplankton, zooplankton, and nutrient systems in one volume can be interpreted in terms of the growth and death rates of the phytoplankton and zooplankton. The equations are as before

$$\frac{dP}{dt} = (G_P - D_P)P + \frac{Q}{V}(P_0 - P) \tag{54}$$

$$\frac{dZ}{dt} = (G_Z - D_Z)Z + \frac{Q}{V}(Z_0 - Z) \tag{55}$$

where P_0 and Z_0 are the concentrations of phytoplankton and zooplankton carbon in the influent flow, Q. A more useful form for these equations is

$$\frac{dP}{dt} = [G_P - (D_P + \frac{Q}{V})]P + \frac{Q}{V}P_0 \tag{56}$$

$$\frac{dZ}{dt} = [G_Z - (D_Z + \frac{Q}{V})]Z + \frac{Q}{V}Z_0 \tag{57}$$

A complete analysis of the properties of these equations is quite difficult since the coefficients of P and Z are time variables and also functions of P and Z. However, the behavior of the solution becomes more accessible if the variation of these coefficients is studied as a function of time. The expressions $G_P - (D_P + Q/V)$ and $G_Z - (D_Z + Q/V)$ can be considered the net growth rates for phytoplankton and zooplankton. The advective or flushing rate, Q/V, is included in these expressions since it acts as a death rate in one segment system.

The sign and magnitude of the net growth rate controls the behavior of the solution. For a linear equation, for which the net growth rate is not a function of the dependent variable (*i.e.*, P or Z), the type of solution obtained depends on the sign and magnitude of the net growth rate. That is, for the equation

$$\frac{dP}{dt} = \alpha P + \frac{Q}{V}P_0 \tag{58}$$

where α, Q, and V are constant, the solution is

$$P(t) = P(o) \, e^{\alpha t} + P_0 \, \frac{Q}{\alpha V} \, (e^{\alpha t} - 1) \qquad (59)$$

For α negative, that is, for a negative net growth rate, the solution tends to the steady state value $P_0 \, Q/|\alpha|V$. However, for α positive, the solution grows exponentially without limit. Thus, for α negative but $|\alpha|$ small, or for α positive, the solution becomes large; whereas for α negative but $|\alpha|$ large, the solution stays small. Hence, the behavior of the solution can be inferred from the plots of the net growth rates. Figure 12a is a plot of the following terms from the 1966 Mossdale calculation: G_P without the Michaelis–Menton multiplicative factor included—*i.e.*, the growth rate at nutrient saturation denoted by $G_P \, (I,T)$; G_P itself denoted by $G_P \, (N,I,T)$—*i.e.*, the growth rate considering the nutrient effects. The net growth rate $G_P - (D_P + Q/V)$ is also plotted. Similarly, in Figure 12b, the growth rate of zooplankton G_Z, the mortality rate D_Z, the flushing rate $Q(t)/V$, and the net growth rate $G_Z - (D_Z + Q/V)$ are plotted.

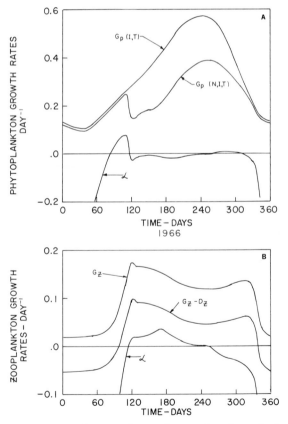

Figure 12.   Theoretical growth rates for phytoplankton and zooplankton populations

The analysis of the 1966 model calculations can now be made by inspecting these figures. The net growth rate for the phytoplankton $G_P - (D_P + Q/V)$ becomes positive at $t = 85$ days owing to an increase in G_P, the result of rising temperature and light intensity, and a decrease in Q/V as the advective flow decreases. The positive net growth rate of the population causes their numbers to increase exponentially fast until the nutrient begins to be in short supply. This is evidenced by the departure of the G_P curve from the G_P at nutrient saturation curve. At the same time, the D_P curve is showing a marked increase because of the increased zooplankton population and their grazing. The result is that the net growth rate becomes zero and then negative as the zooplankton reduce the phytoplankton population by grazing. The growth of the zooplankton can be analyzed in a similar fashion using Figure 12b. The net growth rate becomes positive when the phytoplankton population is large enough to sustain the zooplankters. Then the zooplankton grow until they have reduced the phytoplankton population to a level where they are no longer numerous enough to sustain the zooplankton. The net zooplankton growth rate then becomes negative and the population diminishes in size. This small zooplankton population no longer exerts a significant effect on the death rate of the phytoplankton, D_P, and its value decreases, causing the net phytoplankton growth rate to become positive again, and the smaller autumn bloom results. The decreasing temperature and light intensity and the increasing advective flow then effectively terminate the bloom as the year ends.

Summary and Conclusions

A model of the dynamics of phytoplankton populations based on the principle of conservation of mass has been presented. The growth and death kinetic formulations of the phytoplankton and zooplankton have been empirically determined by an analysis of existing experimental data. Mathematical expressions which are approximations to the biological mechanisms controlling the population are added to the mass transport terms of the conservation equation for phytoplankton, zooplankton, and nutrient mass in order to obtain the equations for the phytoplankton model. The resulting equations are compared with two years' data from the tidal portion of the San Joaquin River, California. Similar comparisons have been made for the lower portion of Delta and are reported elsewhere (62).

It is recognized that certain parameters in the model have been estimated from the data which are then used to demonstrate the veracity of the model. The parameters used in the verification were either obtained from prototype measurements or estimated from the range of values reported in the literature. The refinement of the later set of parameters was made by comparing the observed 1966 data and calculated results. The model was further verified by applying the parameters obtained in the 1966 analysis to the data of the following year. The parameter values finally used were all within the ranges of reported literature values. The agreement achieved between the available data and the model calculations is sufficiently encouraging to prompt further effort in this direction.

The primary aim of this investigation, presenting a useful model as a component in solution of the eutrophication problems, in our opinion, has been achieved. The resulting equations are admittedly complex and require numerical methods for solution. It is anticipated as with all modeling activities that the structure presented herein will be expanded and modified in the future to incorporate additional features of the eutrophication phenomena. In particular, the model as it is presently structured does not address the species changes that might result as the environment is changed. This is a problem of some consequence, and refinements and expansion of the number of species considered is an area for future work (12). However, the initial application of these equations to an actual problem area with specific eutrophication problems has been sufficiently successful to support its engineering use as a preliminary step in the assessment of a potential or actual eutrophication problem.

Acknowledgment

The authors are pleased to acknowledge the participation of John L. Mancini of Hydroscience Inc. in the research reported herein, as well as the assistance of Gerald Cox and Jack Hodges of the Department of Water Resources and Harold Chadwick of the Department of Fish and Game, State of California.

The research was sponsored in part by a research grant to Manhattan College from the Federal Water Quality Administration and in part from research funds made available by Hydroscience, Inc.

The application of the phytoplankton model to the San Joaquin River was sponsored by the California Water Resources Commission and carried out by Hydroscience Inc.

Literature Cited

(1) Riley, G. A., *Theory of Food-Chain Relations in the Ocean*, "The Sea," p. 438–63, M. N. Hill, Ed., Interscience, New York, 1963.
(2) Riley, G. A., "Factors Controlling Phytoplankton Populations on Georges Bank," *J. Marine Res.* (1946) **6** (1), 54–73.
(3) Riley, G. A., "Seasonal Fluctuations of the Phytoplankton Populations in New England Coastal Waters," *J. Marine Res.* (1947) **6** (2), 114–25.

(4) Riley, G. A., Von Arx, R., "Theoretical Analysis of Seasonal Changes in the Phytoplankton of Husan Harbor, Korea," *J. Marine Res.* (1949) **8** (1), 60–72.

(5) Riley, G. A., Stommel, H., Bumpus, D. F., "Quantitative Ecology of the Plankton of the Western North Atlantic," *Bull. Bingham Oceanog. Coll.* (1949) **12** (3), 1–169.

(6) Steele, J. H., "Plant Production on Fladen Ground," *J. Marine Biol. Assoc. U.K.* (1956) **35**, 1–33.

(7) Steele, J. H., "A Study of Production in the Gulf of Mexico," *J. Marine Res.* (1964) **22**, 211–22.

(8) Steele, J. H., *Notes on Some Theoretical Problems in Production Ecology*, "Primary Production in Aquatic Environments," C. R. Goldman, Ed., *Mem. Inst. Idrobiol.*, p. 383–98, 18 Suppl., University of California Press, Berkeley, 1965.

(9) Davidson, R. S., Clymer, A. B., "The Desirability and Applicability of Simulating Ecosystems," *Ann. N. Y. Acad. Sci.* (1966) **128** (3), 790–4.

(10) Cole, C. R., "A Look at Simulation through a Study on Plankton Population Dynamics," Report BNWL-485, p. 1–19, Battelle Northwest Laboratory, Richland, Washington, 1967.

(11) Parker, R. A., "Simulation of an Aquatic Ecosystem," *Biometrics* (1968) **24** (4), 803–22.

(12) Chen, C. W., "Concepts and Utilities of Ecologic Model," *J. Sanit. Eng. Div., Am. Soc. Civil Eng.* (Oct. 1970) **96**, 1085–97.

(13) Thomann, R. V., O'Connor, D. J., DiToro, D. M., "Modeling of the Nitrogen and Algal Cycles in Estuaries," *5th Intern. Water Pollution Res. Conf., San Francisco, Calif., July 1970.*

(14) Thomann, R. V., "Mathematical Model for Dissolved Oxygen," *J. Sanit. Eng. Div., Proc. Am. Soc. Civil Eng.* (October 1963) **83**, SA5, 1–30.

(15) O'Connor, D. J., Thomann, R. V., "Stream Modeling for Pollution Control," *Proc. IBM Sci. Computing Symp. Environ. Sci., November 14–16, 1966,* p. 269.

(16) Hutchinson, G. E., "A Treatise on Limnology Vol. II. Introduction to Lake Biology and the Limnoplankton," p. 306–54, Wiley, New York, 1967.

(17) Strickland, J. D. H., *Production of Organic Matter in the Primary Stages of the Marine Food Chain*, "Chemical Oceanography," Vol. 1, p. 503, J. P. Riley and G. Skivow, Eds., Academic, New York, 1965.

(18) Lund, J. W. G., "The Ecology of the Freshwater Phytoplankton," *Biol. Rev.* (1965) **40**, 231–93.

(19) Raymont, J. E. G., "Plankton and Productivity in the Oceans," p. 93–466, Pergamon, New York, 1963.

(20) Vollenweider, R. A., Ed., "Manual on Methods for Measuring Primary Production in Aquatic Environments," Ch. 2, p. 4–24, Blackwell Scientific Publications, Oxford, England, 1969.

(21) Tamiya, H., Hase, E., Shibata, K., Mituya, A., Iwamura, T., Nihei, T., Sasa, T., *Kinetics of Growth of Chlorella with Special Reference to Its Dependence on Quantity of Available Light and on Temperature*, "Algal Culture From Laboratory to Pilot Plant," p. 204–34, J. S. Burlew, Ed., Publ. 600, Carnegie Institute of Washington, D. C., 1964.

(22) Yentsch, C. S., Lew, R. W., "A Study of Photosynthetic Light Reactions," *J. Marine Res.* (1966) **24** (3).

(23) Spencer, C. P., "Studies on the Culture of a Marine Diatom," *J. Marine Biol. Assoc. U.K.* (1954) **28**; *quoted by* Harvey, H. W., "The Chemistry and Fertility of Sea Waters," p. 94, Cambridge University Press, 1966.

(24) Myers, J., *Growth Characteristics of Algae in Relation to the Problem of Moss Culture*, "Algal Culture From Laboratory to Pilot Plant," p. 37–54, Publ. 600, J. S. Burlew, Ed., Carnegie Institute of Washington, D. C., 1964.

(25) Sorokin, C., Krauss, R. W., "The Effects of Light Intensity on the Growth Rates of Green Algae," *Plant Physiol.* (1958) **33**, 109–13.

(26) Sorokin, C., Krauss, R. W., "Effects of Temperatures and Illuminance on Chlorella Growth Uncoupled from Cell Division," *Plant Physiol.* (1962) **37**, 37–42.

(27) Fogg, G. E., "Algal Cultures and Phytoplankton Ecology," p. 20, University of Wisconsin Press, Madison, Wis., 1965.

(28) Ryther, J. H., "Photosynthesis in the Ocean as a Function of Light Intensity," *Limnol. Oceanog.* (1956) **1**, 61–70.

(29) Shelef, G., Oswald, W. J., McGauhey, P. H., "Algal Reactor for Life Support Systems," *J. Sanit. Eng. Div., Proc. Am. Soc. Civil Eng.* (February 1970) **96**, SA1.

(30) Vollenweider, R. A., *Calculation Models of Photosynthesis—Depth Curves and Some Implications Regarding Day Rate Estimates in Primary Production Measurements*, "Primary Production in Aquatic Environments," C. R. Goldman, Ed., *Mem. Inst. Idrobiol.*, p. 425–57, 18 Suppl., University of California Press, Berkeley, 1965.

(31) Riley, G. A., "Oceanography of Long Island Sound 1952–1954. II. Physical Oceanography," *Bull. Bingham Oceanog. Coll.* (1956) **15**, 15–46.

(32) Small, L. F., Curl, H., Jr., "The Relative Contribution of Particulate Chlorophyll and   to the Extinction of Light off the Coast of Oregon," *Limnol. Oceanog.* (1968) **13** (1), 84.

(33) Oswald, W. J., Gotaas, H. B., Ludwig, H. F., Lynch, V., "Photosynthetic Oxygenation," *Sewage Ind. Wastes* (1953) **25** (6), 692.

(34) Azad, H. S., Borchardt, J. A., "A Method for Predicting the Effects of Light Intensity on Algal Growth and Phosphorus Assimilation," *J. Water Pollution Control Federation* (1969) **41** (11), part 2.

(35) Dugdale, R. C., "Nutrient Limitation in the Sea: Dynamics, Identification, and Significance," *Limnol. Oceanog.* (1967) **12** (4), 685–95.

(36) Eppley, R. W., Rogers, J. N., McCarthy, J. J., "Half Saturation Constants for Uptake of Nitrate and Ammonium by Marine Phytoplankton," *Limnol. Oceanog.* (1969) **14** (6), 912–20.

(37) Ketchum, B. H., "The Absorption of Phosphate and Nitrate by Illuminated Cultures of Nitzschia Closterium," *Am. J. Botany* (June 1939) 26.

(38) Thomas, W. H., Dodson, A. N., "Effects of Phosphate Concentration on Cell Division Rates and Yield of a Tropical Oceanic Diatom," *Biol. Bull.* (1968) **134** (1), 199–208.

(39) Lake Tahoe Area Council, South Lake Tahoe, Calif., "Eutrophication of Surface Waters—Lake Tahoe, May 1969," p. 33, 95705, Report for Federal Water Pollution Control Adm., Dept. of Interior, WPD 48-02.

(40) Riley, G. A., "Mathematical Model of Regional Variations in Plankton," *Limnol. Oceanog.* (1965) **10** (Suppl.), R202–R215.

(41) Gerloff, Skoog, "Nitrogen as a Limiting Factor for the Growth of Microcystis Aerugmosa in Southern Wisconsin Lakes," *Ecology* (1957) **38**, 556–61.

(42) MacIsaac, J. J., Dugsdale, R. C., "The Kinetics of Nitrate and Ammonia Uptake by Natural Populations of Marine Phytoplankton," *Deep Sea Res.* (1969) **16**, 415–22.

(43) Kuentzel, L. E., "Bacteria, Carbon Dioxide, and Algal Blooms," *J. Water Pollution Control Federation* (October 1969) **41** (10).

(44) Droop, M. R., *Organic Micronutrients*, "Physiology and Biochemistry of Algae," p. 141–60, R. A. Lewin, Ed., Academic, New York, 1962.

(45) Adams, J. A., Steele, J. H., *Shipboard Experiments on the Feeding of Calanus Finmarchicus,* "Some Contemporary Studies in Marine Science," p. 19–35, H. Barnes, Ed., G. Allen and Unwin Ltd., London, 1966.

(46) Mullin, M. M., Brooks, E. R., "Laboratory Culture, Growth Rate, and Feeding Behavior of a Planktonic Marine Copepod," *Limnol. Oceanog.* (1967) **12** (4), 657–66.

(47) Anraku, M., Omori, M., "Preliminary Survey of the Relationship Between the Feeding Habits and the Structure of the Mouth Parts of Marine Copepods," *Limnol. Oceanog.* (1963) **8** (1), 116–26.

(48) Wright, J. C., "The Limnology of Canyon Ferry Reservoir, I. Phytoplankton–Zooplankton Relationships," *Limnol. Oceanog.* (1958) **3** (2), 150–9.

(49) Burns, C. W., "Relation between Filtering Rate Temperature and Body Size in Four Species of Daphnia," *Limnol. Oceanog.* (1969) **14** (5), 693–700.

(50) Ryther, J. H., "Inhibitory Effects of Phytoplankton upon the Feeding of Daphnia Magna with Reference to Growth, Reproduction and Survival," *Ecology* (1954) **35**, 522–33.

(51) McMahon, J. W., Rigler, F. H., "Feeding Rate of Daphnia Magna Straus in Different Foods Labeled with Radioactive Phosphorus," *Limnol. Oceanog.* (1965) **10** (1), 105–13.

(52) Conover, R. J., "Oceanography of Long Island Sound, 1952–1954. VI. Biology of *Acartia Clausi* and *A. tonsa*," *Bull. Bingham Oceanog. Coll.* (1956) **15**, 156–233.

(53) Mullin, M. M., "Some Factors Affecting the Feeding of Marine Copepods of the Genus Calanus," *Limnol. Oceanog.* (1963) **8** (2), 239–50.

(54) Burns, C. W., Rigler, F. H., "Comparison of Filtering Rates of Daphnia in Lake Water and in Suspensions of Yeast," *Limnol. Oceanog.* (1967) **12** (3), 492–402.

(55) Riley, G. A., "A Theoretical Analysis of the Zooplankton Population of Georges Bank," *J. Marine Res.* (1947) **6** (2), 104–13.

(56) Conover, R. J., "Assimilation of Organic Matter by Zooplankton," *Limnol. Oceanog.* (1966) **11**, 338–45.

(57) Lotka, A. J., "Elements of Mathematical Biology," p. 88–94, Dover, New York, 1956.

(58) Bishop, J. W., "Respiration Rates of Migrating Zooplankton in the Natural Habitat," *Limnol. Oceanog.* (1968) **13** (1), 58–62.

(59) Comita, G. W., "Oxygen Consumption in Diaptomus," *Limnol. Oceanog.* (1968) **13** (1), 51–7.

(60) Martin, J. H., "Phytoplankton Zooplankton Relationships in Narragansett Bay III," *Limnol. Oceanog.* (1968) **13** (1).

(61) Vollenweider, R. A., "Scientific Fundamentals of the Eutrophication of Lakes and Flowing Waters, with Particular Reference to Nitrogen and Phosphorus as Factors in Eutrophication," p. 117, Organization for Economic Cooperation and Development Directorate for Scientific Affairs, Paris, France, 1968.

(62) O'Connor, D. J., DiToro, D. M., Thomann, R. V., Mancini, J. L., "Phytoplankton Population Model of the Sacramento–San Joaquin Delta Bay," Technical Report to Department of Water Resources, State of California, by Hydroscience, Inc., Westwood, N. J., 1971.

(63) Davis, H. T., "Introduction to Nonlinear Differential and Integral Equations," p. 99–109, Dover, New York, 1962.

(64) Department of Water Resources, State of California, private communication, 1966.

(65) Stiefel, E. L., "An Introduction to Numerical Mathematics," p. 163, Academic, New York, 1966.
(66) Hamming, R. W., "Numerical Methods for Scientists and Engineers," p. 194–210, McGraw-Hill, New York, 1962.
(67) Levy, H., Baggott, E. A., "Numerical Solutions of Differential Equations," p. 91–110, Dover, New York, 1950.

RECEIVED May 27, 1970.

6

Transport of Organic and Inorganic Materials in Small-Scale Ecosystems

EARNEST F. GLOYNA, YOUSEF A. YOUSEF,[1] and THOMAS J. PADDEN

Center for Research in Water Resources, The University of Texas at Austin, Austin, Tex. 78712

This paper describes some studies on the transport of bio-degradable and nondegradable materials through a system of flowing and nonflowing research flumes. A variety of radionuclides and dyes were used as tracers. The uptake and release of pollutants and biomass of various varieties of typical bottom sediments, rooted plants, were monitored in time and space along a model river (200-ft-long flume) which simulated the ecological environment. Laboratory and aquaria studies simulated closed ecosystems. Responses of the total ecosystem and its components to instantaneous and continuous releases of pollutants were investigated. Generalized mathematical relationships have been developed to describe the transport of various materials along a model flume.

A small-scale ecosystem (research flume) has been used successfully to investigate the transport phenomena of radionuclides added as dissolved inorganics and to study and evaluate the parameters affecting a stream when it is subjected to organic stresses.

The research flume (Figure 1) is 200 ft long, 2 ft deep, and 2.5 ft wide with a center removable partition providing two 1.25-ft-wide channels. The intake can be provided either with potable water or with water from a control reservoir rich in phytoplankton. The phytoplankton population can be controlled by dilution with potable water as system simulation may require. Flow can be controlled from 0 to approximately 200 liters per minute per channel and depths controlled from 0 to 2 ft. Bottom sediments were obtained from nearby Lake Austin, and plants, when used

---

[1] Present address: Department of Civil Engineering, Florida Technological University, Orlando, Fla.

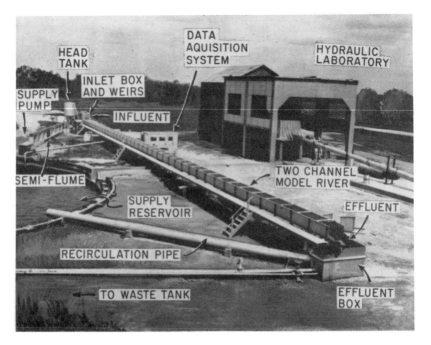

Figure 1.   Model river

in the experimentation, were species common to local streams and ponds. The predominant plants were *Myriophyllum, Chara, Potamogeton, Clado-phora,* and *Spirogyra.* Figure 2 indicates the hydraulic characteristics of the flume. The shaded portion of Figure 2 portrays the general area in which experimentation has been conducted.

Instrumentation on the flume consisted of pH meters, dissolved oxygen probes (galvanic), pyrheliometers, an anemometer, and a paper-punch tape data acquisition system.

The studies on the flume were conducted to determine the effects of specific environmental factors on transport and to specifically determine the physicochemical characteristics of the ecological system in order to better understand the factors that react to pollutional stress. The procedure used was to estimate the uptake and release rates of radionuclides, to establish the effects of organic and inorganic pollution loads on radionuclide movement, and to relate all of these factors into a basic prediction model. Radionuclides including ^{65}Zn, ^{58}Co, ^{137}Cs, ^{85}Sr, ^{106}Ru, and ^{51}Cr were introduced into the flume under different environmental conditions. A close follow-up of the radioactive flume was maintained as the radionuclides dispersed longitudinally and transversely. Continuous monitoring of radioactivity in the water phase, sediment, and biomass was maintained by sampling and use of radionuclide detection equipment.

Transport of Radionuclides

The transport characteristics of radionuclides in flowing ecosystems were established through a series of step-wise operations:

(*a*) logically enumerating the various processes which occur in a model river

(*b*) isolating important processes by using aquaria and model river studies

(*c*) synthesizing logical mathematical models using the above information in conjunction with parametric studies

(*d*) programming of the prediction model in computer language to evaluate the relative importance of factors affecting radionuclide transport.

The basic transport equation developed by Shih and Gloyna (*1*) was used.

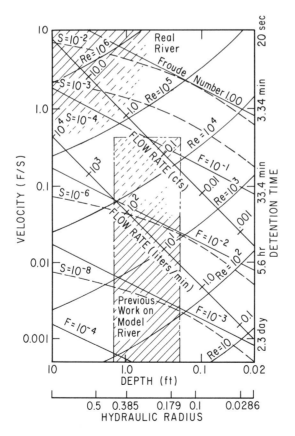

Figure 2.   Hydraulic characteristics of model river

$$\frac{\partial c}{\partial t} = D_x \frac{\partial^2 c}{\partial x^2} - u \frac{\partial c}{\partial x} - \sum_{i=1}^{n} f_i K_i [G_i(c) - C_i] \qquad (1)$$

where

c	=	concentration of a particular radionuclide in the flowing stream at any point (x) at time (t)
x	=	distance in direction of flow
u	=	average velocity of flow
f_i	=	mass or surface area of the ith sorbent per unit volume of the zone of flow
K_i	=	mass or area transfer coefficient for phase i
$G_i(c)$	=	a transfer function relating the concentration of activity in the water to the equilibrium level of activity in phase i
C_i	=	the specific activity in the ith position of the n-sorption phases

The first two terms in the equation define the mixing characteristics and dilution, while the third term establishes the uptake and release by the various aquatic masses.

Longitudinal dispersion coefficients, D_x values, in the absence of bio-mass were calculated for velocities ranging from 0.33 to 3.30 ft/min. The D_x values calculated from the time–concentration curves of the Rhodamine B dye studies, under various flow conditions in the flume, follow an empirical relationship in the form

$$D_x = 3.26 u^{0.607} \qquad (2)$$

where

u	=	the average velocity in ft/min
D_x	=	the longitudinal dispersion coefficient in ft^2/min

Discrepancies between nonsorptive dye and radionuclide movement were observed. In some experiments where 90–100% of Rhodamine B passed through the flume after one day from release, only 5–25% of ^{65}Zn, ^{58}Co, ^{137}Cs, and ^{85}Sr added at the inlet was detected in the exit flume water. The retention of radionuclides was primarily owing to surface concentration by sediments and biomass (1, 2, 3, 4, 5, 6, 7, 8).

Surface Concentration by Bed Sediments

Dissolved radioactive ions in aquatic systems are extracted from the water by bed sediments. These radionuclides are concentrated at the sediment surface and then diffused into the lower layers of bottom sediment. The rate of transfer is dependent on contact time, environmental conditions, and physicochemical characteristics of the sediments. The

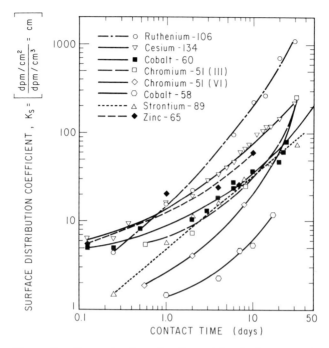

Figure 3. Concentration of radionuclides by bottom sediments in nonflowing ecosystems

surface distribution coefficients, K_s values [the specific activity per unit area of bed sediment divided by the specific activity per unit volume of water (*i.e.*, dpm/cm² ÷ dpm/cm³)], are useful parameters to compare the relative concentration of radionuclides by sediment surface area. The K_s values were measured in aquaria studies for various radionuclides as shown in Figure 3. It was obvious that K_s values increased with contact time; certain radionuclides such as ^{137}Cs and [85]Sr appear to follow an exponential relationship (linear on log–log coordinates). Previous investigations (7) indicated that the concentration of [85]Sr by sediments similar to those used in experiments described herein followed the equation

$$K_s = 5.0 T^{0.64} \tag{3}$$

where

T = contact time (days)
K_s = (dpm/cm²) ÷ (dpm/ml)

The affinity of bottom sediments obtained from Lake Austin to radionuclides released instantaneously in the flume followed the order [137]Cs > [65]Zn > [58]Co > [85]Sr. The maximum uptake of ^{137}Cs did not exceed 24% of released activity.

As radionuclides concentrate on the sediment surface, they tend to diffuse to deeper depths. The surface concentration and migration relationships of radionuclides follow an empirical relationship

$$C_{sd} = C_{so}10^{-pd} \tag{4}$$

where

C_{sd} = specific activity of radionuclide per gram at depth d in core
C_{so} = specific activity of radionuclide at surface
p = penetration coefficient

The penetration coefficient varies inversely with contact time. As the contact time increases, the value of p decreases. Penetration depth indicated in Table I is that depth in which 99.9% of the activity in the sediment core was concentrated during the indicated contact time. This depth has varied between 2 cm for ^{65}Zn and 13.5 cm for ^{137}Cs.

Table I.   Penetration Depth for Various Radionuclides in Bottom Sediments

Radionuclide	Contact Time, Days	Penetration Coefficient p, Cm-1	Penetration, Depth
^{65}Zn	17	1.38	2.0
^{58}Co	17	0.70	4.0
^{137}Cs	30	0.22	13.5
^{85}Sr	32	0.43	7.3
^{106}Ru	30	0.60	5.2

Concentration of Radionuclides by Plants

Some previous studies, such as those conducted on the Clinch River, have indicated that biomass may be neglected in a mass balance analysis of radionuclides (1, 4). However, when the biomass was extensive in the flume, a significant portion of the radionuclide was retained, at least temporarily (Figure 4). Plants sorbed radionuclides in the order indicated.

$$^{58}Co > \,^{65}Zn > \,^{85}Sr > \,^{137}Cs$$

Figure 4 also shows that plants released ^{137}Cs and ^{85}Sr at a faster rate than ^{58}Co and ^{65}Zn.

Radionuclide transport in an aqueous environment probably is related to community metabolism (9). The concentration of radionuclides increases in the plant biomass, with an accompanying loss from water or sediment, when the photosynthesis-to-respiration (P/R) ratio exceeds one (Figure 5). When this ratio fell below one, a net gain of radionuclides in the sediment and water accompanied by a loss in plants was observed.

However, the role of community structure in the transport of radionuclides is not as thoroughly understood.

Mathematical Modeling

Shih and Gloyna (7) presented an analytical solution for Equation 1 describing instantaneous releases of radionuclides, assuming dispersion in the longitudinal direction and sorption by sediments and biomass. A linear relationship between specific activity in sediments and plants as compared with the specific activity in water seemed to exist, providing Equation 5.

$$\frac{\partial c}{\partial t} = K[G(c) - C] \tag{5}$$

The solution of Equation 5 for sediments is

$$K_1 = \frac{1}{t} \ln \frac{K_s(C_w) - C_s}{K_s(C_w)} \tag{6}$$

whereas the solution of Equation 5 for plants is

$$K_2 = \frac{1}{t} \ln \frac{K_p(C_w) - C_p}{K_p(C_w)} \tag{7}$$

where

K_1 and K_2 = mass transfer coefficients for sediments and plants $(\text{time})^{-1}$

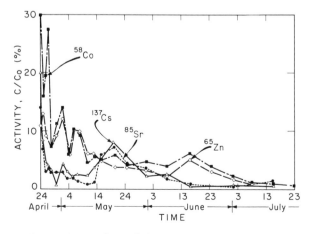

Figure 4.   Radionuclide uptake by aquatic plants

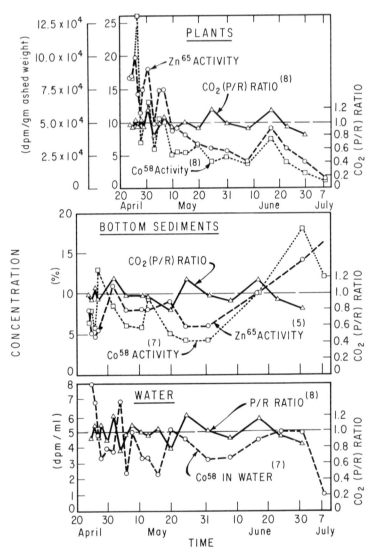

Figure 5. *Co⁵⁸ and Zn⁶⁵ activity in plants, sediment, and water as related to the P/R ratio (CO₂ metabolism)*

K_s and K_p = equilibrium concentration coefficients for sediments and plants

C_s and C_p = specific activity in sediments and plants at time t

C_w = specific activity in water phase

Values of K_s for various isotopes are presented in Figure 3. Values of K_p derived from the flume experiments agree with the published concentration coefficients of radionuclides by edible plants (10).

Equations 6 and 7 were used to determine the mass transfer coefficient K_1 and K_2 for sediments and plants. Detailed studies on [85]Sr showed that K_1 values ranged from 0.0045 and 0.011 hr^{-1} for flow conditions having a Reynolds number of 2440 and 3700, respectively (7). Flume studies also indicated that K_p for [85]Sr ranged from 664 to 683 dpm/gram per dpm/ml and values of K_2 varied from 0.0126 to 0.0322 hr^{-1} with an average of 0.022 hr^{-1} (1).

By determining D_x, K_1, and K_2, it was possible to program a solution for Equation 1 and verify it through repeated flume experiments.

Influence of Organic Pollutional Stresses on Transport

An organic load may influence the transport of other potential pollutants. For example, the reducing environment created by organic pollution may cause the reduction of hexivalent ^{51}Cr to trivalent [51]Cr (3), thus affecting transport of these radioactive ions, since sediments and biomass have different capacities for sorbing Cr VI and Cr III. In this connection, it is possible to evaluate the influence of surface aeration, without wind effects, and photosynthetic oxygenation.

Prior to simulating stream conditions and determining responses to imposed organic loads, it is desirable to investigate the individual factors that influence oxygen balance, namely atmospheric reaeration, photosynthetic production, respiration, and benthic uptake of oxygen. The aerobic or anaerobic conditions of a stream may drastically influence transport of ions, but the effect of reaeration may be difficult to monitor in a stream.

In these investigations, the first step was to determine atmospheric reaeration over various ranges of depths and velocities. This was accomplished in a clean channel devoid of sediment using potable water. Inlet dissolved oxygen was controlled by a solution of Na_2SO_3 with cobalt catalyst dispensed from carboys in which a nitrogen environment was maintained. The nitrogen prevented assimilation of oxygen by the solution, thereby providing reasonably consistent initial dissolved oxygen conditions. Dissolved oxygen profiles along the length of the flume were then obtained. A brisk wind increased atmospheric reaeration rate in a shallow system as much as 20 times the expected rate. To eliminate wind effect, a series of baffles were installed above the flowing surface, and data obtained during winds exceeding 5 knots were discarded. However, even with these precautions, wind affected the aeration, and it was necessary to completely cover the flume. In a natural system, the magnitude and direction of the wind would be a major factor in stream reaeration, but first the minimum surface transfer condition must be met.

Figure 6 shows the relationship between the reaeration coefficient (k_2) and $V/H^{1.5}$ where V is stream velocity in feet per second and H is depth in feet. Each plotted point is a mean value of k_2 for a group of test runs for a given velocity and depth. An average of 17 test runs were made to establish the value of each plotted k_2. Equations 8 and 9 describe the best fit determined by the least square method.

$$\text{Log } k_2 = 0.703 \text{ Log } \frac{V}{H^{1.5}} + \text{Log } 2.98 \tag{8}$$

$$k_2 = 2.98 \left[\frac{V}{H^{1.5}} \right]^{0.703} \tag{9}$$

While these equations are descriptive of the flume, they are reasonably close to the prediction model derived by Churchill, Elmore, and Buckingham from their analysis of atmospheric reaeration of streams in the Tennessee Valley (11). The Churchill et al. formula for prediction of k_2 at 20°C is

$$k_2 = 5.026 \frac{V^{0.969}}{H^{1.673}} \tag{10}$$

As an example of comparison, Table II shows k_2 predictions by the

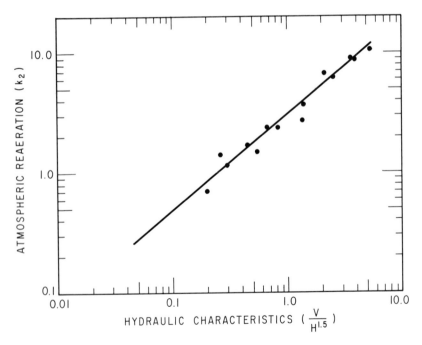

Figure 6.   Atmospheric reaeration as related to velocity and depth (in flume)

Churchill formula and the flume formula for hydraulic conditions occurring in the Jackson River (*12*).

Table II.   Predicted Atmospheric Reaeration Coefficients (k_2) as a Function of Stream Flow in the Jackson River[a]

Subreach	Length, Miles	Mean Velocity, Ft/Sec	Mean Depth, Ft	(k_2) Day^{-1} ($20°C$) Churchill	Flume
0–1	0.757	0.636	1.76	1.25	1.20
1–2	1.290	0.419	1.94	0.71	0.81
2–3	1.579	0.359	2.09	0.54	0.67
3–4	1.408	0.458	3.11	0.36	0.52
4–5	1.157	0.570	1.67	1.24	1.17
5–6	1.069	0.346	2.36	0.44	0.57
6–7	0.746	0.335	1.99	0.55	0.67
7–8	1.477	0.436	2.26	0.57	0.71
8–9	0.826	0.286	2.83	0.28	0.41
9–10	1.034	0.563	2.30	0.72	0.83
10–11	0.955	0.529	1.95	0.89	0.92
11–12	1.472	0.441	2.42	0.52	0.67
12–13	1.053	0.539	2.13	0.78	0.87
13–14	1.038	0.425	2.06	0.66	0.76
14–15	1.163	0.457	1.66	1.00	1.01

[a] Taken from Tsivoglou *et al.*, Ref. *12*.

Several experiments have been conducted also using the water from the control reservoir containing phytoplankton to determine the contribution to reoxygenation by photosynthesis (*P*) and the oxygen consumption or respiration (*R*) during periods of darkness. Since the effects of atmospheric aeration were known, the net production (*P* − *R*) could be determined by direct observation. While (*P* − *R*) in the flume on a 24-hour basis was a positive value, the net effect is to superimpose an oxygen demand on any existing demand during hours of darkness. Notably, the beneficial effect of photosynthesis in a stream must always be questioned since diminished light during daylight hours might result in respiration exceeding production (*P* − *R* negative).

Experiments in aquaria containing 4 inches of lake sediment were conducted to determine rate of oxygen consumption by the benthic load. These experiments indicated that the five-day biochemical oxygen demand (BOD_5) of the sediment would be satisfied in 350 to 450 days. This slow rate of biodegradation indicates anaerobic decomposition and the presence of a reducing environment.

With the preceeding knowledge and using one flume as a control, the influence of an organic pollutional stress on transport of radionuclides can be measured.

Conclusions

Small-scale ecosystems are useful tools for simulating stream processes and evaluating the complex factors which interact with various pollutional stresses.

A transport equation has been developed and verified which describes the mode of radionuclide movement. For dissolved wastewaters containing radionuclides, the principal mechanism of transport involves hydrodynamic action. However, the rate of movement is influenced by other factors such as suspended sediment loads and surface concentrations on both the bed sediments and biomass.

By simulating the physicochemical characteristics of a waterway system, the flume can be used effectively in determining the relationships of flow conditions to nuclide transport, concentration of nuclides by plants and bottom sediment, and distribution of individual nuclide species between aqueous solution and plants and between aqueous solution and sediments. Measurement of responses of the system to stress can aid in the development of, or verification of, prediction models for oxygen balance.

The surface concentration distribution coefficients for sediments K_s and for plants K_p in contact with radioactive contaminated water were formulated and predicted for various radionuclides through aquaria studies and flume experiments. These values reflect the affinity of sediments and plants for various radionuclides.

Acknowledgment

This research was partially supported by the U. S. Atomic Energy Commission and the Office of Water Resources Research, U. S. Department of the Interior.

Literature Cited

(1) Armstrong, N. E., Gloyna, E. F., *Radioactivity Transport in Water—Numerical Solutions of Radionuclide Transport Equations and Role of Plants in Sr-85 Transport*, EHE-12-6703, Technical Report No. 14 to the U.S. Atomic Energy Commission, December 1967.
(2) Bhagat, S. K., Gloyna, E. F., *Radioactivity Transport in Water—Transport of Nitrosylruthenium in an Aquatic Environment*, EHE-11-6502, Technical Report No. 9 to the U.S. Atomic Energy Commission, November 1965.
(3) Canter, L. W., Gloyna, E. F., *Radioactivity Transport in Water—Transport of Cr-51 in an Aqueous Environment*, EHE-04-6701, Technical Report No. 11 to the U.S. Atomic Energy Commission, May 1967.
(4) Gloyna, E. F., Bhagat, S. K., Yousef, H., Shih, C., "Transport of Radionuclides in a Model River," *Proc. Intern. Atomic Energy Agency Symp. Disposal of Radioactive Wastes into Seas, Oceans, and Surface Waters, Vienna, Austria, May 16–20, 1966.*

(5) Kudo, Akira, Gloyna, E. F., *Radioactivity Transport in Water—Interaction Between Flowing Water and Bed Sediment*, EHE-69-03, CRWR-36, Technical Report No. 17 to the U.S. Atomic Energy Commission, January 1969.
(6) Rowe, D. R., Gloyna, E. F., *Radioactivity Transport in Water—The Transport of Zn^{65} in an Aqueous Environment*, EHE-09-6403, Technical Report No. 5 to the U.S. Atomic Energy Commission, September 1964.
(7) Shih, C. S., Gloyna, E. F., *Radioactivity Transport in Water—Mathematical Model for the Transport of Radionuclides*, EHE-04-6702, Technical Report No. 12 to the U.S. Atomic Energy Commission, June 1967.
(8) Yousef, Y. A., Gloyna, E. F., *Radioactivity Transport in Water—The Transport of Co^{58} in an Aqueous Environment*, EHE-12-6405, Technical Report No. 7 to the U.S. Atomic Energy Commission, December 1964.
(9) Copeland, B. J., Gloyna, E. F., *Radioactivity Transport in Water—Structure and Metabolism of a Lotic Community*, Part I (April–July 1964), EHE-02-6501, Technical Report No. 8 to the U.S. Atomic Energy Commission, February 1965.
(10) Chapman, W. H., Fisher, H. L., Pratt, M. W., "Concentration Factors of Chemical Elements of Edible Aquatic Organisms," UCRL-50564, University of California, Lawrence Radiation Laboratory, TJD-4500, UC-48, December 30, 1968.
(11) Churchill, M. A., Buckingham, R. A., Elmore, H. L., "The Prediction of Stream Reaeration Rates," Tennessee Valley Authority, Chattanooga, 1962.
(12) Tsivoglou, E. C., Cohen, J. B., Shearer, S. D., Godsil, P. J., "Tracer Measurement of Stream Reaeration. II. Field Studies," *J. Water Pollution Control Federation* (February 1968) **40**, 2, 285–305.

RECEIVED May 14, 1970.

7

Use of Inert Tracer Data to Estimate the Course of Reaction in Geometrically Complex Natural Flows

LLOYD A. SPIELMAN and FRANCOIS BRIERE

Division of Engineering and Applied Physics, Harvard University, Cambridge, Mass. 02138

The dispersion model, while suitable for interpreting tracer data from geometrically simple flows, is not suited to complex geometries. Here the residence time interpretation, originally developed by Danckwerts and Zwietering for application to steady state industrial reactors, is described and extended to treat transient reactant loadings which occur in modeling for pollution control. Sample computer calculations are based on tracer responses of two flows: one theoretical, the other a natural stream. Results are presented for both slug and step inputs of reactants disappearing according to both first and second order rate expressions. With second order reaction, the downstream response is bracketed by upper and lower bounds corresponding to minimum and complete segregation. For first order reaction, the bounds merge to give a single response curve.

In modeling rivers, streams, and other natural flows for pollution control, it is often necessary to estimate downstream concentration profiles, in distance and time, of substances which simultaneously undergo reaction and convection after their introduction upstream. These substances might be dissolved or suspended pollutants, microorganisms, oxygen, or other relevant constituents of the aqueous environment.

To predict concentration profiles, it is desired to combine independent laboratory reaction rate data with empirical data on the hydrodynamics of the flow. The problem of interpreting laboratory reaction data to obtain rate expressions which are valid within the complex natural system is in itself a difficult task beyond the scope of this work; it is here assumed that valid concentration-dependent expressions for rates of reaction are available.

With the given rate expression, detailed theoretical or empirical knowledge of the spatial velocity distribution would promote use of the equations of convective transport (*1*) to evaluate concentration profiles and give a comprehensive treatment, but direct in-stream measurements required to obtain such velocity data are prohibitively costly and not generally available. Inert tracer studies for the flow are far more practical and economical but do not convey complete hydrodynamic information. Tracer studies are normally carried out by introducing a conveniently monitored solute (such as a radioactive isotope), usually as a slug, and measuring its concentration response with time at stations downstream (*2*). For reaches of the flow which are not geometrically complex, the longitudinal dispersion model can often be applied, using measured tracer response curves to extract numerical dispersion coefficients. These can then be inserted back into the dispersion equation with a reaction rate term to estimate the concentrations of reacting species which are introduced upstream in an assigned fashion (*3, 4*). For regions of the flow which are geometrically complex, however, tracer response curves can differ in shape greatly from those expected from the dispersion model and an alternate method of interpreting them must be sought. Illustrations of such possible complexities are shown in Figure 1.

Interpreting tracer response curves as residence time distributions, along with considerations of complete and minimum segregation in the flow, provides a rigorous basis for treating geometrically complex flows. This approach was pioneered by Danckwerts (*5*) and Zwietering (*6*) for application to nonideal industrial reactors. Industrial reactors over which the engineer has considerable geometrical control behave more predictably than most natural flows and often can be treated theoretically; thus, application of the semiempirical residence time approach to natural systems could be even more significant than its originally intended use.

While residence time theory should be applicable to a wide variety of geometrically complex flows, it possesses inherent theoretical limitations which should be discussed. An important one is that input and monitoring stations should correspond to constrictions of the flow (Figure 1) where concentrations across the stream width may be considered as uniform (although much can also be said about the distribution of concentrations between stations). This is seen to be no great handicap, however, when compared with the one-dimensional dispersion model which assumes a uniform concentration across the stream width for all points along the flow. Another limitation occurs if the reaction dependence on concentration is greatly different from first order; then, with knowledge of the residence time distribution, one can only bracket the concentration

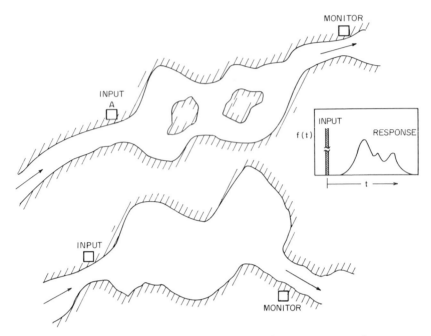

Figure 1. Possible natural flows having complex geometry and tracer response. Input and monitoring stations are at constrictions. Upstream of station A, the dispersion model might apply.

with least upper and greatest lower bounds. Only for the important special case of first order (linear) reaction is the downstream concentration unambiguously determined for a specified inert tracer response. How tightly concentration is bracketed for a given system depends on the particular form of the residence time (tracer response) curve, the degree of nonlinearity of the reaction, and the extent of reaction occurring between stations.

That the tracer response curve does not in general fix system behavior unambiguously for nonlinear reactions is readily demonstrated by considering the particular system consisting of an ideal plug flow vessel in alternate sequence with an ideally-stirred vessel. As illustrated in Figure 2, interchanging the order of the two vessels leaves the shifted exponential response to a slug input of tracer unchanged; yet for reactions with order greater than unity, straightforward calculations confirm that the configuration having the plug flow vessel first yields more extensive reaction. The situation is reversed for reaction orders less than unity. Both arrangements give the same response if the reaction is first order.

Because of the ambiguity of behavior which remains with knowledge of a system's tracer response, it can be concluded that close agreement of inert tracer response curves to the form expected from the dispersion model does not conclusively confirm the validity of that model between two monitoring stations; thus, application of the dispersion model to non-linear reactions should be made with caution even if inert tracer response appears ideal. It is subsequently shown how any given inert tracer response of a system still permits a range of possible responses for nonlinear reaction and how the minimal bounds on that range can be computed.

Elements of Residence Time Theory

Since the foundations of residence time theory are rigorously given elsewhere (5, 6, 7), only those features which are essential to the present treatment will be given here. The residence time distribution (residence time frequency function; exit age distribution), $f(t)$, is defined such that $f(t)dt$ is the fraction of fluid at any instant leaving the system, having spent time between t and $t + dt$ within the system. The cumulative residence time distribution is

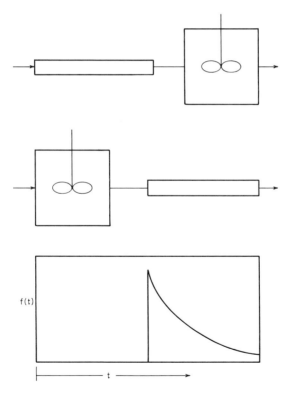

Figure 2. Ideal plug flow and perfectly mixed vessels in alternate sequence. Both arrangements exhibit the same response for inert tracers and first order reactions but not for nonlinear reactions.

$$F(t) = \int_0^t f(t')dt' \qquad (1)$$

$F(t)$ corresponds to the fraction of fluid at any instant leaving the system, having spent time shorter than t within the system. A relation of further importance is

$$V/Q = \tau = \int_0^\infty tf(t)dt \qquad (2)$$

That is, the first moment of the residence time frequency function (mean residence time) equals the quotient of the system volume V and the constant volume flow rate Q through the system.

The function $f(t)$ is straightforwardly related to the experimental tracer response curve; for a slug input, the downstream concentration vs. time curve is proportional to the function $f(t)$. Response curves for different modes of input theoretically convey equivalent information, although certain inputs are experimentally convenient to carry out. Thus, redoing tracer experiments for different inputs gives no added information about the flow.

Extremes of Mixing for an Arbitrary Specified
Residence Time Distribution

Zwietering was able to prove that the two conceptual reactor configurations shown in Figure 3, each having the same specified arbitrary residence time distribution, exhibit opposite extremes of mixing and thus give bounds on chemical conversion. Each configuration is an ideal plug flow vessel having continuous distribution of respective side exits or entrances which are governed to give the arbitrary specified residence time distribution. Zwietering showed that the lower configuration in Figure 3 corresponds to the minimum degree of segregation for a given residence time distribution, while the upper diagram corresponds to complete segregation or degree of segregation unity. The degree of segregation was introduced by Danckwerts and is a measure of the extent to which molecules entering the system at a particular time remain together, not mixing with molecules which enter at other times. While this concept arises in the fundamentals of the theory, its precise definition or numerical evaluation is not needed for application of the theory and will not be gone into here. Zwietering alternately interprets complete and minimum segregation for the specified residence time function as "latest" and "earliest" mixing, respectively. The time coordinate α shown in Figure 3 corresponds to the age (time interval after entering) of a fluid element in the system, and the coordinate λ is the life expectation (time interval before leaving) of an element.

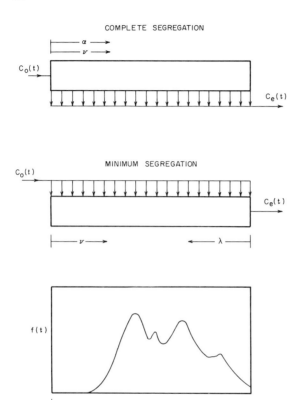

Figure 3. *Conceptual flows corresponding to extremes of complete and minimum segregation. The distributed entrances and exits are governed to give the same arbitrary specified residence time distribution for each system.*

Analysis for Complete Segregation

In the case of complete segregation, the volume dv of an element of fluid with age between α and $\alpha + d\alpha$ is

$$\frac{dv}{V} = \frac{1}{\tau} [1 - F(\alpha)]d\alpha \tag{3a}$$

The relation between the volume coordinate v (Figure 3) and α is then

$$\frac{v}{V} = \frac{1}{\tau} \int_0^\alpha [1 - F(\alpha')]d\alpha' \tag{3b}$$

in which F is given by Equation 1.

The equations describing reaction in the system were derived previously only for the steady state with a constant incoming concentration. For industrial reactors that case is of utmost importance, but for natural systems transient situations are equally important, and the treatment is extended here to include an accumulation term. Following Zwietering, let $c(\alpha,t)$ denote the concentration of reacting substance in the system as it depends on both position (through Equation 3b) and time t, and let $R(c)$ be the concentration-dependent rate of disappearance of reactant per unit volume. Considering now a volume element of the system and using Equation 3b, the quantity of reactant leaving through side exits is $Qc(\alpha,t)f(\alpha)d\alpha$. A detailed material balance on the volume element gives the partial differential equation for the concentration profile as

$$\frac{\partial c}{\partial \alpha} + \frac{\partial c}{\partial t} = -R(c) \tag{4}$$

The time-dependent concentration in the combined exit stream is

$$c_e(t) = \int_0^\infty c(\alpha,t)f(\alpha)d\alpha \tag{5}$$

in which $c(\alpha,t)$ is the solution of Equation 4. Equation 4 is to be solved under the conditions: before time $t = 0$, no reactant is present in the system, and after $t = 0$, the incoming concentration c_o is an arbitrary specified function of time. That is,

$$c(\alpha,t) = 0 \text{ for } t \leq 0 \tag{6a}$$

$$c_o = 0 \text{ for } t \leq 0 \tag{6b}$$

$$c_o = c_o(t) \text{ for } t > 0 \tag{6c}$$

Equation 4 is quasilinear and can be solved readily by the method of characteristics (8). Its solution under conditions 6a, 6b, and 6c is given by the implicit relation

$$-\int_{c_o(t-\alpha)}^{c(\alpha,t)} \frac{dc}{R(c)} = \alpha \tag{7}$$

For a disappearance reaction of order $n \neq 1$, $R(c) = kc^n$ and Equation 7 gives

$$c(\alpha,t) = c_o(t - \alpha)/[1 + (n - 1) \cdot k \cdot c_o^{n-1}(t - \alpha) \cdot \alpha] \qquad (8)$$

For first order reaction, $R(c) = kc$, and Equation 7 gives

$$c(\alpha,t) = c_o(t - \alpha)e^{-k\alpha} \qquad \text{for } n = 1 \qquad (9)$$

Equations 5, 9, 6b, and 6c then give the exit concentration as

$$c_e(t) = \int_0^t c_o(t - \alpha)e^{-k\alpha}f(\alpha)d\alpha \qquad \text{for } n = 1 \qquad (10)$$

For an initial constant step increase of concentration c_o and infinite time, Equation 10 becomes the well-known steady state result

$$c_e = c_o \int_0^\infty e^{-k\alpha}f(\alpha)d\alpha = \text{constant} \qquad (11)$$

Analysis for Minimum Segregation

In the case of minimum segregation, the equation which relates the volume element dv to the life expectation λ, analogous to Equation 3a, is

$$\frac{dv}{V} = -\frac{1}{\tau}[1 - F(\lambda)]d\lambda \qquad (12)$$

The rate of reactant now entering at time t through side entrances in the volume element is $Qc_o(t)f(\lambda)d\lambda$. An unsteady material balance on the element gives the partial differential equation describing the concentration profile $c(\lambda,t)$.

$$\frac{\partial c}{\partial \lambda} - \frac{\partial c}{\partial t} = R(c) + \frac{f(\lambda)}{1 - F(\lambda)} \cdot [c - c_o(t)] \qquad (13)$$

The time-dependent exit concentration in this case corresponds to zero life expectation ($\lambda = 0$), thus

$$c_e(t) = c(0,t) \qquad (14)$$

Equation 13 is now solved under similar conditions as previously, *viz.*, before $t = 0$ no reactant is present in the system, and after $t = 0$ the incoming concentration is an arbitrary specified function of time. That is,

$$c(\lambda,t) = 0 \text{ for } t \leq 0 \tag{15a}$$

$$c_o = 0 \text{ for } t \leq 0 \tag{15b}$$

$$c_o = c_o(t) \text{ for } t > 0 \tag{15c}$$

As was the case with Equation 4, Equation 13 is quasilinear and can be solved by the method of characteristics (8). The problem is thereby simplified to that of solving the ordinary first order differential equation

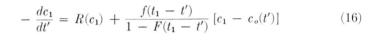

$$-\frac{dc_1}{dt'} = R(c_1) + \frac{f(t_1 - t')}{1 - F(t_1 - t')} [c_1 - c_o(t')] \tag{16}$$

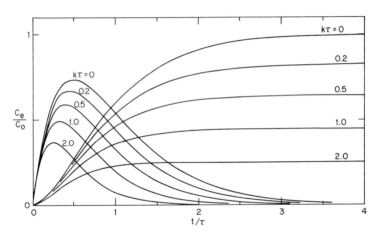

Figure 4.   Transient responses to slug (left) and step (right) inputs of reactant for first order reaction and selected $k\tau$ values. The flow has the same residence time distribution as two perfectly mixed vessels in sequence. For slug input, $c_o = m/V$.

In Equation 16, $c_1(t';t_1)$ is a one-place function of the independent variable t' for assigned values of the constant parameter t_1 and is related to $c(\lambda,t)$ by

$$c_1(t'; t_1) = c(t_1 - t', t') \tag{17}$$

The integration of Equation 16 is started at

$$c_1 = 0, t' = 0 \tag{18}$$

since from Equations 15a and 17, $c_1(0;t_1) = c(t_1,0) = 0$. Thus, $c(\lambda,t)$ can be obtained from Equation 17 by integrating Equation 16 for any assigned $\lambda = t_1 - t'$ and $t = t'$ by appropriate selection of t_1 and t'. In particular, the exit concentration given by Equation 14 corresponding to $\lambda = 0$ is obtained by choosing $t_1 = t$ and integrating Equation 16 to $t' = t$, thus

$$c_e(t) = c_1(t:t) \tag{19}$$

In the special case of first order reaction, Equation 16 becomes linear and after some manipulation gives for the exit concentration a result identical to that of Equation 10; thus, for first order reaction, calculations for complete and minimum segregation yield a single result in the transient case. Inputting the reactant as a slug of amount m, $c_o(t) = (m/V)\delta(t)$, and Equation 10 gives for the output of either system

$$c_e(t) = \frac{m}{V} e^{-kt} f(t) \quad \text{for } n = 1 \tag{20}$$

With a slug of inert tracer, $k = 0$, and Equation 20 confirms the response to be simply proportional to the residence time distribution.

Theoretical considerations of residence time theory imply the following important trends:

The closer the reaction order is to unity, the more closely the calculated bounds will approach one another.

The more peaked the residence time distribution, the closer the extremes of possible behavior. This is because the limit of plug flow, for which $f(t) = \delta(t - \tau)$, yields unambiguous response for nonlinear reactions.

The smaller the extent of reaction, the closer the extremes of possible behavior. This is because the reactant behavior then approaches that of an inert tracer, which exhibits an unambigous response.

Sample Calculations

Numerical calculations to demonstrate application of the preceding equations to specific tracer responses were performed using an IBM 360 computer. The results, in dimensionless form, are given in Figures 4 through 8. In general, calculations were quite rapid, typically requiring seconds to compute a response curve. Both first and second order disappearance reactions were investigated. Responses to slug and step inputs of reactant were computed for two flow systems: one theoretical and the other a segment of a real natural flow.

   Results for a Theoretical Flow. The results shown in Figures 4 and
5 correspond to a theoretical flow having a residence time distribution
identical to that of two equal-sized ideally-stirred vessels in sequence.
For this flow, the residence time distribution is given by

$$f(t) = (4t/\tau^2)e^{-2t/\tau} \tag{21}$$

Calculations were first performed for this situation to provide comparison
with Zwietering's steady state calculations for the same system, as well as
for convenient analytical checks on the computer program. Figure 4
shows the unambiguous system response for first order reaction with slug
and step inputs. The response curve for an inert slug ($k\tau = 0$) corresponds to Equation 21.

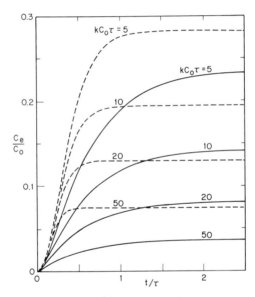

*Figure 5. Possible extremes of transient re-
sponse to step change of reactant input concentration for second order reaction and
selected values of $kc_o\tau$. The flow has the
same residence time distribution as two perfectly mixed vessels in sequence. Solid
curves: complete segregation. Broken
curves: minimum segregation.*

   Figure 5 shows predicted response of the same system for the extremes of complete and minimum segregation, second order reaction,
and step inputs. Results are shown for various selected values of $kc_o\tau$.
The concentrations after long times are in substantial agreement with
Zwietering's steady state results (the titles of the two tables in Ref. 6

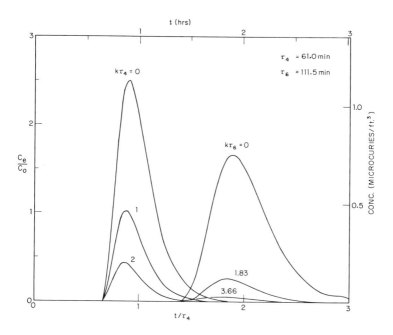

Figure 6.   Transient responses at two consecutive downstream stations 4 and 6 to the same slug inputs of reactant undergoing first order reaction. The flow is a stretch of natural stream for which the responses to inert tracer ($k = 0$) are empirical (2). $c_o = m/V$.

should be interchanged). For second order reaction, complete segregation predicts most extensive reaction (lowest concentrations).

Results for a Natural Stream. The results shown in Figures 6, 7, and 8 were calculated from measured tracer data given by Fischer (2). The source of these data was cited as the U.S. Geological Survey of 1959–1961. The flow is Copper Creek, Va. The experimental tracer curves are given in Figure 6 for $k\tau = 0$ and correspond to points 7870 and 13,550 ft downstream of the point of release. These are respectively referred to as stations 4 and 6. Figure 6 shows the unambiguously predicted response for first order reaction with slug inputs at the point of release. The rate constants k were arbitrarily chosen to give convenient values of dimensionless $k\tau$ at station 4 and retained at station 6 in order to observe the progress of reaction in passing from station 4 to station 6. Corresponding results for step input are shown in Figure 7.

Figure 8 shows predicted response at station 4 in the extremes of complete and minimum segregation for second order reaction and step increase in reactant concentration at the point of release. Results are shown for various chosen $kc_o\tau$ values.

Summary and Conclusions

Interpretation of tracer data by means of residence time theory, in the extremes of complete and minimum segregation, has been reviewed and extended to treat transient response under reacting conditions. While residence time theory was initially developed for industrial application to nonideal steady state reactors, its transient extension seems especially well suited for describing segments of natural flows which are too complex to interpret using simpler models, such as dispersion.

The closer the reaction is to first order, the closer the predicted extremes of response. The more peaked the residence time distribution, the closer the extremes, and the smaller the extent of reaction, the closer the extremes.

Sample calculations have been presented for first and second order reactions, slug and step reactant inputs upstream, and two flow configurations: one theoretical, the other a segment of a natural stream. For the second order reaction, lower and upper bounds on downstream reactant concentration, bracketing system response and respectively corresponding to complete and minimum segregation were calculated (for $n < 1$, the situation is reversed). For first order reaction, the bounds merge to yield unambiguous predictions of downstream profiles.

The theory was presented here for a single entrance, single exit, and a single reactant. Modeling for pollution control can require more com-

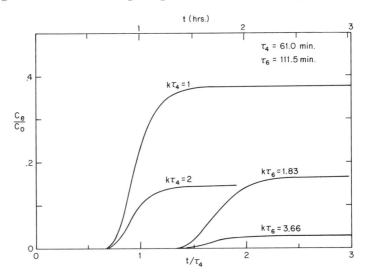

Figure 7. Transient responses at downstream stations 4 and 6 to the same step change of reactant input concentration and first order reaction. The flow is the stretch of natural stream having measured tracer responses shown in Figure 6.

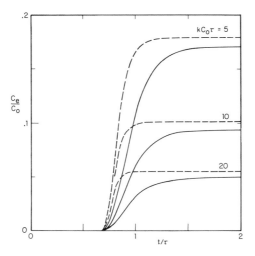

Figure 8. Extremes of transient response at downstream station 4 to step change of reactant input concentration, second order reaction, and selected values of $kc_o\tau$. The flow is the stretch of stream up to station 4, having measured tracer response shown in Figure 6. Solid curves: complete segregation. Broken curves: minimum segregation.

plicated possibilities, such as multiple entrances and exits, sources, sinks, etc. Whereas some complications can be dealt with by straightforward analysis into treatable subsystems, others may require significant theoretical extensions which should be sought.

Experiments should be performed in natural flows to test the validity of the theory and to indicate improvements. It might be the case that certain kinds of flows closely approximate one of the extremes of segregation; this would promote narrowing of the bounds for nonlinear reactions.

List of Symbols

English symbols:

c	= concentration of reactant
$f(t)$	= residence time frequency distribution
$F(t)$	= cumulative residence time distribution
k	= reaction rate constant
m	= amount in slug input
n	= reaction order
Q	= volume flow rate through system
$R(c)$	= rate of disappearance of reactant per unit volume
v	= volume coordinate
V	= total volume of system

Greek symbols:

α = age of fluid element
λ = life expectation of fluid element
τ = mean residence time

Subscripts:

0   = upstream inlet
4   = station 4
6   = station 6
e = downstream exit

Acknowledgment

This work was supported in part by the Federal Water Pollution Control Administration, U.S. Department of Interior, and in part by the Ministry of Natural Resources, Provincial Government of Quebec.

Literature Cited

(1) Bird, R. B., Stewart, W. E., Lightfoot, E. N., "Transport Phenomena," Wiley, New York, 1960.
(2) Fisher, H. B., J. Sanit. Eng. Div., Am. Soc. Civil Engrs. (Oct. 1968) 94, 927.
(3) Bella, D. A., Dobbins, W. E., J. Sanit. Eng. Div., Am. Soc. Civil Engrs. (Oct. 1968) 94, 995.
(4) Harleman, D. R. F., Lee, Chok-Hung, Hall, L. C., J. Sanit. Eng. Div., Am. Soc. Civil Engrs. (Oct. 1968) 94, 897.
(5) Danckwerts, P. V., Chem. Eng. Sci. (1958b) 8, 93.
(6) Zwietering, Th. N., Chem. Eng. Sci. (1959) 11, 1.
(7) Levenspiel, O., Bischoff, K. B., "Advances in Chemical Engineering," Vol. 4, p. 95, T. B. Drew et al., Eds., Academic, New York, 1963.
(8) Hildebrand, F. B., "Advanced Calculus for Applications," p. 394–400, Prentice-Hall, Englewood Cliffs, N. J., 1962.

RECEIVED July 20, 1970.

Variations in the Stability of Precipitated Ferric Oxyhydroxides

DONALD LANGMUIR and DONALD O. WHITTEMORE

Department of Geosciences, and Mineral Conservation Section,
The Pennsylvania State University, University Park, Pa. 16802

The apparent thermodynamic stability of ferric oxyhydroxide precipitates may be described by the activity product $pQ = -log[Fe^{3+}][OH^-]^3$ in solution. Such precipitates in natural waters usually begin as mixed amorphous material and goethite, sometimes including lepidocrite. In laboratory solutions initially 10^{-2} molar in Fe^{2+} or Fe^{3+}, pQ was 37.3 to 43.3. In 24 well waters containing $10^{-3.33}$ to $10^{-5.40}$ molar Fe^{2+} and suspended oxyhydroxides, pQ was 37.1 to 43.5. Variations in particle size are probably the major control on absolute and relative stabilities of the oxyhydroxides. H^+ leaching of precipitates raises pQ by removing the most soluble material. Measurable crystallization of precipitates occurs within a few hours in 10^{-2} molar Fe^{2+} solutions, but may take thousands of years in waters 10^{-6} molar in Fe^{2+}.

It has generally been assumed that the thermodynamic stability of precipitated ferric oxyhydroxides increases continuously with aging in or out of aqueous solution (*1, 2*). Little thought has been given to the role of natural water chemistry in changes in the stability of the oxyhydroxides with time. This report first examines the effects of changes in solution composition on the stability of ferric oxyhydroxides precipitated in ferric and ferrous sulfate solutions in the laboratory. These solutions are similar in composition to some iron-rich acid mine discharges. Results of the laboratory studies and of work by others are then used to explain observed variations in the stability of suspended ferric oxyhydroxides in some ground waters of coastal plain New Jersey and Maryland.

Naturally-Occurring Ferric Oxyhydroxides

Ferric oxyhydroxide minerals or phases which occur naturally are listed in Table I, and except for akaganéite, their occurrences have been described by Palache *et al.* (3). The oxyhydroxides found as precipitates in natural waters are usually goethite and x-ray amorphous material. Amorphous material, which comprises a relatively large proportion of most fresh precipitates, is formed under conditions of substantial supersaturation with respect to the crystalline oxyhydroxides. An amorphous phase develops by rapid hydrolysis of dissolved ferric species, particularly at pH's below 4–5 where the total concentration of such species can exceed 0.01 ppm. Amorphous material is also produced during the rapid oxidation and hydrolysis of ferrous iron-rich solutions.

Table I.   Naturally Occurring Ferric Oxyhydroxides

Ferric Oxyhydroxide	Ideal Formula
Amorphous	indefinite
Goethite	α–FeOOH
Akaganéite	β–FeOOH
Lepidocrocite	γ–FeOOH
Hematite	α–Fe$_2$O$_3$
Maghemite	γ–Fe$_2$O$_3$

Goethite precipitates by the relatively slow oxidation and hydrolysis of aqueous or solid ferrous iron species. Under sterile laboratory conditions, goethite forms most rapidly at pH's greater than 3.5 because the oxidation rate of dissolved ferrous iron is extremely slow under more acid conditions (4). Goethite is also formed by the long-term aging of amorphous material precipitated during hydrolysis of ferric salt solutions. Under natural conditions, the oxyhydroxides which precipitate on and coat bottom and bank sediments in streams polluted by acid mine discharges are usually mixtures of goethite and amorphous material.

Lepidocrocite has most often been precipitated in the laboratory at pH's between 4 and 7 by the oxidation of aqueous or solid ferrous iron species. The mineral is usually accompanied by goethite. Akaganéite is the rarest of the oxyhydroxides under natural conditions. Perhaps the only published descriptions of natural occurrences are by Van Tassel (5) and Chandy (6). The mineral has been precipitated in the laboratory at room temperatures by the slow hydrolysis of 0.02–0.08M FeCl$_3$ solutions (7).

Maghemite is rare as a direct precipitate, and usually forms by the oxidation of magnetite (Fe$_3$O$_4$) or the dehydration of lepidocrocite. Hematite and goethite are probably the most stable ferric oxyhydroxides under earth-surface conditions. Hematite is rare as a direct precipitate but slowly crystallizes from amorphous material by dehydration or long-term aging out of solution.

In most cases, the first oxyhydroxide precipitate is amorphous, having been formed under conditions of substantial supersaturation. Early in precipitation, poorly crystalline goethite and poorly crystalline lepido-crocite may also form. Other oxyhydroxides than these three are rarely precipitated from natural waters.

Oxyhydroxide precipitates are generally mixtures of different phases. The apparent thermodynamic stability or solubility of such mixtures depends in large part on the solubility of the least stable phase present. This phase is of course often amorphous. Oxyhydroxide precipitates usually contain substantial amounts of collodial-sized material. Sizes from molecular $Fe(OH)_3{}^\circ$ to micron-size crystals or crystal aggregates are possible (8). Both crystalline and amorphous precipitates remain colloidal indefinitely in some solutions. Particle size is probably the major control on the thermodynamic stability of the oxyhydroxides. Other important controls on stability are mineralogy, crystallinity, the degree of hydration of the precipitate, and the presence of impurities.

Solution reactions for all the oxyhydroxides can be written in the same form, as shown by expressions 1, 2, and 3.

$$Fe(OH)_3 = Fe^{3+} + 3OH^- \qquad (1)$$
$$\text{amorphous}$$

$$\alpha-FeOOH + H_2O = Fe^{3+} + 3OH^- \qquad (2)$$

$$1/2\,(\alpha-Fe_2O_3) + 3/2\,H_2O = Fe^{3+} + 3OH^- \qquad (3)$$

Here and henceforth, the composition of amorphous material is given for simplicity as $Fe(OH)_3$. The negative logarithm of the thermodynamic activity product expressions for these reactions may be written in each case as

$$pK = -\log [Fe^{3+}][OH^-]^3 \qquad (4)$$

when the activity of water is close to unity. Values of pK range from about 37.1 for fresh, rapidly precipitated amorphous material (8) to probably 44 or more for crystalline goethite and hematite, based on other work (9) and solubility measurements described in this paper.

Particle Size Effect

It may be shown that the decrease in pK ($-\delta pK$) of an oxyhydroxide relative to the pK of its thermodynamically most stable form and owing to the particle size effect equals

$$-\delta pK = 10.5\ S(gfw)/\rho x \qquad (5)$$

In this expression, S is the surface energy of the oxyhydroxide in ergs per square centimeter, gfw is its gram formula weight, ρ its density in grams per square centimeter, and assuming that particles of the oxyhydroxide are cubes, x is the cube edge length in Angstroms. Densities of hematite and goethite are 5.26 and 4.28 gm/cm^3, respectively (3). Based on heat of solution measurements at 70°C, Ferrier (10) found the surface enthalpies of hematite (as $1/2\ \alpha\text{-}Fe_2O_3$) and goethite (as $\alpha\text{-}FeOOH$) to be 770 and 1250 ergs/cm^2, respectively. Assuming as an approximation that surface enthalpies and Gibbs free energies are equal, and constant from 25° to 70°C, equations for the variation in δpK with particle size of hematite and goethite are, respectively

$$-\delta pK = 123/x \qquad (6)$$

and

$$-\delta pK = 272/x \qquad (7)$$

and have been plotted in Figure 1. Hematite and goethite particles of 100 Å and less in average dimension are common in laboratory solutions and natural waters. Based on Figure 1, such particles are at least 1.2 and 2.7 pK units less stable than well-crystallized samples of hematite and goethite, respectively. Ferrier's studies further show that two coexisting oxyhydroxides can reverse their relative stabilities because of particle size effects. Thus, for example, if we consider equal-sized hematite and goethite particles, at 70°C the enthalpy of the reaction

$$1/2\ \alpha-Fe_2O_3 + 1/2\ H_2O = \alpha-FeOOH \qquad (8)$$

equals

$$\Delta H°_{343}\ (cal/mole) = (-\ 1720 \pm 250) + 1.90 \times 10^5/x \qquad (9)$$

(10), where x is the edge length of cubes of hematite or goethite. This expression is roughly equivalent to

$$\Delta G°_{298}\ (cal/mole) = (-\ 250 \pm 300) + 1.90 \times 10^5/x \qquad (10)$$

for Reaction 8 $(9, 11)$. Expression 10 is plotted in Figure 2 and shows that for equal-sized hematite and goethite crystals, goethite is more stable than hematite when x exceeds 760 Å, but less stable than hematite at smaller particle sizes. Variations in $\Delta G°_{298}$ can be considerably greater than this when unequal particle sizes of hematite and goethite are involved (11).

Eh–pH Relations

The behavior of precipitated ferric oxyhydroxides in solution can best be described in terms of Eh and pH. Figure 3 is an Eh–pH diagram showing fields of solids and predominant ionic species in the system Fe–H_2O–O_2. Detailed procedures for the construction of Eh–pH diagrams are given by Hem and Cropper (*13*) and Garrels and Christ (*14*). In Figure 3, ion–ion boundaries are drawn at equal ion activities. Ion-solid boundaries are drawn for $10^{-4}M$ dissolved iron species (5.6 ppm iron), a typical concentration in the iron-rich ground waters of New Jersey and Maryland described below. The ferric and ferrous species–ferric oxyhydroxide boundaries have been drawn for pK values of 37.1 and 44. This difference in pK means that for a given pH and Eh, a water can contain roughly seven orders of magnitude more dissolved iron in equilibrium with fresh, amorphous material than in equilibrium with well-crystallized goethite or hematite.

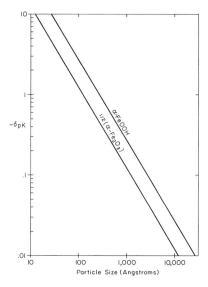

Figure 1.   Decrease in pK ($-\delta pK$) with decreasing particle size of goethite (α-FeOOH) and hematite ($1/2\alpha$-Fe_2O_3)

The Eh and pH of iron-rich natural waters, most of which contain suspended ferric oxyhydroxides, places these waters on the ferric or ferrous species–ferric oxyhydroxide boundary. Acid-mine discharges usually have pH's between 2 and 5 and Eh's from within the Fe^{3+} field down to zero volts. Iron-rich ground waters are usually on the same boundary, with pH's between 5 and 8 and Eh's from +0.3 to −0.1 volt.

Calculation of Apparent Stabilities

It is useful to compute what may be called the apparent stability of a mixture of oxyhydroxides. This apparent stability may be defined in terms of pQ where in the solution

$$pQ = - \log [Fe^{3+}][OH^-]^3 \qquad (11)$$

For solutions having a constant aqueous chemical composition and constant composition of associated ferric oxyhydroxides, the measured pQ will equal the stability of all the ferric oxyhydroxides present. For solutions in which one ferric oxyhydroxide is altering or dissolving to form another, the measured pQ will represent a stability greater than that of the disappearing phase and less than that of the phase which is forming.

In ferric sulfate-rich solutions, pQ may be calculated from a knowledge of pH and $[Fe^{3+}]$, the activity of uncomplexed ferric ion. The complexes which must be considered to evaluate ionic strength (I) are HSO_4^-, $FeSO_4^+$, and $FeOH^{2+}$. Dissociation constants for these complexes have been measured at 25°C by Nair and Nancollas (15), Willix (16), and Milburn (12), respectively, and are

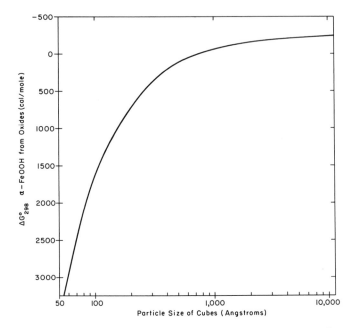

Figure 2. Gibbs free energy of formation of goethite (α-FeOOH) from hematite (1/2α-Fe₂O₃) and water as a function of particle size, assuming equal particle sizes of goethite and hematite

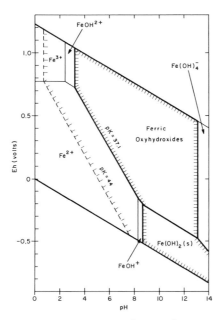

Figure 3. Eh–pH relations between solids and predominant ionic species in the system Fe–H$_2$O–O$_2$. The position of the Fe^{2+}–ferric oxyhydroxide boundary is shown for pK's of 37.1 and 44. The position of the Fe^{3+}–FeOH^{2+} boundary is based on Milburn (12). Other thermochemical data used to construct the figure are from Langmuir (8).

$$K_{HSO_4^-} = [H^+][SO_4^{2-}]/[HSO_4^-] = 10^{-1.96} \tag{12}$$

$$K_{FeSO_4^+} = [Fe^{3+}][SO_4^{2-}]/[FeSO_4^+] = 10^{-4.15} \tag{13}$$

$$K_{FeOH^{2+}} = [Fe^{3+}][OH^-]/[FeOH^{2+}] = 10^{-11.83} \tag{14}$$

When significant ferrous iron is present, this species must also be considered. Transposing and taking the antilog of Equation 25 gives the relationship between [Fe^{2+}], [Fe^{3+}], and the measured Eh at 25°C, which is

$$[Fe^{2+}]/[Fe^{3+}] = 10^{(E^\circ - Eh)/0.05961} \tag{15}$$

where E° is the standard electrode potential of the ferric–ferrous iron couple (*see* below).

Known concentrations of total dissolved iron, ferrous iron, and sulfate species yield the following mass balance equations.

$$m\mathrm{Fe(total)} = m\mathrm{Fe^{3+}} + m\mathrm{FeOH^{2+}} + m\mathrm{FeSO_4^+} + m\mathrm{Fe^{2+}} \qquad (16)$$

$$m\mathrm{Fe(II)} = m\mathrm{Fe^{2+}} \qquad (17)$$

$$m\mathrm{SO_4^{2-}} \text{ (total)} = m\mathrm{FeSO_4^+} + m\mathrm{HSO_4^-} + m\mathrm{SO_4^{2-}} \qquad (18)$$

If NaOH is the base used to precipitate the oxyhydroxides, then

$$m\mathrm{Na^+} \text{ (total)} = m\mathrm{Na^+} \qquad (19)$$

The over-all charge balance equation for acid solutions is

$$m\mathrm{Na^+} + m\mathrm{H^+} + 3m\mathrm{Fe^{3+}} + 2m\mathrm{FeOH^{2+}} + m\mathrm{FeSO_4^+} + 2m\mathrm{Fe^{2+}} =$$
$$2m\mathrm{SO_4^{2-}} + m\mathrm{HSO_4^-} \qquad (20)$$

The definition of ionic strength is

$$I = 1/2 \sum_i m_i z_i^2 \qquad (21)$$

where m_i and z_i are the molality and valence of ionic species i, respectively. The full expression for I is then

$$I = 1/2\,[m\mathrm{Na^+} + m\mathrm{H^+} + 9m\mathrm{Fe^{3+}} + 4m\mathrm{FeOH^{2+}} + m\mathrm{FeSO_4^+} +$$
$$4m\mathrm{Fe^{2+}} + 4m\mathrm{SO_4^{2-}} + m\mathrm{HSO_4^-}] \qquad (22)$$

Because significant concentrations of complex ions are present, I and $[\mathrm{Fe^{3+}}]$ must be calculated by a method of successive approximations. The first step is to establish which dissolved species are the major ones. At this point, differences between ion activities and molalities may be ignored. The pH is used to estimate the relative importance of $m\mathrm{SO_4^{2-}}$ and $m\mathrm{HSO_4^-}$ through $K_{\mathrm{HSO_4^-}}$, and $m\mathrm{Fe^{3+}}$ vs. $m\mathrm{FeOH^{2+}}$ through $K_{\mathrm{FeOH^{2+}}}$. As a first approximation in most waters with pH > 2.5, $m\mathrm{SO_4^{2-}} = m\mathrm{SO_4^{2-}}$ (total), which permits an estimate of $m\mathrm{FeSO_4^+}$ vs. $m\mathrm{Fe^{3+}}$ through $K_{\mathrm{FeSO_4^+}}$. With a rough estimate of I at values below 0.1, approximate individual ion activity coefficients may be calculated using the extended Debye–Hückel equation (*17*). To a good approximation, the following equalities may be assumed: $\gamma\mathrm{FeOH^{2+}} = \gamma\mathrm{Fe^{2+}}$, and $\gamma\mathrm{FeSO_4^+} = \gamma\mathrm{HSO_4^-} = \gamma\mathrm{HCO_3^-}$. Substituting values for these coefficients into Expressions 12 through 15, we may refine the molal relationships of free and complex ion species. Combining these relationships with mass balance Equations 16 and 18 permits calculation of approximate molalities of all the ionic species present and of a more accurate ionic strength through Expression 22. The cycle is repeated about two or three times until no further change in ionic strength can be detected. The final value of $[\mathrm{Fe^{3+}}]$ is then used to compute pQ.

The preceding calculations are possible using a total dissolved iron and Eh analysis with or without an analysis for mFe (II), although the most accurate results are obtained with both total and ferrous iron values. Because mFe^{3+} is a small percentage of mFe (III), errors in $K_{FeSO_4^+}$ and $K_{FeOH^{2+}}$ are magnified in the final value of $[Fe^{3+}]$. Also, the uncertainty in γFe^{3+} at moderate ionic strengths is likely to be significant. In view of such uncertainties, calculated pQ values based on the above approach are considered accurate to ± 0.4 unit when only a total iron analysis is available, and to ± 0.2 unit when mFe (II) has also been measured.

The most reliable values of pQ are based on the mFe (II) analysis in solutions of known ionic strength. This approach is applicable in mixed ferric–ferrous salt solutions as above, or in ferrous salt solutions. An equation relating pQ to solution composition may be derived as follows. For the reduction reaction

$$Fe^{3+} + e^- = Fe^{2+} \qquad (23)$$

$$Eh = E° + 1.9842 \times 10^{-4}\, T \times \log\,([Fe^{3+}]/[Fe^{2+}]) \qquad (24)$$

where $E°$ is the standard electrode potential of the reaction and T the temperature in degrees Kelvin. Expanding the log term and transposing gives

$$-\log\,[Fe^{3+}] = (E° - Eh)/1.9842 \times 10^{-4} T - \log\,[Fe^{2+}] \qquad (25)$$

Adding $-3 \log\,[OH^-]$ to both sides, the left hand side then equals pQ and the final expression may be written

$$pQ = (E° - Eh)/1.9842 \times 10^{-4} T - \log\,[Fe^{2+}] - 3(\log K_w + pH) \qquad (26)$$

where K_w is the activity product of water. This expression was used to calculate pQ in both ferric–ferrous sulfate and ferrous sulfate laboratory solutions and in coastal-plain ground waters of New Jersey and Maryland. Ionic strength was computed for the laboratory solutions from the detailed ion content of the water through Equation 21. The New Jersey ground waters studied were chiefly of the calcium bicarbonate type with ionic strengths less than 4×10^{-3} so that I could be computed accurately with the empirical equation

$$I = 1.5 \times 10^{-5} \mu \qquad (27)$$

where μ is the specific conductance in micromhos at $25°C$ (*18*). Ionic strengths of the Maryland ground waters studied, which are all less than 5×10^{-3}, were provided by W. Back (*19*).

Because pQ is quite sensitive to $E°$ and values of $E°$ at other temperatures than $25°C$ have not been published, careful laboratory measurements of this quantity were made from 5° to 35°C at a pH of 1.5 in mixed ferrous–ferric perchlorate solutions. A detailed description of these measurements will be given elsewhere (20). Resultant smoothed and rounded values of $E°$ at 5°, 10°, 15°, 20°, 25°, 30°, and 35°C are 0.746, 0.752, 0.758, 0.764, 0.770, 0.775, and 0.780 volts, respectively, and are probably accurate to ± 0.001 volt. The equation

$$E° = -1.226 \times 10^{-2} + 4.147 \times 10^{-3}T - 5.111 \times 10^{-6}T^2 \qquad (28)$$

fits both measured and smoothed values of $E°$ well within the uncertainty of the measurements.

FeOH$^+$ is the only complex ion in the ferrous-rich laboratory solutions or the ground waters examined. The equilibrium constant, K_{eq}, for the reaction

$$Fe^{2+} + H_2O = FeOH^+ + H^+ \qquad (29)$$

equals

$$K_{eq} = [FeOH^+][H^+]/[Fe^{2+}] = 10^{-8.30} \qquad (30)$$

at 25°C. At temperatures near 25°C

$$K_{eq} = K_w \cdot 10^{5.70} \qquad (31)$$

This expression is probably accurate to about ± 0.05 log unit at 15°C (18). The FeOH$^+$ complex is clearly negligible relative to Fe^{2+} at pH's below about 6.3, but amounts to a maximum of about 10% of the total dissolved Fe^{2+} content in ground water sample 172 at a pH of 7.75 (Table III). Values of K_w as a function of temperature used in Expressions 26 and 31 are from Ackerman (21).

In the laboratory studies, the largest uncertainties in terms of Equation 26 are in the measured values of Eh and pH, which lead to an uncertainty of about ± 0.2 unit in pQ. Measurement uncertainties in Eh, pH, and mFe (II) are significant for the ground waters examined. The resultant uncertainty in pQ for ground waters highest in mFe (II) is about ± 0.3 unit; for ground waters lowest in mFe (II), about ± 0.4 unit.

Laboratory Studies

Methods of Chemical Analysis. During oxidation and hydrolysis studies of ferrous sulfate solutions, ferrous iron concentrations which ranged from 570 to 157 ppm were analyzed by titration with potassium dichromate solution using sodium diphenylamine sulfonate as an indicator (22). The same method was used to analyze for ferrous iron in mixed ferric–ferrous sulfate solutions. Total dissolved iron concentrations, which ranged from 560 to 100 ppm during hydrolysis of mixed ferric–ferrous sulfate solutions, were similiarly titrated after acidification with HCl and boiling with reduction by $SnCl_2$ (22). Samples for dissolved total iron analysis were withdrawn from the cleared supernatant solution in the reaction vessels after the solution had stood undisturbed for 48 hours. Upon withdrawal, samples were passed through 0.45-micron filter paper prior to analysis. Although some suspended ferric oxyhydroxides could have been present and have passed through the filter paper, the amount of such material must have been negligible for purposes of this study in that pQ values calculated using the mFe (total) analyses were identical to such values calculated from mFe (II) concentrations measured in the same samples. Analyzed ferrous iron values are probably accurate to $\pm 0.5\%$; total dissolved iron values to $\pm 1\%$.

A combination glass, silver–silver chloride reference electrode was used for pH measurement, a platinum thimble electrode and silver–silver chloride reference electrode for Eh measurement. Eh and pH measurements were made with a Coleman Medallion Model 37 battery- or line-operated pH-millivolt meter and/or a Corning 12 Research pH–millivolt meter. The pH measurements were calibrated with nominal pH 4 and 7 buffers. Double buffer checks were within ± 0.02 pH unit. Eh measurements were calibrated with Zobell solution [$0.0003M$ $K_3Fe(CN)_6$, $0.0003M$ $K_4Fe(CN)_6 \cdot 3H_2O$, and $0.1M$ KCl]. From $0°$ to $25°C$, the Eh of Zobell solution obeys the equation

$$Eh(\text{volts}) = 0.429 + 0.0024(25 - t) \tag{32}$$

where t is in degrees Celsius (23). Laboratory Eh and pH measurements are probably accurate to ± 0.002 volt and 0.02 pH unit, respectively.

Precipitates were smear-mounted and dried on a glass slide for x-ray diffraction analysis using a General Electric x-ray diffractometer and $FeK\alpha$ radiation. An estimate of the size of crystalline particles in the precipitates was made by measurement of the width of the major diffraction peak for the mineral of interest at half-maximum intensity (24).

Precipitates for examination by electron microscopy were first disaggregated with an ultrasonic vibrator so that single crystals could be studied. A drop of the dispersion was then placed on a collodion-covered grid in a Zeiss electron microscope, Model 9S. Electron micrographs of the precipitates show preferred orientation of goethite and lepidocrocite crystals (Figures 5 and 7). Geothite commonly occurs as accicular and narrow prismatic crystals elongated in the [001] or c-axis direction. Lepidocrocite is typically present in bladed prismatic or micaceous forms flattened on {010} (3). Thus, sizes based on the x-ray diffraction method represent the mean thickness of goethite crystals in the [110] direction, and of lepidocrocite crystals in the b-axis or [020] direction. The accuracy of x-ray determined values for these mean dimensions is probably ± 20–40% for the sizes encountered in this study.

Experimental Methods. Studies were made of the precipitation and aging of ferric oxyhydroxides formed in sulfate solutions initially about $10^{-2}M$ in ferrous iron. Three representative runs are described below. Oxidation of ferrous iron was accomplished by bubbling the solutions with air. Hydrolysis was with 0.1M NaOH or 0.5M $NaHCO_3$. Runs were made in a constant-temperature bath at $25° \pm 0.2°C$ with about 3.5

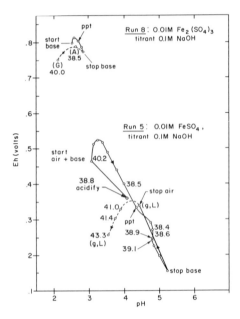

Figure 4. Observed changes in Eh, pH, and calculated pQ values during Runs 5 and 8. Bracketed letters A, G, and L indicate that the precipitate is mostly amorphous, goethite, or lepidocrocite, respectively. Bracketed letters g and l indicate the presence of minor amounts of goethite or lepidocrocite, respectively. Runs 5 and 8 were begun at points labeled "acidify" and "start base," respectively.

liters of solution in 4-liter borosilicate glass beakers. The beakers were fitted with airtight Lucite covers which had holes to accommodate stopper-fitted Eh and pH electrodes, a specific conductance probe, gas intake and discharge tubes, a thermometer, and a buret tip for base addition. Solutions were stirred with a teflon-coated stirring bar powered from beneath the beaker by a water-driven magnetic rotor. Periodically during the runs, which lasted from 4 to 10 hours, and later during aging, measurements were made of Eh, pH, and specific conductance, samples of solution were withdrawn for total dissolved iron or ferrous iron analysis, and precipitate was collected for x-ray and electron microscope analysis.

Hydrolysis of $Fe_2(SO_4)_3$ Solutions. Run 8 involved the hydrolysis with $0.1M$ NaOH of a solution initially $10^{-2}M$ in total iron. The solution was prepared with reagent-grade ferric sulfate which contained about 1% of ferrous iron. The base was added periodically during four hours of the run so that the final solution contained about 100 ppm of dissolved total iron species. During the run, over 90% of $mFe(III)$ was present as the $FeSO_4^+$ complex. Eh and pH changes during the run and subsequent aging are shown in Figure 4. Other results evaluated during aging are listed in Table II. The solution was allowed to evaporate to about one third its original volume during 109 days following the run. The goethite which had appeared as of 112 days' aging evidently formed by crystallization of the amorphous material. Electron microscopic examination of the precipitate after 198 days showed rod-like crystals of goethite surrounded by amorphous-appearing material. The goethite crystals were about 50 to 100 Å thick and 300 to 600 Å long.

Oxidation and Hydrolysis of FeSO₄ Solutions. Initial solution of reagent grade $FeSO_4$ in Run 5 (Figure 4) gave a hydrolysis pH of 4.12. The small amounts of suspended ferric oxyhydroxides present had a pQ value of 38.8. Addition of a few drops of concentrated H_2SO_4 brought the pH to 3.08 and leached away the most soluble oxyhydroxide material present, leaving behind relatively more crystalline particles with a pQ of 40.2. Leaching of the ferric oxyhydroxides increased the $[Fe^{3+}]/[Fe^{2+}]$ ratio and so raised the Eh. With the beginning of aeration and base addition, Eh dropped because of preferential hydrolysis and precipitation of Fe^{3+}, whereas the rate of Fe^{2+} oxidation at these pH's is extremely slow (4). Solution composition thus moved along the ferrous ion–ferric oxyhydroxide boundary. Substantial precipitation began at a pH of 4.34, when pQ = 38.4, indicating that the first precipitate was chiefly amorphous. The value of pQ subsequently increased to 39.1 and probably higher for two reasons. First, initial precipitation occurs in the vicinity of drops of added base and thus develops under highly supersaturated conditions. With mixing, much of this relatively unstable precipitate redissolves in the bulk of solution, leaving behind a more stable (less soluble) fraction. Also as the degree of supersaturation decreases, precipitation occurs more slowly with the consequent development of better crystallized material. At pQ = 39.1, the solution contained 335 ppm of dissolved ferrous iron.

With the end of base addition, oxidation of Fe^{2+} to ferric oxyhydroxides by the air became the important reaction, causing an increase in Eh and decrease in pH. Two days after cessation of bubbling with air, pQ = 41.0, and x-rays showed the crystalline fraction of the precipitate to be predominantly lepidocrocite (crystal thickness 64 ± 15 Å) with a few percent goethite (crystal thickness 54 ± 15Å). After 93 days (last point shown in Figure 4), pQ = 43.3, and the goethite to lepidocrocite ratio had doubled. The thickness of goethite crystals remained unchanged, while that of lepidocrocite had increased to 75 ± 15 Å. X-ray diffraction

Table II.   Some Results During Aging of Run 8

Days After Run	pH	Fe(II), ppm	Fe(III), ppm	pQ	X-ray Analysis and Crystal Thickness, Å
3	2.65	–	100	38.5	Amorphous
21	–	–	–	–	Amorphous
112	2.15	–	–	–	Goethite, 67 ± 20
147	2.14	16.1	309	40.0	Goethite, 58 ± 20
192	2.11	16.1	–	40.0	Goethite, 61 ± 20

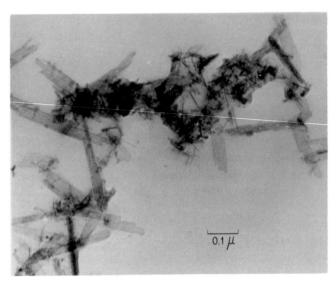

Figure 5.   Electron micrograph taken after 298 days' aging of Run 5, showing lath-like lepidocrocite crystals and smaller needle-like goethite crystals

after 293 days showed the crystal thickness of goethite fairly constant at 63 ± 15 Å, while that of lepidocrocite had increased to 93 ± 20 Å. At this time, the goethite to lepidocrocite ratio had again doubled. After 258 and 298 days, the electron microscope showed lepidocrocite aggregates and lath-like single crystals, and relatively smaller goethite needles or rods (Figure 5). The goethite crystals were elongated in the c-axis direction, and the lepidocrocite laths lay with their b-axis vertical. This orientation was confirmed by the relative intensities of reflections in the x-ray diffraction patterns. As shown in Figure 5, the width of the lepidocrocite laths averaged about 400 Å, their length about 2300 Å. The goethite rods were 40 to 60 Å in width, and averaged about 500 Å in length.

Run 4 (Figure 6) was also begun with a $10^{-2}M$ FeSO$_4$ solution. Acid leaching of small amounts of suspended oxyhydroxides initially present raised pQ from 40.1 to 41.8. Increments of $0.5M$ NaHCO$_3$ were then introduced without aeration. Base was added more rapidly than in Run 5 so that the solution became more highly supersaturated with respect to crystalline oxyhydroxides, and precipitation of amorphous material began with pQ = 37.3. As in Run 5, pQ increased because of resolution of the most soluble material and a decreasing rate of precipitation. X-ray analysis at pQ = 38.4 showed the crystalline fraction of the precipitate to be chiefly lepidocrocite (crystal thickness 115 ± 35 Å) with minor goethite (crystal thickness 100 ± 30 Å). Subsequent x-ray analyses during the run at pQ = 39.1 and 39.2 showed a rapidly increasing proportion of goethite relative to lepidocrocite with roughly equal amounts of the two phases present at pQ = 39.2. The mean crystal thickness of goethite remained roughly constant during this time; however, that of lepidocrocite increased to about 200 ± 60 Å, probably because smaller particle sizes of this mineral were redissolved or converted to goethite.

Aeration was then begun, with the result that Fe^{2+} concentrations dropped to negligible values (< 0.005 ppm). After one day and also after 216 days' aging (last point in Figure 6), the crystalline fraction of the precipitate contained roughly twice as much goethite as lepidocrocite, while crystal thicknesses remained about 100 and 200 Å, respectively. In 296 days, the goethite-to-lepidocrocite ratio had increased to greater than 3, although crystal thickness of the two minerals remained unchanged. Electron micrographs taken at 262 and 302 days (Figure 7) show compact aggregates with single and intergrown crystals of goethite and lepidocrocite projecting radially from their surfaces. Isolated needle-shaped goethite crystals had an average width and length of roughly 100 and 2000 Å, respectively. Lepidocrocite occurred only in the aggregates where its crystalline dimensions could not be adequately measured. X-ray diffraction intensities indicated fairly random orientation of crystals in the aggregates.

Discussion. Observations of other workers provide further insights into the behavior of precipitated ferric oxyhydroxides. Lamb and Jacques (*25*) found that in pure ferric salt solutions the particle size and stability of oxyhydroxides increases with the concentration of dissolved ferric iron present during precipitation. Feitknecht and Michaelis (*26*) and Watson *et al.* (*7*) noted that crystalline oxyhydroxides such as goethite and

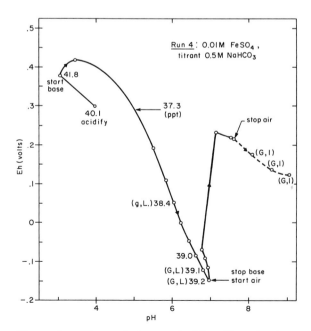

Figure 6.   Observed changes in Eh, pH, and calcu-lated pQ values during Run 4; bracketed letters have the same significance as in Figure 4

akaganéite grow rapidly with aging in ferric salt solutions, but may form or crystallize extremely slowly if at all in solutions which contain only trace amounts of dissolved ferric iron. Feitknecht and Michaelis (26) observed the complete crystallization of amorphous material to goethite when the amorphous phase was precipitated in the presence of Fe^{2+} ions. This study similarly shows that the rate of formation of crystalline particles of goethite or lepidocrocite is proportional to the dissolved ferrous or ferric iron concentration and is faster in ferrous than in ferric iron solutions of the same molarity.

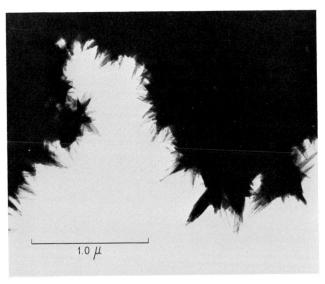

1.0 μ

Figure 7.   Electron micrograph taken after 302 days' aging at Run 4, showing compact crystalline intergrowth of goethite and lepidocrocite

Other generalizations based on results of the study are possible. These are:

1. The initial precipitate of ferric oxyhydroxide is usually low in pQ, reflecting the presence of relatively large amounts of amorphous material. The faster the rate of precipitation by oxidation and/or hydrolysis, the greater the supersaturation with respect to crystalline oxyhydroxides, and the lower the initial pQ.

2. Because of the effect of particle size on relative stabilities, minerals such as lepidocrocite and goethite (Run 4) or goethite and hematite may reverse in stability.

3. Partial solution or leaching of precipitates removes the least stable (most soluble) material first, so that pQ values in solution increase.

As was evident from Runs 4, 5, and 8, after long periods of aging, the average thickness of goethite rods or needles may remain in the 50 to 100 Å range, with average lengths of from 500 to 2000 Å. Lepidocrocite crystals similarly can remain as laths averaging about 80–200 Å thick, 400 Å wide, and 2300 Å in length. If the entire surfaces of such crystals were in contact with the solution, one could predict a large particle size effect on thermodynamic stability. For goethite, Equation 5 may be modified to give

$$-\delta pK = 45.3\ A/V \tag{33}$$

where A is the surface area of a crystal in Angstroms squared, and V is its volume in Angstroms cubed. At the close of Run 8, the mean diameter of the rodlike goethite crystals was 60 ± 15 Å, their mean length 500 ± 100 Å, and $pQ = 40.0$. Based on Equation 33, δpK is 3.2 for the goethite, and the pK of macroscopic goethite must then exceed 43.2 because the goethite is forming by crystallization of less-stable amorphous material.

Preliminary estimates for goethite crystals in Run 5 (Figure 5) similarly yield $\delta pK = 3.2$. If the lepidocrocite crystals in this run are considered rectangular prisms with a surface energy the same as that of goethite, then δpK for the lepidocrocite equals 1.3. The above estimates of δpK are calculated from the geometry of isolated crystals and may be in error for several reasons, including nonuniformity of crystal shapes and sizes for a particular mineral, uncertainty in the surface energy of lepidocrocite, and differences in the state and composition of crystal aggregates. When aggregates or intergrowths of a mineral are present, as in Figure 7 (*see also* Watson *et al.*, Ref. 27), δpK values based on the geometry of isolated crystals will be too large.

Suspended Ferric Oxyhydroxides in Some Ground Waters

Methods of Chemical Analysis. Analyses of the iron content in ground waters from the Camden, New Jersey area (Table III) were made in a trailer modified as a mobile chemical laboratory. Ferrous iron values were measured within one hour of collection using a modified version of the bathophenanthroline spectrophotometric method of Lee and Stumm (28) which involves extraction of the ferrous iron–bathophenanthroline complex from aqueous solution into *n*-hexyl alcohol. The extraction is desirable in that colloidal-sized ferric oxyhydroxides present in the sample are excluded from the extractant. A second advantage is that low sample iron concentrations may be determined by extracting the iron from a large sample volume into a relatively small volume of *n*-hexyl alcohol. Lee and Stumm recommend acidifying and boiling the sample prior to ferrous iron analysis. Boiling and acidification were avoided in this study because Shapiro (29) has shown that such treatment reduces ferric to ferrous iron, in amounts dependent on the pH and time of boiling.

Table III.　Description of Wells and Chemical Data for
and Magothy Formations

Well No.	Altitude above Sea Level Ft	Screen Setting, Interval in Ft	Date	Temp., °C
103	19	118–148	12–17–69	13.0
104	19	204–224	12–17–69	12.9
123	59	298–338	12–15–69	13.8
125	10	211–272	12–14–69	13.1
127	45	248–288	12–13–69	13.0
131	44	309–367	12–16–69	14.2
162	65	452–473, 541–594	12–15–69	15.0
171	100	425–445	12–14–69	16.2
172	100	369–389	12–14–69	15.9
187	40	325–375	12–15–69	13.2
188	45	258–293	12–13–69	13.8
189	65	220–272	12–15–69	14.0

[a] Specific conductance (μ) is in micromhos at 25°C.

Total iron concentrations given in Table III were determined with the unmodified bathophenanthroline method of Lee and Stumm (28).

All spectrophotometric measurements were made with a Bausch and Lomb Spectronic 20, using 1-inch sample test tubes (light path 2.235 cm). The accuracy and reproducibility of ferrous and total iron analyses with the bathophenanthroline method was about ±0.005 ppm at concentrations near 0.02 ppm, and about ±0.1 ppm at concentrations near 10 ppm.

The same electrodes were used for Eh and pH measurement, both in the laboratory and the field. Measurements of pH were made with the Coleman Medallion Model 37 meter after calibration in nominal pH 4 and 7 buffers brought to within 0.5°C of ground water temperature. Double buffer checks were always within ±0.05 pH unit. Eh values were read with the Coleman meter or an Orion Model 407 battery-powered pH–millivolt meter. The Eh response of the electrodes and meters was checked daily with Zobell solution. Ground water for Eh analysis was forced by well-pump pressure through one side arm of a glass U tube and out the opposite side arm. The platinum thimble and reference electrodes were inserted through rubber stoppers into the larger U tube openings to give a water-tight seal. Eh was recorded with time during constant flow through the U tube until values changed by less than 0.005 volt in 20 minutes. At this point, flow was stopped to eliminate the flowing potential (usually −0.010 to −0.030 volt) and the final Eh reading taken. Eh measurements typically required from one to two hours for completion because of relatively low redox capacity of the ground water and the need to flush all extraneous oxygen from plumbing between the ground water aquifer and electrodes. A detailed explanation of the methods, theory, and limitations of Eh measurement are given elsewhere (23). Field Eh values listed in Table III are considered accurate to ±0.020 volt.

Well-Water Samples from the Potomac Group and Raritan near Camden, New Jersey [a]

pH	Eh, Volts	Fe(total), ppm	Fe(II), ppm	μ	pQ
4.92	+0.220	0.74	0.65	64	43.0
5.64	+0.246	0.17	0.15	66	41.0
6.52	−0.092	9.4	8.6	173	42.5
6.15	−0.049	11.2	10.2	128	42.8
6.22	−0.046	11.5	10.5	146	42.6
6.71	−0.088	2.6	2.5	226	42.3
7.36	−0.090	0.67	0.51	186	41.0
7.65	−0.034	0.22	0.07	253	39.9
7.75	−0.040	0.21	0.11	253	39.5
5.00	+0.295	0.024	0.022	40	42.9
6.21	−0.053	10.3	9.3	170	42.7
7.25	−0.085	0.61	0.21	212	41.8

Specific conductance was measured with a Beckman conductivity bridge, Model RC-1682, and a dipping glass conductivity cell with a cell constant near 1 cm^{-1}. This equipment was calibrated periodically in a standard KCl solution. Conductance values were corrected to 25°C and are probably accurate to within ±2%.

Results of the chemical analyses of ground waters from the twelve wells near Camden, New Jersey, are given in Table III.

Ground Waters Near Camden, New Jersey. Results of the laboratory studies and of related work by others may be used to explain the behavior of suspended ferric oxyhydroxides in ground water. The ground waters to be considered are pumped from the Potomac Group and Raritan and Magothy Formations of Cretaceous age as they occur near Camden, New Jersey, and in southern Maryland. A detailed description of the iron content of these ground waters in New Jersey has been published by Langmuir (*18, 30*), while the iron content of the Maryland ground waters has been examined by Back and Barnes (*31*).

The Potomac Group and Raritan and Magothy Formations, which may be considered a single aquifer system, are made up of sands, silts, and gravels with a combined thickness of 200–250 feet in the outcrop area near Camden, New Jersey (Figure 8). The formations dip 40 to 100 ft/mile to the southeast where they are overlain by silts and clays of the Merchantville and Woodbury Formations. Ground waters present in artesian parts of the aquifer system have entered as recharge in the outcrop area and by vertical movement through overlying sediments within a few miles southeast of the outcrop area.

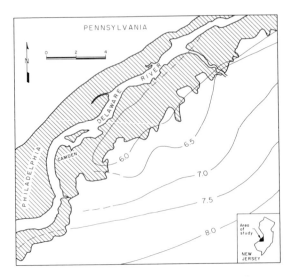

*Figure 8.   Location of the New Jersey study area
and generalized pH values of ground water in the
Potomac Group and Raritan and Magothy Formations as shown by contours, 1966–67; crosshatch-
ing denotes the outcrop area of the formations*

Where fresh, the ground water is of the calcium bicarbonate type. Its dissolved solids content ranges from less than 100 ppm where unpolluted in the outcrop area to 500 ppm downdip ten or more miles, where fresh ground waters mix with residual saline ground waters.

In and near the outcrop area, pH's are usually between 5 and 6 (Figure 8) because of H^+ ion production resulting from the oxidation of traces of the FeS_2 minerals pyrite and marcasite and from solution of soil-zone CO_2. Ground waters in this area are probably at most a few months in age. The chief H^+ ion producing reactions require oxygen. Without fresh sources of oxygen in artesian parts of the aquifer system, pH values rise to about 8 as H^+ ions are depleted with time by hydrolysis reactions with silicate minerals and traces of carbonate shell materials present in the formations. Arguments presented elsewhere (30) support an age of several thousand years or more for ground waters in the area having a pH of about 8. Thus, the pH contours shown in Figure 8 are a measure of the relative age of the ground water.

A map of total iron concentrations based on analyses of 180 well waters (30) is shown in Figure 9. Also indicated in the figure are locations of 12 wells which were sampled and their waters chemically analyzed in December 1969 (Table III). A comparison of Figures 8 and 9 shows that total iron contours parallel pH contours. The latter figure shows that total iron concentrations increase rapidly to a maximum and then gradually decrease downdip. Total iron is present chiefly as ferrous iron species which ranged from 0.022 to 10.2 ppm in the 12 well waters. The differences between total and ferrous iron values in Table III are the concentration of suspended ferric oxyhydroxides which represented from 5 to 70% of the total iron content in the 12 well waters. Filtration studies (*18*) have shown that the suspended material ranges from less than 0.01 micron (100 Å) to greater than 5 microns and averages 1–2 microns. Unfortunately, it was not possible to collect enough of this material on a membrane filter for x-ray analysis. However, laboratory studies by others described elsewhere (*18*) support a mixture of amorphous material and goethite for the suspension.

Taking pH as a measure of relative age, Figure 10 is a plot of pH *vs.* other calculated and measured parameters in the ground water. Wells 103, 104, and 187, which are at or adjacent to the outcrop area, plot to the left of the diagram. The other wells tap artesian ground waters down-

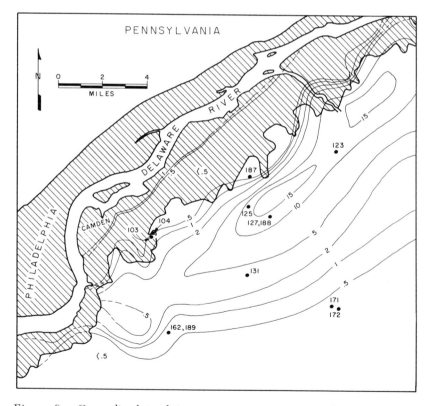

Figure 9.   Generalized total iron concentration in ground waters of the formations in parts per million, 1965 (18); total iron values used to construct the map were measured in the laboratory by the spectrophotometric bipyridine method (32)

dip. The diagram shows that specific conductances increase with age of
the ground water from 50 micromhos in the outcrop area to 250 micromhos downdip. In the outcrop of the formations, ground water recharge, which is chiefly from precipitation, is low in dissolved solids but
high in dissolved oxygen so that Eh is high and ferrous iron concentrations are less than 1 ppm ($10^{-4.7}M$). The Fe^{2+} maximum just southeast
of the outcrop area reflects the absence of sources of oxygen and mixing
of the ground water with iron-rich recharge which enters the aquifer
system from overlying sediments. Further southeast, Fe^{2+} concentrations
decrease by hydrolysis and precipitation as pH rises. There are no other
ferrous iron sources downdip.

Variations in pQ may be explained in light of the behavior of ferric
oxyhydroxides in laboratory studies. The two ground waters in and
adjacent to the outcrop area with pH's near 5 (wells 103 and 187) are
supplied with H^+ ions by reactions already noted. The acidity tends to
leach away the less stable oxyhydroxides, raising pQ values close to 43.
In the water from well 104 (pH = 5.64), precipitation is occurring at
relatively low Fe^{2+} concentrations so that pQ is rather low (41.0). Com-
paratively high pQ values (42.5–42.9) are found with maximum ferrous

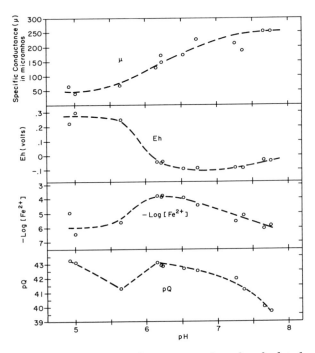

*Figure 10.   pH vs. other measured and calculated
chemical parameters for the 12 New Jersey well waters*

iron concentrations of about 10 ppm. Thus, as in the laboratory, the relatively stable oxyhydroxides develop in the presence of high Fe^{2+} concentrations, under which conditions fresh precipitates can recrystallize to more stable forms. Precipitation continues as the ground water moves further downdip. However, pQ values gradually decrease because the oxyhydroxides recrystallize more slowly in the presence of decreasing amounts of ferrous iron.

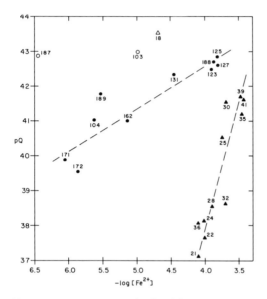

Figure 11.   pQ vs. $-log[Fe^{2+}]$ for 12 New Jersey well waters (circles) and 12 Maryland well waters (triangles); open symbols denote waters with pH values of 5.00 or less

Comparison with Ground Waters in Southern Maryland. The principles described above also apply to the behavior of suspended ferric oxyhydroxides in ground waters from the Raritan and Magothy formations in southern Maryland. Figure 11 is a plot of pQ *vs.* $-log[Fe^{2+}]$ for the 12 New Jersey well waters and for 12 well waters from Maryland which were studied by Back and Barnes (*31*). In the Maryland waters, Fe^{2+} concentrations were generally higher (1.3 to 26 ppm), Eh's slightly higher (-0.020 to $+0.384$ volt) and pH's lower (3.66 to 6.87) than in the New Jersey waters. These differences reflect relatively greater abundances of H^+ and Fe^{2+}-producing minerals (FeS_2, pyrite, and marcasite) in lignite within the Maryland sediments and the fact that all the Maryland wells tap ground waters which are relatively oxygen-rich, being under water table or at most semiconfined conditions. Waters from Maryland well 18 have a pH of 3.66 because of oxidation of FeS_2 minerals so that oxyhydroxide precipitates have been leached, and pQ is a high 43.5. For the other Maryland well waters, the plot shows a good correlation between pQ and $-\log[Fe^{2+}]$, indicating that the stability of suspended oxyhydroxides increases with the Fe^{2+} concentration.

Summary

Ferric oxyhydroxides precipitated in natural waters are usually mixtures of x-ray amorphous material and goethite (α-FeOOH). The apparent thermodynamic stability of oxyhydroxide precipitates may be described in terms of the ion activity product $pQ = -\log[Fe^{3+}][OH^-]^3$ in solution. Values of pQ are calculable from measurements of total dissolved iron and (or) ferrous iron, pH, Eh, and a knowledge of ionic strength. Ferric oxyhydroxides precipitated in the laboratory from solutions initially $10^{-2}M$ in iron as ferric sulfate or ferrous sulfate had pQ values ranging from 37.3 to 43.3. The lowest values were found with freshly precipitated amorphous material, the highest values with aged mixtures of goethite and lepidocrocite formed in $FeSO_4$ solutions. Based on x-ray diffraction and electron microscopy, goethite occurred as rod or needle-like crystals elongated parallel to the c-axis direction. The crystals typically ranged from 50 to 100 Å in diameter and 500 to 2000 Å in length. Lepidocrocite appeared as laths ranging from 60 to 200 Å thick along the b-axis, averaging 400 Å wide in the c-axis direction and 2300 Å long in the a-axis direction.

Chemical analyses were made of 24 well waters from coastal plain New Jersey and Maryland in which the iron is present in solution chiefly as aqueous ferrous species ($10^{-3.33}$ to $10^{-5.40}M$ Fe) and suspended ferric oxyhydroxides ($10^{-4.13}$ to $10^{-7.4}M$ Fe, or 5 to 70% of the total iron content). The suspended oxyhydroxides are probably mixtures of amorphous material and goethite. Based on filtration studies, they have a particle size from <100 to $>50,000$ Å and averaging 10,000 to to 20,000 Å. Values of pQ ranged from 37.1 in a shallow, oxidized ground water with active precipitation to 43.5 in a shallow ground water with active leaching of oxyhydroxides by H^+ ions (pH = 3.66) from oxidation of FeS_2 minerals. The highest pQ values for deeper, artesian ground waters ranged from about 41 to 43 and were found in relatively young waters at the highest measured ferrous iron concentrations. Ground waters at even greater depths in coastal plain New Jersey, which are probably several thousand years old, had pQ values as low as 39.5, having aged along with low ferrous iron concentrations ($10^{-5.90}M$ Fe).

Some general conclusions of the study are:

1. The greater the supersaturation with respect to crystalline oxyhydroxides, the faster the precipitation and the lower the initial pQ.

2. High pQ values can result from H^+ ion leaching of oxyhydroxide precipitates which preferentially dissolves the least stable material.

3. The rate of crystal formation and pQ increase of a precipitate (usually goethite) is proportional to the dissolved iron concentration and is faster in ferrous than in ferric iron solutions. Measurable crystallization of amorphous material occurs within minutes in $10^{-2}M$ ferrous iron solutions, but may take thousands of years in waters which contain about $10^{-6}M$ ferrous iron.

4. Ferric oxyhydroxides with crystal thicknesses less than 100 Å can remain indefinitely in some waters.

5. The relative stabilities of two crystalline oxyhydroxides such as goethite and hematite or goethite and lepidocrocite can reverse with time owing to particle size effects.

Acknowledgment

Financial support for this study was provided by the Mineral Conservation Section and the Institute for Research on Land and Water Resources, both of The Pennsylvania State University, University Park, Pa.

Literature Cited

(1) Garrels, R. M., in "Researches in Geochemistry," p. 25, Philip H. Abelson, Ed., Wiley, New York, 1959.
(2) Welo, L. A., Baudisch, O., *Chem. Rev.* (1934) **15**, 1–43.
(3) Palache, C., Berman, H., Frondel, C., "The System of Mineralogy," Vol. I, p. 527, 642, 680, 708, Wiley, New York, 1944.
(4) Singer, P. C., Stumm, W., *Science* (1970) **167**, 1121–3.
(5) Van Tassel, R., *Bull. Soc. Belge Geol.* (1959) **68**, 360–7.
(6) Chandy, K. C., *Indian J. Phys.* (1962) **36**, 484–9.
(7) Watson, J. H. L., Heller, W., Poplawski, L. E., *Third European Regional Conf. Elect. Microscopy* (1964) 315–6.
(8) Langmuir, D., *U. S. Geol. Surv. Profess. Paper* (1969) **650-B**, B180–B184.
(9) Langmuir, D., *Am. J. Sci.* (1971), in press.
(10) Ferrier, A., *Rev. Chim. Minerale* (1966) **3**, 587–615.
(11) Langmuir, D., *Geol. Soc. Am. Ann. Mtg., Milwaukee, 1970*, Program with Abstracts.
(12) Milburn, R. M., *J. Am. Chem. Soc.* (1957) **79**, 537–40.
(13) Hem, J. D., Cropper, W. H., *U. S. Geol. Surv. Water Supply Paper* (1959) **1459A**, 4.
(14) Garrels, R. M., Christ, C. L., "Solutions, Minerals, and Equilibria," p. 172, Harper and Row, New York, 1965.
(15) Nair, R. L. S., Nancollas, G. H., *J. Chem. Soc.* (1958) 4144–7.
(16) Willix, R. L. S., *Trans. Faraday Soc.* (1963) **59**, 1315–24.
(17) Klotz, I. M., "Chemical Thermodynamics," revised ed., p. 419, Benjamin, New York, 1964.
(18) Langmuir, D., *U. S. Geol. Surv. Profess. Paper* (1969) **650-C**, C224–C235.
(19) Back, W., U. S. Geological Survey, Washington, D. C., written communication, 1970.

(20) Whittemore, D. O., Langmuir, D., in preparation.
(21) Ackerman, T., Z. *Electrochem.* (1958) **62**, 411–9.
(22) Kolthoff, I. M., Sandell, E. B., "Textbook of Quantitative Analysis," 3rd ed., p. 571, 579, MacMillan, New York, 1952.
(23) Langmuir, D., "Procedures in Sedimentary Petrology," Ch. 26, p. 597–635, Robert E. Carver, Ed., Wiley, New York, 1971.
(24) Cullity, B. D., "Elements of X-ray Diffraction," p. 97, 261, Addison-Wesley, Reading, Mass., 1956.
(25) Lamb, A. B., Jacques, A. G., *J. Am. Chem. Soc.* (1938) **60**, 1215–25.
(26) Feitknecht, W., Michaelis, W., *Helv. Chim. Acta* (1962) **45**, 212–24.
(27) Watson, J. H. L., Cardel, R. R., Jr., Heller, W., *J. Phys. Chem.* (1963) **66**, 1757–63.
(28) Lee, G. F., Stumm, W., *J. Am. Water Works Assoc.* (1960) **52**, 1767–74.
(29) Shapiro, J., *Limnol. Oceanog.* (1966) **11**, 293–8.
(30) Langmuir, D., N. J. Department of Conservation and Economic Development, *N. J. Water Resources Circ.* (1969) **19**, 43 p.
(31) Back, W., Barnes, L., *U. S. Geol. Surv. Profess. Papers* (1965) **498-C**, C1–C16.
(32) Rainwater, F. H., Thatcher, L. L., *U. S. Geol. Surv. Water Supply Papers* (1960) **1454**, 184.

RECEIVED July 30, 1970.

9

The Rate of Hydrolysis of Hafnium in 1M NaCl

HALKA BILINSKI and S. Y. TYREE, JR.

College of William and Mary, Williamsburg, Va. 23185

The behavior of hafnium chloride in aqueous 1M NaCl has been studied over the temperature interval 25°–73°C. Total hafnium concentration was varied from 0.60 to 30 mM, and hydroxyl number was varied from 0 to 3. Rayleigh turbidities and pH values were measured as functions of time. The existence of polynuclear species was confirmed, but the measured properties change slowly with time, the most rapidly at the highest temperature. The rate of hydrolysis is related analytically to total hafnium concentration, hydroxyl number, and temperature.

The hydrolysis of hafnium and zirconium appear to be among the most difficult to understand; *i.e.*, their hydrolytic behavior has proved to be exceedingly complicated. Furthermore, the elements zirconium and hafnium offer an ideal pair with which to search for small differences in hydrolytic behavior between two elements. Consequently, the results reported here are the complement of a similar study on zirconium (*1*).

The hydrolysis of hafnium has been studied less extensively than that of zirconium. Larsen and Gamill (*2*) carried out pH titrations of dilute aqueous solutions of hafnium chloride, nitrate, and perchlorate at 25°C, varying total hafnium concentration from 0.0047 to 0.038M. They concluded, using the points at the solubility boundary, that an ion of average formula $[\mathrm{Hf(OH)}_{3.4}{}^{0.6+}]_n$ is the hafnium-containing solution species in equilibrium with the solid phase. Larsen and Wang (*3*) studied the ion exchange behavior of hafnium at the same temperature but in 0.50, 1.00, and 2.00M perchloric acid solution, varying the hafnium concentration in the aqueous phase from 0.01 to 2.5 × $10^{-5}M$. They found that three days were needed to reach equilibrium. Using equilibrium ultracentrifugation, Johnson and Kraus (*4*) found that the hafnium-bearing solute species in 0.5–2.0M acid at room temperature are trimers and tetramers, principally, where total hafnium was 0.05M. They did not attain equilibrium, however, and expressed the view that hydrolytic reactions become more complicated at low acidities. Deshpande and coworkers (5) measured diffusion coefficients of hafnium solutions as a function of acid molarity, using solutions which had been equilibrated for two hours at 27°C. The results indicated continuous polymerization of hafnium ions. Peshkova and Ang (6) used the TTA–solvent extraction technique to study hydrolysis and polymerization of hafnium in 1 and 2M HClO$_4$. They actually calculated hydrolysis constants of hafnium monomeric, trimeric, and tetrameric hydrolysis products at 25°C. Matijevic and coworkers (7) measured the charge on hydrolyzed hafnium ions in very dilute (10^{-5}–$10^{-6}M$) solutions, using colloid chemical techniques. They found no "aging" effects on keeping hafnium tetrachloride solutions for six months, but did not state what measurements were used to detect "aging" effects. Stryker and Matijevic (8) measured the adsorption of hydrolyzed hafnium species on glass surfaces. From the fraction of adsorbed hafnium as functions of total hafnium and different equilibration times, they concluded that approximately 70 hours are required to reach equilibrium saturation. Their experiments were in the range of 10^{-4}–$10^{-5}M$ total hafnium. Savenko and Scheka (9) have studied polymerization of hafnium in solution by dialysis. They have found that the degree of polymerization of hafnium depends on total hafnium and pH but not on the nature of the anion; polymerization increases as the degree of hydrolysis increases. The same conclusion was reached by Copley and Tyree (10), who used the light-scattering method to determine whether or not a steady state had been reached. However, the latter experiments were preliminary only. We now wish to report the results of a somewhat more lengthy series of experiments. This series was designed to see if hafnium solutions, within the range of variables available, do or do not come to equilibrium, or at least some steady state, in measurable time. We hoped to establish what the effect of several variables is upon the rate at which hafnium solutions approach a steady state. One variable was eliminated early in the study, that of ionic strength, by conducting the entire study in 1M NaCl. Preliminary experiments indicated that ionic strength has a very large effect on the rate of change of hafnium solutions in the $10^{-3}M$ total hafnium range. A subsequent study will deal with that variable. The three variables, total hafnium concentration, hydroxyl number of hafnium, and temperature, were considered in this study. It is our opinion that the results are of significance to this symposium topic since little attention has been given by water chemists to the possibility that simple inorganic solutes may take long (in the geological sense) times to approach equilibrium. No attempts have been made in this work to elucidate the presumably very fast pre-equilibrium steps in aqueous metal ion systems of the sort reported by Eyring and his coworkers (*11*).

Experimental

A stock solution of hafnium oxide dichloride octahydrate was prepared in the manner described previously (*10*). The stock solution was analyzed for hafnium gravimetrically, as the oxide, and for chloride by the modified Volhard method. Other stock solutions—*i.e.*, NaCl, NaHCO$_3$, and HCl—were prepared from reagent grade chemicals and standardized by the usual methods.

Experimental solutions were prepared from stock solutions in 250-ml volumetric flasks, each to simulate a point in a potentiometric titration. The solutions were prepared as several series, each series representing a total hafnium concentration. Each member of a series was made up to a particular ratio of analytical concentration of acid added to total hafnium, designated as Z*. Individual solutions were prepared by adding the calculated volumes of hafnium stock, acid or base stock, and NaCl stock in that order to volumetric flasks. Mixing and dilution to the mark completed the preparation of solutions. All solutions were prepared using borosilicate volumetric ware and were stored in polyethylene bottles at 25°C.

Volumes of stock solution were calculated as shown in the following example. H_o was calculated for each experimental solution to be prepared from the relationship, $H_o = [Hf]_{TOT} \times \dfrac{[H^+]_{x's}}{[Hf]_{stock}}$, where

$[Hf]_{TOT}$ = hafnium molarity of experimental solution

$[Hf]_{stock}$ = hafnium molarity of stock solution

$[H^+]_{x's}$ = chloride molarity of stock solution minus twice its hafnium molarity. This assumes an initial ideal solution of Hf(OH)$_2^{2+}$ 2Cl$^-$ and x's H$^+$Cl$^-$.

To prepare solutions of lower Z* values, sufficient additional HCl was added.

$$Z^* = 2 - \frac{[H^+]_{added} + H_o}{[Hf]_{TOT}}$$

To prepare solutions of higher Z* values, sufficient NaHCO$_3$ solution was added.

$$Z^* = 2 + \frac{[HCO_3^-]_{added} - H_o}{[Hf]_{TOT}}$$

In some few cases, OH$^-$ was added coulometrically, with no detectable difference in result.

Table I. Turbidities and pH Values at 25°C

Solution No.	$[Hf]_{TOT}$	Z^*	Time, Days	$\tau^* \times 10^5 Cm^{-1}$	pH
1	1×10^{-2}	0.013	2	0	1.60
			30	2.0	1.61
			67	–	1.58
			267	4.5	1.53
2	1×10^{-2}	0.513	2	0	1.69
			30	3.0	1.68
			67	–	1.65
			267	4.2	1.59
3	1×10^{-2}	0.913	2	0	1.77
			30	3.0	1.77
			67	–	1.73
			267	5.0	1.67
4	1×10^{-2}	1.313	2	0	1.87
			30	4.0	1.85
			67	3.6	1.81
			267	5.5	1.69
5	1×10^{-2}	2.213	2	0	2.17
			30	7.4	2.15
			67	9.1	2.11
			267	15.0	1.95
6	1×10^{-2}	2.513	2	5.0	2.35
			30	15.0	2.33
			67	19.0	2.29
			267	57.0	2.11
7	1×10^{-2}	2.713	2	9.2	2.49
			30	33.0	2.46
			67	58.0	2.42
			267	633.0	2.25
8	1×10^{-2}	2.913	2	32	2.68
			30	760	2.68
			67	ppt	2.63
			267	ppt	2.48
9	1×10^{-3}	0.013	2	0	2.56
			30	1.0	2.54
			60	–	2.55
			260	14.0	2.47
11	1×10^{-3}	0.913	2	1.0	2.70
			30	1.8	2.69
			60	2.5	2.70
			260	14.2	2.57
12	1×10^{-3}	1.313	2	1.0	2.81
			30	2.8	2.79
			60	4.9	2.78
			260	38.5	2.65

Table I. Continued

Solution No.	$[Hf]_{TOT}$	Z^*	Time, Days	$\tau^* \times 10^5 Cm^{-1}$	pH
13	1×10^{-3}	2.213	2	1.1	3.09
			30	47	3.06
			60	174	3.06
			183	275	3.03
			260	ppt	3.11
14	1×10^{-3}	2.513	2	10.7	3.23
			30	685	3.22
			60	ppt	3.22
			260	ppt	3.53
49	6×10^{-4}	−0.004	2	0	2.80
			323	8.0	−
50	6×10^{-4}	0.313	2	0	2.84
			323	17	−
51	6×10^{-4}	0.979	2	0	2.95
			323	93	−
52	6×10^{-4}	1.313	2	1.7	3.20
			27	5.7	−
			323	289	−
53	6×10^{-4}	2.396	2	9.8	3.41
			27	ppt	−
54	6×10^{-4}	2.596	2	59	3.53
			27	ppt	−
55	6×10^{-4}	2.796	2	227	3.69
			27	ppt	−
56	6×10^{-4}	2.963	2	ppt	3.85
65	3×10^{-2}	0.013	2	3.9	1.17
			346	6.3	−
66	3×10^{-2}	0.979	2	4.9	1.33
			346	8.4	−
70	3×10^{-2}	1.313	2	5	1.42
			37	15	−
			346	16	−
67	3×10^{-2}	2.313	2	7	1.82
			346	17	−
68	3×10^{-2}	2.613	2	12	2.03
			37	28	−
			346	35	−
69	3×10^{-2}	2.779	2	22	2.20
			37	22	−
			346	123	−

Aliquots of experimental solutions were taken at appropriate time intervals for measurements. For room temperature measurements, the aliquots were of just sufficient size for one immediate measurement of each parameter, pH and Rayleigh turbidity. For measurements at the higher temperatures, aliquots were transferred to glass bottles and held in thermostated baths at temperature. At appropriate time intervals, the bottles were removed from the baths, cooled to room temperature, and aliquots taken for measurements. The measurements of hydrogen ion concentration (pH) were accomplished using the cell: + glass electrode / experimental solution / calomel electrode −. The emf of the cell can be written as $E = E_o + 59.15 \log [H^+] + j_{ac}[H^+]$. E of the cell in $1M$ NaCl (containing ca. 8×10^{-4} mole/liter HCl) was measured as a function of HCl added to the solution (added by coulometer). An extrapolation of the values of E plotted against HCl added gives $E°$. Variations in E_o over a year were less than 0.5 mV. Since we are not interested in accuracy of absolute values of pH to greater than ±0.01 pH (the significant observation being changes in pH), we feel justified in neglecting $j_{ac}[H^+]$. This is true even though the total ionic strength was not kept precisely at 1; rather, all solutions were made up in $1M$ NaCl as the solvent, such that the very small contributions to the ionic strength from the hafnium chloride and hydrochloric acid were neglected. Values of pH ranged upward from 2; i.e., $[HCl] \leqslant 10^{-2}M$. Rayleigh turbidities and and refractive indices were determined using the same method and instrument as described previously (12). From these data, it is possible to calculate the degree of aggregation of hafnium (12), which was done for many of the solutions. τ^* values are the turbidities owing to the hafnium solute only.

Results and Discussion

Series were prepared for $[Hf]_{TOT}$ values of 10^{-4}, 6.00×10^{-4}, 1.00×10^{-3}, 1.00×10^{-2}, and $3.00 \times 10^{-2}M$. It did not prove possible to prepare stable solutions over all values of Z^* for each of the series. Furthermore, the solutions at 10^{-4} mole/liter did not yield consistent or reproducible data. The series at $10^{-3}M$ gave the largest amount of data prior to precipitation from which to observe the effects of Z^* and temperature upon the rate of change. The meaningful data at 25°C are shown in Table I.

As expected and in agreement with previous work, for a given value of Z^*, solution acidity increases regularly with $[Hf]_{TOT}$, and for a given $[Hf]_{TOT}$, acidity decreases with Z^*. However, pH increases with time, except following the incidence of precipitation, and changes in most solutions take months to reach a steady state. An example of the latter trend is seen in Figure 1. More significantly, the changes in τ^* values do not follow pH changes and appear to be changing much more slowly. The rate of increase in τ^* is a measure of the rate of increase in average size of hafnium-containing solute species and is most rapid for low $[Hf]_{TOT}$; cf. solutions numbered 49–56 in Table I. Within each series—

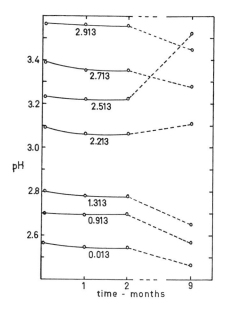

Figure 1.   pH vs. time (months) at
$25°C$ $[Hf]_{TOT} = 1 \times 10^{-3}$ M; Z^* *as*
parameter

i.e., at constant $[Hf]_{TOT}$—the rate of change of τ^* increases with Z^*. The latter effect is seen clearly in Figure 2 for $[Hf]_{TOT} = 1 \times 10^{-2}$. Note that differences do not appear to be significant among solutions 1 through 4 at 25°C or, for example, between solutions 9 and 11 at 25°C.

Data at higher temperatures are tabulated in Tables II and III.

The acidity changes at high temperatures occur so rapidly that only the steady state pH is recorded in some instances.

Clearly, all changes in pH have taken place during the first few hours at 73°C, while aliquots of the same experimental solutions have only begun to change after two months at 25°C. Thus, for example, solution No. 9 (Table I) after *ca.* eight months at 25°C has reached a value approximately the same as the one it reached in *ca.* five hours at 50°C. Clearly, the rate of hydrolysis increases dramatically with temperature. Equally clearly, hafnium solutions in the 10^{-2}–$10^{-4}M$ and pH 2 range require longer than nine months to approach a steady state as measured by the attainment of steady pH values. Thus, solutions of hafnium chlorides in this range of concentration and acidities at 25°C are hardly susceptible to calculations of distribution of hafnium among solute species from tabulated equilibrium constants. The constants of Peshkova and Ang (6) are for $[Hf]_{TOT}$ no more than 10^{-3} and acidities no less than $1M$ strong acid. Obviously, their values must not be con-

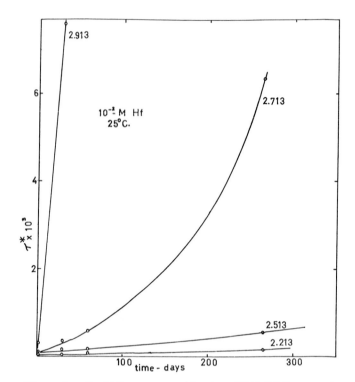

Figure 2.   Turbidity vs. time (days) at 25°C $[Hf]_{TOT} = 1 \times 10^{-2}$ M; Z^* as parameter

sidered as applicable under conditions encountered in natural water systems. In fact, they hardly can be considered as applicable in solutions less acid than $1M$ H^+.

From the pH data alone, it can be concluded that solutions of hafnium salts for $[Hf]_{TOT} = 10^{-1}-10^{-4}M$, pH 1, and at 25°C require a year or more to reach a steady state. Thus, the solutions are not at equilibrium. Nonetheless, it is possible to calculate much about the nature of the hafnium species from the data at any particular time, in accordance with Sillen's model (*13*). In order to make such calculations, it is necessary to calculate Z from pH data for each solution. Values of Z define the actual stoichiometry of the hafnium species, $[Hf(OH)_Z]_N^{(4-Z)N+}$. For experimental solutions at 25°C after two days, the Z vs. pH plot (with $[Hf]_{TOT}$ as the parameter) is shown in Figure 3. Clearly, the hydrolysis products are polynuclear.

Once having calculated the average stoichiometry of the solute species, the turbidity data permit the estimation of N, the degree of aggregation. For several series, the results are shown in Table IV. Even

Table II.   Turbidities and pH Values at 50°C

Solution No.	$[Hf]_{TOT}$	Z^*	Time, Hours	$\tau^* \times 10^5 Cm^{-1}$	pH
9	1×10^{-3}	0.013	0	–	–
			5	1.8	2.47
			9	3.2	2.46
			23	6.7	2.48
			34	23	2.48
11	1×10^{-3}	0.913	0	0.4	2.77
			5	4.9	2.63
			9	10	2.64
			23	46	2.64
			28	82	2.63
			34	116	2.63
12	1×10^{-3}	1.313	0	0.4	2.85
			4	8	2.71
			9	25	2.73
			23	224	2.74
			25	250	2.73
			28	358	2.74
13	1×10^{-3}	2.213	0	1.1	3.16
			2	52	3.00
			3	149	3.00
			4	313	3.01
			23	ppt	2.99

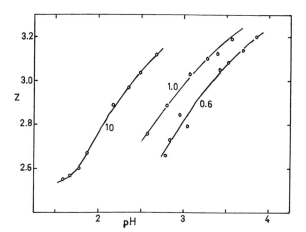

Figure 3.   Z vs. pH after two days at 25°C $[Hf]_{TOT}$ in mM as parameter

Table III. Turbidities and pH Values at 73°C

Solution No.	$[Hf]_{TOT}$	Z^*	Time, Hours	$\tau^* \times 10^5 Cm^{-1}$	pH
1	1×10^{-2}	0.013	7	20	1.50
			25	12	1.50
			219	20	–
			288	21	1.51
3	1×10^{-2}	0.913	97	52	1.63
			164	57	–
			234	69	1.64
4	1×10^{-2}	1.313	1	27	1.72
			3.5	76	1.68
			7.0	76	1.75
			14.5	240	1.73
			20	500	1.69
5	1×10^{-2}	2.213	1	202	1.97
			2	820	2.02
9	1×10^{-3}	0.013	2	118	2.46
			3	344	2.47
			4	560	2.44
			5	ppt	–
11	1×10^{-3}	0.913	1	222	2.58
			2	647	2.60
			3	ppt	–
12	1×10^{-3}	1.313	1	529	2.66
			2	ppt	2.69
70	3×10^{-2}	1.313	0	5.0	1.42
			2	21	1.32
			6.5	36	1.30
			18	36	1.28
67	3×10^{-2}	2.313	0	7	1.82
			4.8	137	1.63
			21.5	1240	1.57
68	3×10^{-2}	2.613	2	ppt	–

though the solutions are not at equilibrium, it is evident that the degree of aggregation, N, increases with time, and the rate of increase, $\dfrac{dN}{dt}$, increases rapidly with both Z^* and temperature. Table III indicates that Solution 11 had a τ^* value of 647×10^{-5} cm^{-1} after two hours at 73°C corresponding to $N = ca.$ 15000. After three hours, a precipitate was visible. Such behavior is typical, in that precipitation occurred in all solutions soon after they had reached a τ^* value of $ca.$ 10^{-2} cm^{-1}. All values of N were calculated for $Z' = 0$, assuming sufficient ion-pairing of chloride ion to the polymeric cations to reduce the apparent charge to zero (*14*). Such an assumption gives minimum values for N.

The point has already been made that the rates of change of pH and of τ^* do not appear to be related. The effect of temperature on the rate of change of τ^*, $\partial\tau^*/\partial\tau$, can be seen by comparing Figures 4 and 2, both for $[Hf]_{TOT} = 10^{-2}M$. The ordinate scales are the same, but the abscissa scales are hours and days, respectively. Also, the rate of change for $Z^* = 2.213$ at 73°C is comparable with that for $Z^* = 2.913$ at 25°C. Figure 5 shows most clearly the effect of Z^* at an intermediate temperature. In particular, Solution 9, which did not exhibit very much change at 25°C over eight months, changes substantially in a day at 50°C. Table III shows that Solution 9 yields a precipitate in five hours at 73°C.

The effect of concentration is seen in Figure 6. It is but typical of all such plots of τ^* *vs.* time, constant Z^* and temperature, with $[Hf]_{TOT}$ as the parameter. This result is the only one to emerge from this study which is unexpected.

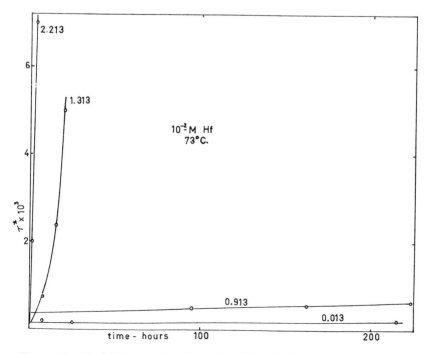

Figure 4.   Turbidity vs. *time (hours) at 73°C* $[Hf]_{TOT} = 1 \times 10^{-2}$ M; Z^*
as parameter

Table IV. Degree of Aggregation

$[Hf]_{TOT}$ (Molarity)	Z^*	Temp., °C.	Time	N
3×10^{-2}	1.313	25	2 days	4
	1.313	73	2 hours	16
	2.313	25	2 days	4
	2.313	73	5 hours	100
1×10^{-3}	0.013	25	30 days	30
	0.013	50	10 hours	100
	0.013	73	2 hours	2500
	0.913	25	30 days	40
	0.913	50	10 hours	250
	0.913	73	2 hours	15000
	1.313	25	30 days	60
	1.313	50	9 hours	600
	1.313	73	2 hours	ppt

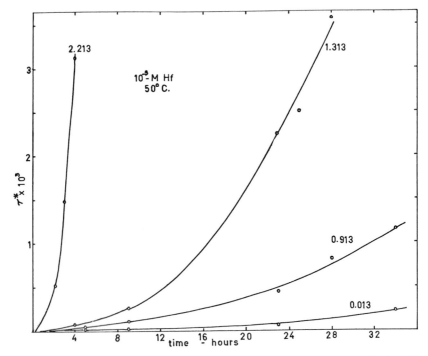

Figure 5.   Turbidity vs. time (hours) at 50°C $[Hf]_{TOT} = 1 \times 10^{-3}$ M; Z^*
as parameter

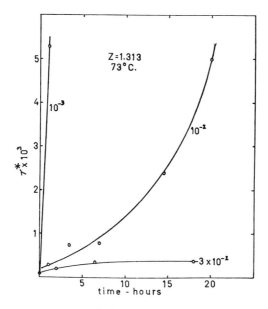

Figure 6.   Turbidity vs. *time (hours) at 73°C*
$Z^* = 1.313$; $[Hf]_{TOT}$ *as parameter*

Conclusions

In general, the behavior of hafnium in aqueous solution for pH values > 2 is very similar to that of zirconium. In the concentration range 10^{-2}–$10^{-4}M$ total hafnium and at ordinary temperatures, we can draw several firm conclusions.

Freshly prepared solutions are not at equilibrium. A detailed description of the way the solution was prepared, how long it has been standing, and at what temperature must be available in order to predict what the solute species are. In all cases, under these conditions the species are aggregated. All of the τ^* data for hafnium in $1M$ NaCl can be fitted to an empirical equation, $\tau^* = At^2$, where t is time and τ^* is the measured turbidity. A is a function of $[Hf]_{TOT}$, Z^*, and temperature, such that:

$d\tau^*/dt$ increases rapidly with increasing temperature

$d\ ^*/dt$ increases with increasing Z^*

$d\tau^*/dt$ decreases with increasing $[Hf]_{TOT}$

For the data herein reported, the best fit is obtained for $A = 1/\exp$, where $\exp = (Z^* - 3)\{[Hf]_{TOT} \cdot 10^{-0.0268\text{temp}+4.17} + 0.026\text{temp} - 2.65\} + 10^{0.0192\text{temp}+1.25}$. A similar relationship, displaced very slightly toward

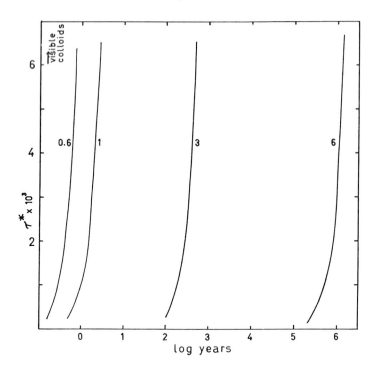

Figure 7.   Turbidity vs. *log time (years) at 25°C* $Z^* = 2.00$;
$[Zr]_{TOT}$ *in mM as parameter*

longer times, was found for zirconium, for which more extensive data
were taken. An estimated function of the rate of change of τ^* of zirconium solution for $Z^* = 2.00$ at 25°C in $1M$ NaCl is plotted as Figure
7, with $[Zr]_{TOT}$ as the parameter. The corresponding curves for hafnium
are displaced very slightly toward shorter times. It is not proposed that
all metals in aqueous solution will behave like hafnium and zirconium.
However, the behavior of these elements shows conclusively that an
"equilibrium model" alone is insufficient to use in describing the behavior
of solutions containing hafnium salts.

Literature Cited

(1) Bilinski, H., Tyree, S. Y., Jr., *ACS Meeting, 158th, New York, September*
   8, 1969.
(2) Larsen, E. M., Gammill, A. M., *J. Am. Chem. Soc.* (1950) **72**, 3615–9.
(3) Larsen, E. M., Wang, P., *J. Am. Chem. Soc.* (1954) **76**, 6223–9.
(4) Johnson, J. S., Kraus, K. A., *J. Am. Chem. Soc.* (1956) **78**, 3937–43.
(5) Deshpande, R. G., Khopkar, P. K., Rao, C. L., Sharma, H. D., *J. Inorg.*
   Nucl. Chem. (1965) **27**, 2171–81.
(6) Peshkova, V. M., Ang, P., *Russ. J. Inorg. Chem.* (1962) **7**, 1091–4.
(7) Matijevic, E., Kratohvil, S., Stryker, L. J., *Disc. Faraday Soc.* (1966) **No.**
   42, 187–96.

(8) Stryker, L., Matijevic, E., ADVAN. CHEM. SER. (1968) **79**, 44.
(9) Savenko, N. F., Scheka, I. A., *Ukr. Khim. Zh.* (1968) **34**, 309–12.
(10) Copley, D. B., Tyree, S. Y., Jr., *Inorg. Chem.* (1968) **7**, 1472–4.
(11) Cole, D. L., Rich, L. D., Owen, J. D., Eyring, E. M., *Inorg. Chem.* (1969) **8**, 682–5.
(12) Angstadt, R. L., Tyree, S. Y., Jr., *J. Inorg. Nucl. Chem.* (1962) **24**, 913–7.
(13) Sillen, L. G., *Acta Chem. Scand.* (1954) **8**, 299–317.
(14) Craig, H. R., Tyree, S. Y., Jr., *Inorg. Chem.* (1969) **8**, 591–4.

RECEIVED April 25, 1970.

10

Relations Among Equilibrium and Nonequilibrium Aqueous Species of Aluminum Hydroxy Complexes

ROSS W. SMITH

University of Nevada, Reno, Nev. 89503

The form of aluminum in acid aqueous media was studied by preparing solutions containing constant concentration of aluminum and constant ionic strength, but with varying ratios of OH⁻ to aluminum (r_n values) and determining the composition and pH of the solutions as a function of aging time. The compositions of the solutions were determined by a timed colorimetric procedure that allowed estimation of three separate types of aluminum that have been designated Al^a, Al^b, and Al^c. Al^a was composed of monomeric species. Al^b was polynuclear material. Al^c was composed of small, solid $Al(OH)_3$ particles. For each r_n value, the concentration of Al^a was constant, Al^b decreased in concentration, and Al^c increased in concentration with aging time. In all cases, equilibrium was only slowly achieved.

The nature of aluminum (III) in aqueous environments has been explored in a number of papers (*1–24*). If the pH of the solution is above neutrality, it appears that the predominant species present is the anion $Al(OH)_4(H_2O)_2^-$ (*8, 9, 10, 13*). Deltombe and Pourbaix (*4*) write this species as AlO_2^- and/or $H_2AlO_3^-$. If the pH is below about 4, most authors agree that the hexaaquo-Al(III) ion $Al(H_2O)_6^{3+}$ dominates. Between pH 4 and 7, there is little agreement as to what species are present.

Schofield and Taylor (*21*), Frink and Peech (*5*), and Raupach (*17*) believe that the system can be handled satisfactorily in this region on the basis of simple monomeric species. The agreement among their determinations of K_1, the equilibrium constant for the reaction

$$Al(H_2O)_6^{3+} + H_2O \rightleftarrows Al(H_2O)_5(OH)^{2+} + H_3O^+$$

(pK values 4.98, 5.02, and 4.97, respectively) is excellent.

In spite of this good agreement and the fact that simple hydrolysis reactions represent their data well, the simple hydrolytic mechanism has been questioned by a number of researchers.

Brosset (2), from measurements of the pH of aluminum perchlorate solutions, postulated that the product of hydrolysis of the aluminum ion is an infinite series of polynuclear complexes with the generalized formula $Al[(OH)_3Al]_n^{3+}$. Brosset, Biedermann, and Sillén (3) later recalculated Brosset's data and concluded that the major hydrolysis product could be either a single complex such as $Al_6(OH)_{15}^{3+}$ or an infinite series of complexes of the type $Al[(OH)_5Al_2]_n^{+3+n}$. Aveston, using ultracentrifugation, found evidence for either $[Al_2(OH)_2]^{4+}$ or $[Al_{13}(OH)_{32}]^{7+}$. The latter species has been favored recently by Sillén (25) and by Johansson (26).

According to Hsu and Bates (11, 12), dissolved aluminum below or near pH 4 is in the form of a six-member ring of approximate composition $Al_6(OH)_{12}^{6+}$. The rings polymerize as pH is increased above 4 with an average equilibrium polymerization number that increases with pH until pH 7 is exceeded. When this happens, bayerite or gibbsite is precipitated. Thus, in their scheme, polynuclear complexes of a size determined by pH exist in solution—*i.e.*, the higher the pH up to 7, the greater the size of the complexes. Also, as pH increases, the average charge per aluminum atom decreases from about 1^+ at pH 4 to 0 near pH 7 where the ratio of OH/Al is 3 or greater. Up to this point the hydroxo-aluminum polymers would repel one another and limit their growth, but slightly above it they should and do rapidly precipitate as $Al(OH)_3$. At still higher pH values, more aluminum goes into solution as increasing amounts of $Al(OH)_4^-$ are formed.

Hem and Roberson (8) believe that polymeric species, probably six-member rings and combinations of these rings, form rapidly in freshly prepared supersaturated aluminum solutions in which the initial pH is between 4 and 7 and the ratio of OH to aluminum in complexes averages between 0.6 and 3. The complex species are not stable, however, and ultimately grow to a size that must be considered a solid phase. Also, the very large polymers are organized and appear to be crystalline with the structure of gibbsite, yet are small enough to pass a 0.45-millipore filter and hence should be considered colloidal. After equilibrium becomes established, which may take years, it may be that the only ionic aluminum species present in significant quantities are the monomeric $Al(OH_2)_6^{3+}$, $Al(OH)(OH_2)_5^{2+}$, $Al(OH)_2(OH_2)_4^+$, and $Al(OH)_4^-$ ions.

It is possible that distribution of Al(III) among hydroxo-complexes is highly variable and is sensitive to ionic strength, total aluminum concentration, total OH available, pH, temperature, the identity of other species present such as NO_3^-, ClO_4^-, SO_4^{2-}, Cl^-, etc., and, perhaps most important, time. The latter factor appears particularly important, considering the experimental work of Hem and Roberson (8) and the statement by Brosset, Biedermann, and Sillén (3) that "Considerable difficulties were met with because equilibrium was attained rather slowly, especially in the region where precipitation occurred. It is, however, thought that the values finally given were not far from those at real equilibrium." From the present work, it would appear that the data used by these authors were considerably further from equilibrium than they had thought. At any rate, it appears that the nature of Al(III) in aqueous media is not well known, particularly at mildly acid pH values.

Experimental

The form of aluminum in acid aqueous media was studied by preparing a series of solutions containing the same total concentration of aluminum, but with varying amounts of added base and determining the composition and pH of the solutions after various periods of aging. For convenience, these solutions will be designated "aging study solutions." Electron microscopy was used to help determine the nature of colloidal-size material that formed in some of the solutions.

The solutions studied contained 4.54×10^{-4} mole/liter aluminum and total ionic strength was 10^{-2}, the remainder of the total ionic strength being made up with sodium and perchlorate ions. The ratio of OH to Al in the solutions as made up (nominal r value or r_n) varied from 0.55 to 3.01. In preparing these aging study solutions, three "stock" solutions were prepared initially and mixed together in correct proportions to achieve the desired r_n value. The procedure for doing this has been described by Hem and Roberson (8). In all cases, the solution containing base, but no aluminum, was added last in solution preparation. Reagent grade chemicals were used.

Analytical Procedure

The compositions of solutions were determined by a timed spectrophotometric method. The technique was a modification of a standard ferron–orthophenanthroline method for aluminum (27, 28) and is identical in principle to a method developed by Turner (22). The method was modified in the following manner.

Standard Procedure	Modification
1) Pipet a volume of sample containing not more than 0.075 mg (25 ml max.) into a 50-ml beaker and adjust the volume to 25.0 ml.	1) Same as standard procedure.

2) Prepare a 25-ml metal-free water blank and necessary standards.
3) Add 2.0 ml NH$_2$OH HCl reagent to blank standards and sample and let stand 30 minutes.

4) Add 5.0 ml ferron–orthophenanthroline reagent and stir.
5) Add 2.0 ml NaC$_2$H$_3$O$_2$. Stir and let stand for at least 10 minutes but not more than 30 minutes before taking a reading of color.
6) Determine the absorbancy of the test sample and standards against the blank at 370 mμ.

2) Same as standard procedure.

3) Add 2 ml NaC$_2$H$_3$O$_2$ to 5 ml ferron–orthophenanthroline reagent (reagent is twice as strong in ferron as standard procedure).
4) Add (3) above to both blank and sample and stir.
5) As quickly as possible (at least within 4 minutes), add 2.0 ml NH$_2$OH HCl reagent, stir quickly, and at the same time start timing.
6) As quickly as possible, read absorbency against the blank at 370 mμ. Time the reading and continue to take timed readings (at 3–4 min intervals for the first half hour, then at wider intervals for as long as necessary).

It should be noted in the modified procedure that the last reagent added is the hydroxylamine hydrochloride, which brings the pH to about 5. Thus, a pH lower than this value is not obtained at any stage of adding analytical reagents, unlike in the standard procedure where the hydroxylamine is added at the start of the procedure and usually lowers pH of the sample to about 1.5. Hem and Roberson (8) discuss at some length the possible complex formed between ferron and aluminum.

When optical density of aluminum standards prepared by dissolving either aluminum wire in HCl solution or aluminum sulfate in water is measured as a function of aluminum concentration using the modified procedure, curves of the type shown in Figure 1 are obtained. Optical density values plotted on this figure, which are for duplicate sets of experiments, were read about 30 minutes after having added the 2 ml of hydroxylamine hydrochloride. The straight line curve appears to obey Beer's law at least up to 0.05–0.06 mg of aluminum. The pH of the standard solutions before analysis was in all cases below 3.

The modified procedure allowed for the estimation of three different types of aluminum present in a particular sample at any particular aging time, based on the manner in which the various types of aluminum reacted with ferron. The way in which the estimation is made is illustrated by Figures 2 and 3, which have been calculated from raw analytical data.

Shown on Figure 2 is aluminum recovered as a function of analysis time after having added the hydroxylamine reagent. In this figure, optical absorbance readings have been converted to molarities and ppm's using proper factors to account for aliquots taken, etc. Curves are for

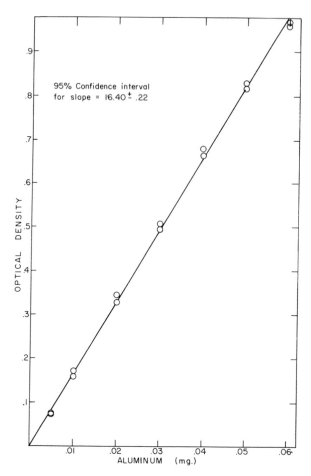

Figure 1.   Optical density as a function of aluminum concentration for aluminum standards using the modified ferron procedure

solutions with r_n values ranging from 0.94 to 2.76 aged 625 hours, and in addition curves are shown for a solution with an r_n value of 2.13 aged from 23 hours to 961 days. All these solutions had a total aluminum concentration of 4.54×10^{-4} mole/liter. Also shown are similar timed data for 25-ml standard solutions containing 2, 6, and 10 ppm aluminum. The fact that the colorimetric readings for the standard solutions changed little during the three hours of analysis time is good evidence that a mere color change of ferron with time is not what is being observed when the amount of aluminum recovered increases with analysis time. Additional evidence is to be found in the way in which the curves for the solution with $r_n = 2.13$ converge when extrapolated to zero analysis time.

Figure 2 shows that the solutions contained three different types of aluminum species, one of which disappeared slowly during long aging. For convenience, the three species will be referred to as Al^a, Al^b, and Al^c, and from the stoichiometry of preparation of the solutions, it is known that in each case $Al = Al^a + Al^b + Al^c = 4.54 \times 10^{-4}$ mole/liter. It is interesting to note that Turner (22), using his similar analytical technique, also concluded that three different types of aluminum can exist in aqueous media.

The fastest reacting form, Al^a, is converted to the ferron complex almost immediately, and for a particular r_n value is present in nearly the same amount regardless of aging time, as shown by the convergency of all the determinations at zero time for the solution with an r_n value of 2.13. The slowest reacting material, Al^c, is represented by the nearly flat

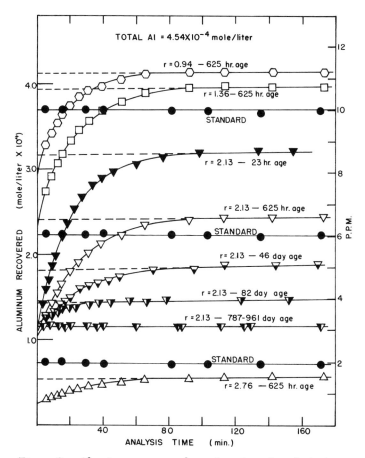

Figure 2.   Aluminum recovered as a function of analysis time for selected aging study solutions plus several standards

slope shown by the sample readings after about 120 minutes. Metastable material, Al^b, is the material which reacts with an intermediate rate, giving the curved portions of the lines of Figure 2.

From these curves, it is a simple matter to measure the amounts present of the three types of aluminum. By extrapolating the essentially straight portions of the curves (Al^c sections) back to zero time, one can estimate the total quantity of $Al^a + Al^b$ present at a particular aging time. Al^a can be obtained directly from the zero time value of aluminum recovered and Al^b obtained by subtracting the value of Al^a from the value of $Al^a + Al^b$. However, it is somewhat difficult to extrapolate these curves accurately, and thus the curves of the type of Figure 3 were constructed to determine more accurately Al^a and Al^b concentrations and to obtain additional information on the ferron reaction with type b aluminum.

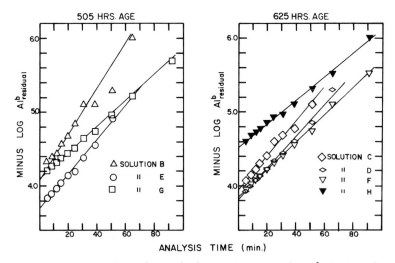

Figure 3.   Minus log Al^b residual as a function of analysis time for age study solutions aged 505–625 hours

Figure 3 shows first order rate plots of the negative log of Al^b residual *vs.* time for solutions aged 505 to 625 hours. The individual Al^b residual points were determined by subtracting data points of Figure 2 from the values of the extrapolated (dotted) lines at the same analysis reaction times.

Because the curves obtained were consistently good, straight lines, even when 90% of the reacting Al^b had been consumed, it was a simple matter to extrapolate them to zero time to obtain more exact values of Al^b and therefore also Al^a. Thus, the kinetic behavior of the aluminum species provided a convenient means of identification. Also of interest is the fact that Al^b reacts with ferron according to a first order rate law.

General Equation for Curves of Figure 2

The following equation should be general for the curves of Figure 2, assuming all of type a aluminum to react with ferron instantaneously

$$^tAl^T = (4.54 \times 10^{-4} - Al_t^T) = Al^a + (Al_0^b - Al_t^b) + (Al_0^c - Al_t^c) \quad (1)$$

where

$^tAl^T$ = total amount of aluminum that has reacted in time t (in mole Al per liter)

Al^a = amount of instantly reacting aluminum present (mole per liter)

Al_0^b = amount of Al^b present at zero analysis time (mole per liter)

Al_t^b = amount of Al^b unreacted at analysis time t (mole per liter)

Al_0^c = amount of Al^c present at zero analysis time (mole per liter)

Al_t^c = amount of Al^c unreacted at analysis time t (mole per liter)

It is assumed that during the analysis reaction there are no interactions between Al^a, Al^b, and Al^c.

If Al^b reacts according to a first order law

$$\frac{-dAl^b}{dt} = k_b Al^b \quad (2a)$$

where k_b is a first order rate constant for type b aluminum.

Integrating between the limits 0 and t and Al_0^b and Al_t^b

$$k_b t = \ln Al_0^b - \ln Al_t^b$$

or

$$Al_t^b = Al_0^b e^{-k_b t} \quad (3)$$

If Al^c reacts according to some rate law of order n where $1 > n > 0$ (and n is probably close to zero)

$$\frac{-dAl^c}{dt} = k_c Al^{cn} \quad (2b)$$

where k_c is a rate constant of some unknown order for type c aluminum.

Table I.   First Order Rate Constants for Al^b

| | | | | Aging | |
Solution	23 Hours	48 Hours	96 Hours	168 Hours	288 Hours
B	9.9	8.5	–	7.4	–
C	7.8	–	5.5	5.8	5.5
D	5.5	–	4.8	5.8	5.5
E	5.8	4.8	–	5.3	–
F	5.1	–	4.6	5.3	5.1
G	5.1	5.1	–	5.1	–
H	4.4	–	4.4	4.4	3.9
J	3.2	–	–	–	–

[a] Average (48 hours–259 days): B 7.0; C 5.3; D 5.2; E 5.3; F 4.2; G 4.2; H 3.8.

Integrating between the limits 0 and t and $Al_0{}^c$ and $Al_t{}^c$

$$Al^{c^{1-n}} = Al_0{}^{c^{1-n}} - \frac{k_c t}{1-n}$$

and

$$Al_t{}^c = \left(Al_0{}^{c^{1-n}} - \frac{k_c t}{1-n} \right)^{\frac{1}{1-n}} \tag{4}$$

Substituting Equations 3 and 4 into Equation 1

$${}^tAl^T = Al^a + [Al_0{}^b - Al_0{}^b e^{-k_b t}] + [Al_0{}^c - \left(Al_0{}^{c^{1-n}} - \frac{k_c t}{1-n} \right)^{\frac{1}{1-n}}] \tag{5}$$

which should be the general equation for the analysis curves. If a zero order rate is assumed for Al^c, then Equation 5 reduces to

$${}^tAl^T = Al^a + [Al_0{}^b - Al_0{}^b e^{-k_b t}] + k_c t$$

or using logarithms to the base ten rather than to the base e

$${}^tAl^T = Al^a + [Al_0{}^b - Al_0{}^b 10^{-k' b t}] + k_c t \tag{6}$$

Aging Studies

Solutions were prepared having the following r_n values: 0.55, 0.94, 1.36, 1.84, 2.13, 2.47, 2.76, and 3.01 and were designated respectively solutions B, C, D, E, F, G, H, J. These solutions were allowed to age in a CO_2-free atmosphere near 25°C for 254–259 days. At various intermediate aging times, aliquots of the solutions were analyzed for amounts of Al^a, Al^b, and Al^c as previously described. Also, first order rate constants

Reaction with Ferron ($-\ln$ Al^b Residual/Min $\times$ 10^2)

Time^a

505–625 Hours	41–46 Days	77–82 Days	116–121 Days	188–193 Days	254–259 Days
7.1	8.5	6.0	5.5	7.4	5.8
5.3	5.1	4.1	4.6	5.8	6.2
5.1	5.3	4.8	4.8	–	–
5.3	4.8	5.3	4.8	6.7	5.8
4.1	4.4	4.8	4.4	–	–
4.1	3.7	3.7	3.7	–	–
3.7	3.7	3.3	3.2	–	–
–	–	–	–	–	–

for the reaction of Al^b with ferron were determined from figures of the type of Figure 3. At the same time, pH of the solutions was measured. Table I shows first order rate constants obtained. No attempt was made to measure rate constants in solutions aged 1–1.5 hours, and, except for solutions B, C, and E, in solutions aged 188 days or more. In the former case, it was thought that the solutions themselves were changing character so rapidly that the measurement would have been meaningless. In the latter case, so little (except for solutions B, C, and possibly E) Al^b was present that small errors in analysis measurement would lead to large errors in the calculation of the rate constants. For solution J, very little Al^b was present except at 23 hours aging, and hence a rate constant was measured only for this one aging time.

Rate constants appear to decrease with increasing r_n values of the solutions. For the case of solution B, rate constants decreased from 23 hours up to at least 254 days aging. For the remaining solutions, rate constants decreased between 23 hours and about 96 hours aging time. At greater aging times, the rate constants appear to remain at about the same magnitude. In all cases, the decrease in rate constant is not great, indicating, perhaps, that Al^b structures probably consist of a limited series of different size particles which increase in size with r_n value and somewhat with aging time. Further, the first order reaction rate followed by Al^b indicates that individual aluminum atoms in the Al^b structures are individually reacting with ferron. It would thus seem that the Al^b structures are not extremely large, and most of the aluminum atoms are in contact with the aqueous solution and not hidden within the structure. However, since it takes a moderate length of time to break down the Al^b structure, these structures must be rather strongly bound together.

Table II lists concentrations of Al^a, Al^b, and Al^c and pH of the solutions at various aging times determined from plots of the types of Figures 2 and 3.

Table II. Concentrations of Various Types of Aluminum at All Age Times

Solution B

$(r_n = 0.55)$

Concentration $\times$ 10^4

Age Time	Al^a	Al^b	Al^c	pH
1.1 hrs	3.90	0.31	0.34	4.46
23 hrs	3.64	0.57	0.34	4.44
48 hrs	3.69	0.40	0.46	4.49
168 hrs	3.57	0.54	0.44	4.45
505 hrs	3.72	0.64	0.19	4.46
41 days	3.84	0.48	0.23	4.43
77 days	3.80	0.50	0.25	4.43
116 days	3.76	0.35	0.44	4.43
188 days	3.77	0.32	0.46	4.42
254 days	3.56	0.59	0.40	4.43

Old solution:
r_n value near B

1001 days	3.75	0.11	0.69	4.43

Solution C

$(r_n = 0.94)$

Concentration $\times$ 10^4

Age Time	Al^a	Al^b	Al^c	pH
1.5 hrs	3.40	0.65	0.49	4.52
23 hrs	3.24	0.94	0.36	4.46
96 hrs	3.20	1.06	0.28	4.43
168 hrs	3.31	1.10	0.13	4.43
288 hrs	3.12	1.00	0.42	4.46
625 hrs	2.94	1.19	0.42	4.48
46 days	3.35	1.14	0.05	4.46
82 days	3.20	0.97	0.37	4.48
121 days	3.23	0.79	0.52	4.47
193 days	3.07	0.65	0.82	4.47
259 days	3.10	0.76	0.68	5.47

Old solution:
r_n value near C

1038 days	3.23	0.43	0.89	4.36

Table II.   Continued

Solution D
$(r_n = 1.36)$

Concentration $\times 10^4$

Age Time	Al^a	Al^b	Al^c	pH
1.3 hrs	2.84	1.06	0.64	4.66
23 hrs	2.50	1.58	0.46	4.56
96 hrs	2.42	1.54	0.58	4.52
168 hrs	2.32	1.62	0.60	4.49
288 hrs	2.35	1.66	0.53	4.49
625 hrs	2.30	1.63	0.61	4.40
46 days	2.31	1.11	1.12	4.32
82 days	2.34	0.34	1.86	4.26
121 days	2.39	0.06	2.09	4.23
193 days	2.43	0.09	2.02	4.14
259 days	2.30	0.02	2.22	4.12

Old solution:
r_n value near D

1038 days	2.46	–	2.08	4.09

Solution E
$(r_n = 1.84)$

Concentration $\times 10^4$

Age Time	Al^a	Al^b	Al^c	pH
1.0 hrs	2.24	1.50	0.80	4.79
23 hrs	1.56	2.10	0.88	4.68
48 hrs	1.82	2.00	0.72	4.66
168 hrs	1.46	1.86	1.22	4.61
505 hrs	1.44	2.14	0.96	4.56
41 days	1.50	1.81	1.23	4.54
77 days	1.53	1.45	1.56	4.54
116 days	1.60	1.20	1.74	4.52
188 days	1.51	0.49	2.54	4.44
254 days	1.50	0.12	2.92	4.32

Old solution:
r_n value near E

1038 days	1.79	–	2.77	4.13

Table II. Continued

Solution F
$(r_n = 2.13)$

Concentration $\times$ 10^4

Age Time	Al^a	Al^b	Al^c	pH
1.2 hrs	1.46	1.78	1.30	4.88
23 hrs	1.11	2.05	1.38	4.75
96 hrs	1.12	1.86	1.56	4.63
168 hrs	1.12	1.86	1.56	4.60
288 hrs	1.05	1.82	1.67	4.52
625 hrs	1.05	1.35	2.04	4.42
46 days	1.02	0.78	2.74	4.37
82 days	1.20	0.26	3.08	4.30
121 days	1.21	0.10	3.23	4.28
193 days	1.07	0.04	3.43	4.22
259 days	1.02	0.02	3.50	4.20

Old solution:
r_n value near F

967 days	1.16	–	3.33	4.19

Solution G
$(r_n = 2.47)$

Concentration $\times$ 10^4

Age Time	Al^a	Al^b	Al^c	pH
1.2 hrs	1.17	1.97	1.40	5.02
23 hrs	0.60	1.20	2.74	4.82
48 hrs	0.60	1.15	2.79	4.77
168 hrs	0.55	1.05	2.94	4.75
505 hrs	0.64	0.81	3.09	4.49
41 days	0.68	0.40	3.46	4.42
77 days	0.68	0.13	3.63	4.40
116 days	0.66	0.043	3.84	4.35
188 days	0.61	0.019	3.91	4.31
254 days	0.62	0.007	3.91	4.28

Table II.   (Continued)

Solution H
$(r_n = 2.76)$

Concentration $\times 10^4$

Age Time	Al^a	Al^b	Al^c	pH
1.1 hrs	0.78	1.36	3.40	5.23
23 hrs	0.20	0.50	3.84	5.16
96 hrs	0.20	0.40	3.94	5.00
168 hrs	0.23	0.39	3.92	4.76
288 hrs	0.24	0.36	3.94	4.72
625 hrs	0.24	0.30	4.00	4.62
46 days	0.28	0.17	4.09	4.53
82 days	0.29	0.05	4.20	4.51
121 days	0.284	0.011	4.24	4.49
193 days	0.232	0.008	4.30	4.45
259 days	0.235	–	4.30	4.45

Solution J
$(r_n = 3.01)$

Concentration $\times 10^4$

Age Time	Al^a	Al^b	Al^c	pH
1.0 hrs	0.56	1.90	2.08	7.15
23 hrs	0.08	0.22	4.24	6.64
48 hrs	0.07	0.16	4.31	6.59
168 hrs	0.05	0.07	4.42	6.52
505 hrs	0.05	0.03	4.46	6.54
41 days	0.065	0.01	4.47	6.43
77 days	0.08	–	4.46	6.28
116 days	0.057	–	4.48	6.40
188 days	0.019	–	4.52	6.21

Of interest from this table is the fact that for most solutions Al^a concentration remains constant after 23 hours of aging, Al^b decreases as a function of aging time, and Al^c increases. The relationship between r_n value and Al^a value is shown graphically in Figure 4.

Experimental Character of Al^a

Since Al^a reacts almost instantly with ferron, it would seem reasonable that it consists of only simple monomeric species—*i.e.*, Al^{3+}, $Al(OH)^{2+}$, $Al(OH)_2^+$, and $Al(OH)_4^-$ (with appropriate coordinated water molecules). Standard Gibbs free energies of formation values (ΔG°) for the species are available in the chemical literature. Table III

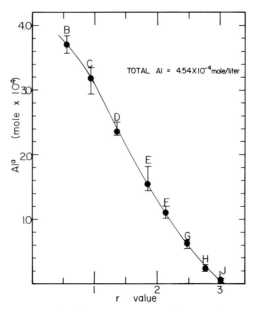

Figure 4. Concentration of Al^a as a function of r_n value for solutions containing 4.54 × 10^{-4} mole/liter total aluminum

lists some of these values plus a Gibbs free energy value for gibbsite [$\alpha Al(OH)_3$].

The values for Al^{3+}, $Al(OH)^{2+}$, $Al(OH)_4^-$, and $\alpha Al(OH)_3$ were selected to be consistent with the work of Hem and Roberson. Few values for $Al(OH)_2^+$ were available, and Raupach's was selected for consistency because of the use of his value for $Al(OH)^{2+}$.

From data of this table, the following equations can be written.

$$Al(OH)_3 \text{ (gibbsite)} + 3H^+ \rightleftarrows Al^{3+} + 3H_2O \qquad (7)$$

$$*K_{s0} = 10^{+8.22}$$

Table III. Gibbs Free Energies of Formation of Monomeric Aluminum Species

Species	Standard Gibbs Free Energies of Formation ($\Delta G° kcal$)	Reference
Al^{3+}	−115.0	(29)
$Al(OH)^{2+}$	−164.9	(18,19)
$Al(OH)_2^+$	−215.1	(18,19)
$Al(OH)_4^-$	−311.7	(8)
$\alpha Al(OH)_3$ (gibbsite)	−273.9	(29)

$$Al(OH)_3 \text{ (gibbsite)} + 2H^+ \rightleftarrows AlOH^{2+} + 2H_2O \qquad (8)$$

$$*K_{s1} = 10^{+3.22}$$

$$Al(OH)_3 \text{ (gibbsite)} + H^+ \rightleftarrows Al(OH)_2^+ + H_2O \qquad (9)$$

$$*K_{s2} = 10^{-1.54}$$

$$Al(OH)_3 \text{ (gibbsite)} + H_2O \rightleftarrows Al(OH)_4^- + H^+ \qquad (10)$$

$$*K_{s4} = 10^{-13.9}$$

$$Al^{3+} + H_2O \rightleftarrows Al(OH)^{2+} + H^+ \qquad (11)$$

$$*K_1 = 10^{-5.00}$$

$$Al(OH)^{2+} + H_2O \rightleftarrows Al(OH)_2^+ + H^+ \qquad (12)$$

$$*K_2 = 10^{-4.76}$$

Using Equations 7–12 plus estimated activity coefficients (listed in Table IV), the theoretical concentrations of the monomeric species can be calculated and then plotted as a function of pH. The activity coefficients used for Al^{3+} and $Al(OH)^{2+}$ were identical to the values used by Hem and Roberson (8) for similar solutions. Further, actual concentrations of Al^a for the various solutions and aging times can be plotted on the same chart. This plot is shown in Figure 5. Increasing aging time is indicated by the arrows.

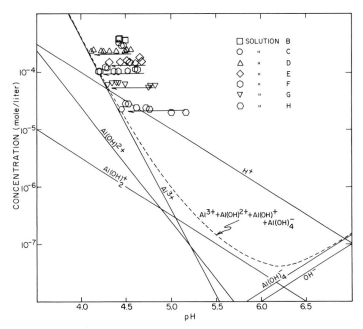

Figure 5.   Concentration of Al^a as a function of pH for solutions B–H at various aging times; arrows indicate increasing aging time

**Table IV.   Activity Coefficients of Monomeric Species
for $\sim$10^{-2} Ionic Strength Solutions**

Species	Coefficient
Al^{3+}	0.427^a
$AlOH^{2+}$	0.678^a
$Al(OH)_2{}^+$	0.907^a
$Al(OH)_4{}^-$	0.890^b

[a] From Debye–Huckel limiting law using $a = 9$.
[b] From Debye–Huckel limiting law using $a = 3$.

Figure 5 shows how Al^a concentration remains nearly constant as pH drops, at least for solutions D–H, as a function of aging time until the theoretical equilibrium line for Al^{3+} + $Al(OH)^{2+}$ + $Al(OH)_2{}^+$ + $Al(OH)_4{}^-$ is reached.

No such constancy of Al^a + Al^b is to be noted in Figure 6, which shows curves similar to those of Figure 5 except Al^a + Al^b concentrations are plotted rather than Al^a concentrations.

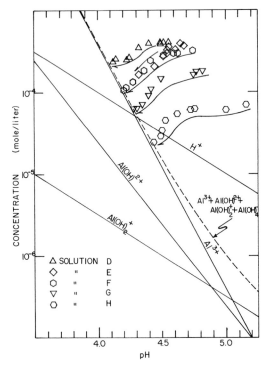

Figure 6.   Concentration of Al^a + Al^b as a function of pH for solutions D–H at various aging times; arrows indicate increasing aging time

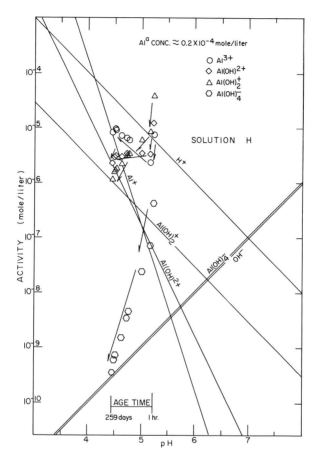

*Figure 7. Aging time effect on activities of mono-
meric species as a function of pH for solution H;
arrows indicate increasing aging time*

From Equations 7–12 and the several activity coefficients, it is possible to calculate activities of the various possible monomeric species at
all aging times using formulae of the type

$$[Al^{3+}] = \frac{Al^a}{234 + \dfrac{1.47 \times 10^{-5}}{H^+} + \dfrac{1.91 \times 10^{-10}}{[H^+]^2} + \dfrac{7.6 \times 10^{-23}}{[H^+]^4}} \quad (13)$$

where $[H^+]$ is hydrogen ion activity.

Activities of monomeric aluminum species calculated in this manner
can then be plotted on a graph of log activity as a function of pH and
aging time. An example (for solution H) of such a plot is illustrated in

Figure 7. Equilibrium lines shown are based on Equations 7–10. For solution H (and similarly, it can be shown for solutions D, E, F, and G), all species move toward the equilibrium lines with increasing aging times.

For each solution, the total concentration of monomeric aluminum remains constant with age. This concentration, strangely enough, is the concentration of Al^a expected at the pH the solution evolves to at long aging times. At intermediate aging times, Al^a concentration is greater than it should be if the solid and observed pH were in equilibrium. Further, all adjustments of the system have to do with pH, Al^b, and Al^c. Changes in monomeric species' activities merely reflect an essentially instantaneous adjustment of these activities to changing pH.

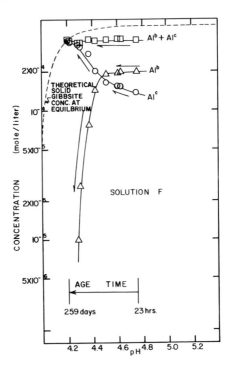

Figure 8. Concentration of Al^b and Al^c as a function of pH and aging time for solution F; arrows indicate increasing aging time

Experimental Characterization of Al^b and Al^c

Figure 8 for solution F has been constructed to show clearly progress of the amounts of Al^b, Al^c, and Al^{b+c} with time and pH. Similar figures can be constructed for the other solutions. Since Al^c appears to be solid material, it is actually incorrect to think of Al^c in terms of concentration "in solution." However, such a designation is convenient and helpful in following the system as a function of time. These curves of minus log concentration of Al^b and Al^c *vs.* pH with appropriate time indicated again show that Al^b decreases in concentration with time and Al^c increases.

As a corollary to constant Al^a, it should be noted that $Al^b + Al^c$ concentration stays constant with aging time. Equilibrium is reached in the systems when the Al^c and $Al^b + Al^c$ lines reach the dotted line indicating the amount of crystalline gibbsite that should be present at equilibrium as a function of pH. After a few days aging, the equilibrium pH for a system can be accurately predicted by extrapolating the $Al^b + Al^c$ curve, which appears to be usually a straight line parallel to the pH axis, to its intersection with the dotted gibbsite concentration curve.

Al^a appears to be composed of monomeric species, and its total concentration is approximately constant after about 23 hours' aging. Al^c, from both analysis and aging data and from electron microscope observations, appears to be composed of clearly solid, colloidal, particles, and its concentration increases as a function of aging time until equilibrium is achieved. Al^b is present in all the solutions at 23 hours' aging and often in considerable concentrations. Its concentration then drops as a function of aging time until it apparently disappears from solution when equilibrium is reached. It is obviously not a stable material and is ultimately converted to Al^c material. It reacts with ferron in the analytical procedure according to a first order rate law which suggests that all aluminum atoms in the material are individually reacting with ferron in a way that does not change as the Al^b structures react and diminish in size in the procedure. This would indicate that the structures formed of Al^b cannot be too large or else we would expect more surface control of the reaction and a rate law of order less than one. However, since it takes ferron a finite length of time to break down the Al^b structures as compared with the ferron reaction with Al^a, it would appear that the structures are rather strongly bound together. A structure of limited size (perhaps containing 20–100 aluminum atoms, as estimated from a study of the pH drop in the solutions as a function of time) composed of coalesced six-membered aluminum hydroxide rings as suggested by either Hem and Roberson or Hsu and Bates would appear reasonable.

The disappearance of Al^b from solution during aging (and its apparent conversion to Al^c) can be noted in Figure 9 which is a plot of minus log concentration Al^b as a function of aging time for solutions D, E, F, G, and H. Except for solution E, which behaved as if the reaction had been blocked for a time, the curves obtained appear to be composed of three parts. The first part, for up to 20–30 days aging, consists of a nonlinear but nearly flat slope region. During this time, judging from the

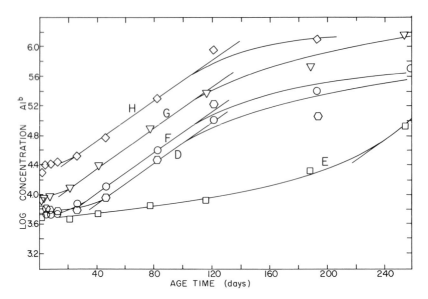

Figure 9. Minus log concentration Al^b as a function of aging time for solutions D, E, F, G, and H

pH drop of the solutions, Al^b particles are growing in size, but relatively few Al^b particles are being converted to Al^c particles.

In the time interval between 20–30 and 120–200 days' aging, there is a steeper slope to the curves and a roughly straight line relation between minus log concentration Al^b and aging time is to be noted. Thus, in this region an approximate first order rate law is obeyed relative to the conversion of Al^b to Al^c (and accompanying reaction of individual Al^b atoms with H_2O with release of H^+ to the solution).

At aging times greater than 120–200 days, the slopes of the curves become nonlinear and flatten. One explanation for this behavior would be that as equilibrium is approached there are relatively few Al^b particles left and the rate-determining step for conversion to Al^c changes to one involving chance for encounter among Al^b particles.

Electron Microscope Observations

Al^c particles are of colloidal size and can be observed under an electron microscope. Figures 10, 11, and 12 show electron micrographs of Al^c from the solutions. X-ray diffraction work has indicated that this material (the hexagonal platelets) is gibbsite (8, 30). Also shown on the micrographs are gold sol particles (small, dark spheres and tetrahedrons).

These figures show much of the detail of the particles, such as their typical crystalline shape and imperfections. The gold sol particles as prepared in this work are negatively charged (*31, 32*). It is interesting to note that these particles tend to adsorb on gibbsite edges, corners, and imperfection points, particularly at the latter two types of sites. The suggestion is, then, that these sites are the most highly positive sites on the gibbsite surface.

The electron micrograph of Figure 10 is of particular interest since it apparently shows edges (rectangular shapes) as well as faces and indicates the relative thickness of the particles. Also, it can be seen that the particles appeared layered much in the manner of microscopic mica particles. The electron micrograph of Figure 11, which is of material from solution C, indicates that microcrystalline gibbsite is ultimately formed even if r_n value is as low as 0.94 and pH is near 4.

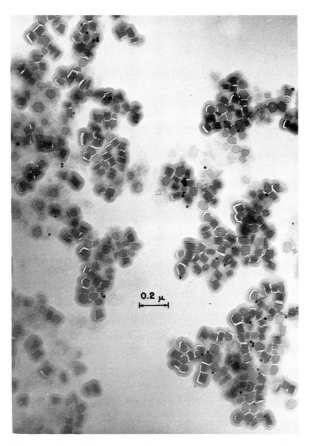

Figure 10.   Electron micrograph of microcrystalline gibbsite from solution F with negative gold particles adsorbed

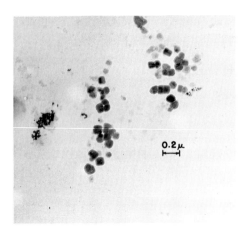

Figure 11. Electron micrograph of microcrystalline gibbsite from solution C with negative gold particles adsorbed

Thus, the electron micrographs show conclusively that colloidal material is formed in the aging study solutions. Further, this material appears to be microcrystalline gibbsite.

Effect of Rate of Addition of Base on Quantity of Al^b Initially Formed

It would seem possible that the rate of adding base in the original mixing of the aging study solutions might in some manner affect the amount of Al^b (and also, therefore, Al^c) initially formed in the aging test solutions. An experiment was performed taking different times to add base in making up an H solution. Results of the experiment are shown in Figures 13 and 14. In all cases, aging time was seven days. Al^a appears independent of rate of base addition but Al^b decreases with increasing speed of addition. The slower the base addition, the higher the solution pH and, thus, the more out of equilibrium is the solution.

Rapid addition probably promotes formation of solid Al(OH)$_3$ particles, possibly originally of a partly amorphous character. These particles appear to start organizing themselves soon into crystalline form (into gibbsite under our experimental conditions). Slow addition gives less opportunity for local excess of base and formation of amorphous solid Al(OH)$_3$. With slower addition of base, more monomeric and polynuclear species are formed initially. The monomeric species very rapidly convert to polynuclear microparticles or macroions in roughly the manner outlined by Hem and Roberson (8). This conversion stops as soon as there are sufficient microparticles present to bring into being the particle size effect. The particle size effect has to do with very small particles being more soluble than larger ones. These microparticles then slowly grow in size with time until they reach the size and character of true crystalline gibbsite particles. The nature of the analysis curves (such as those shown on Figure 2) indicates, however, that there apparently is a distinct separation or "jump" between the polynuclear macroions or microparticles and crystalline gibbsite particles.

That is, apparently, the Al^b polynuclear particles grow in size by combining and "splitting out" protons. If the particle is smaller than some certain size, it reacts in a moderately fast manner with ferron (over a period of a couple of hours) according to a first order rate relative to the Al^b concentration. After a certain size is achieved, the reaction with ferron becomes very much slower, and the rate law approaches zero order. This indicates a rather profound structural change in the particles upon reaching a certain size.

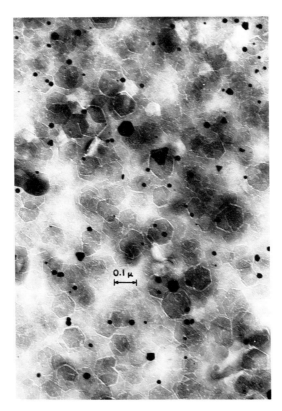

Figure 12. Electron micrograph of microcrystalline gibbsite from a J solution with negative gold particles adsorbed

As soon as a solution is prepared, several things start happening. One is probably a change in structure of rapidly formed $Al(OH)_3$ particles from a more or less amorphous state to crystalline gibbsite. The change is probably complete after a few weeks aging. Secondary evidence from electron micrographs of solutions aged a few months help confirm this statement. The other is the slow growth of polynuclear Al^b particles or macroions accompanied by release of protons from water molecules present in some octahedral positions. Ultimately, the particles grow to such a size that they must be considered crystalline gibbsite. Perhaps these particles also combine with gibbsite particles formed from the rapidly formed $Al(OH)_3$.

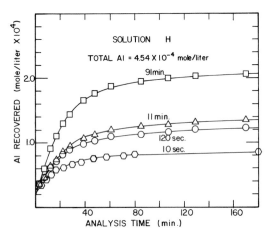

Figure 13. Effect of length of time of adding base in preparing solutions on amount of Al^b formed; Al recovered as a function of analysis time

Al^b Structure and pH Drop

If it is assumed at 23 hours' aging that the average Al^b particle size and makeup is somewhere between $Al_{24}(OH)_{60}{}^{12+}$ and $Al_{96}(OH)_{264}{}^{24+}$— *i.e.*, the OH/Al value for this material is between 2.5 and 2.75—and that the Al^c particles after 254–259 days' aging have the composition $Al(OH)_3$, it is possible to calculate how much H^+ ion concentration should change during the period between 23 hours and 254–259 days. The H^+ ion change can then be compared with actual measured H^+ ion change. This comparison has been made for solutions D through H.

Table V lists results obtained assuming several different average Al^b particle sizes at 23 hours' aging.

Table V.   Actual *vs.* Calculated H⁺ Ion Concentration Changes During the Aging of Solutions from 23 Hours to 254–259 Days (Mole/Liter × 10⁵)

Solution	*Actual Change*	*Calculated Change, Assuming 23-Hour Size to be*		
		$Al_{24}(OH)_{60}^{12+}$	$Al_{45}(OH)_{144}^{18+}$	$Al_{96}(OH)_{264}^{24+}$
D	5.3	7.8	5.2[a]	3.9
E	3.0	9.9	6.6	4.9[a]
F	5.0	10.2	6.8	5.0[a]
G	4.1	5.9	4.0[a]	3.0
H	3.2	2.5[a]	1.7	1.3

[a] Best fit.

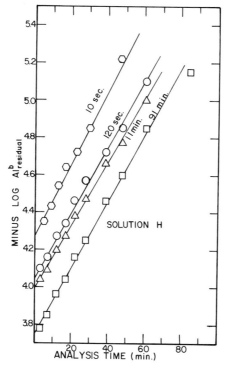

Figure 14.   First order rate plot of Al^b data from Figure 13

Summary on the Approach of the Solutions to Equilibrium

The slower base is added in aging solution preparation, the more Al^a and Al^b material is initially formed. Al^a formed probably starts quickly converting to Al^b material. However, conversion of monomeric material to Al^b is stopped before equilibrium is achieved as soon as sufficient polynuclear material is present to bring the particle size effect into being.

Thus, at any particular age time after 23 hours' aging but prior to the achievement of equilibrium, a solution is held at supersaturation with respect to crystalline $Al(OH)_3$ and in a metastable condition by the particle size effect. Yet slowly the systems head toward equilibrium by growth in size of Al^b material and its conversion to Al^c particles.

Since the concentration of Al^a remains constant, it appears that the monomeric species do not directly contribute to the progress of the reaction although they should adjust their relative amounts according to pH change in the systems. Al^b decreases in amount, Al^c increases, and pH drops with time. Individual Al^b particles must be combining by the bonding of adjacent Al atoms in separate Al^b particles.

All Al atoms are surrounded by some combination of six H_2O molecules plus OH ions. The smaller the Al^b particle, the greater percentage of these coordinated groups will be water molecules. For example, for a single six-membered ring there will be, at least potentially, two water molecules per aluminum. For a structure containing 54 Al ions, there should be about 0.667 water molecules per aluminum. As aluminum atoms, each coordinated with at least one H_2O molecule, in different particles approach each other, they bond together by deprotonation of the coordinated water molecule. As the deprotonation takes place, the pH of the solution drops. There is of course a charged repulsion barrier to the reaction and the reaction should take place more readily at the higher pH values where the individual particles should be less highly charged. However, there is abundant evidence from the present work that such a reaction does take place down to pH values near 4. Ultimately, the Al^b particles coalesce to Al^c particles.

It is curious that for a solution made up to have a particular r_n value one always finds the solution to contain a particular set concentration of Al^a independently of how rapidly the solution was mixed and how long the solution has aged.

Conclusions Relative to Aluminum (III) in Mildly Acid Aqueous Media

From the aluminum hydrolysis work and electron microscopy, several conclusions can be reached relative to the behavior of aluminum in acid aqueous systems. These conclusions are:

1) If aluminum perchlorate solutions are made up to contain 4.54 $\times$ 10^{-4} moles/liter aluminum (total ionic strength about 10^{-2} molar), provided the nominal ratio of OH_{bound}/Al (r_n value) is 3.00 or less, there will initially be present in the system three different types of aluminum which can be designated as Al^a, Al^b, and Al^c. Al^a appears to be composed en-

tirely of monomeric species—*i.e.*, Al^{3+}, $AlOH^{2+}$, $Al(OH)_2^+$, and $Al(OH)_4^-$. Al^b appears to consist of polynuclear aluminum hydroxide species, probably of a general six-membered ring structure in which each aluminum is bonded to its neighbor through shared pairs of OH ions. The individual rings tend to coalesce into larger structures with time until they ultimately become large enough to be filtered out and identified by electron microscopy and x-ray diffraction as gibbsite crystals. The manner in which the rings coalesce appears to be governed by a first order rate law relative to the Al^b material. The Al^b particles appear to range in size from around $Al_{24}(OH)_{60}^{12+}$ to $Al_{96}(OH)_{264}^{24+}$ and perhaps larger. The Al^c is solid material that may be initially all or partly amorphous, but rapidly becomes crystalline and takes on the structure of gibbsite.

2) For a particular r_n value, the amount of Al^a is nearly constant, at least after 23 hours' aging, independent of how long the solution has aged and the rate at which base was added to the solution in the initial solution preparation. A corollary to this statement is that the amount of Al^b plus Al^c also is constant. The significance of these statements is great, for they indicate that the Gibbs free energy of formation of Al^b is not much less than solid Al^c. Although the concentration of Al^a remains constant with aging time, the activities of the individual monomeric species adjust themselves almost instantly to change in pH of the solution.

3) The pH of the solutions decrease with aging time until equilibrium pH values are ultimately achieved. The equilibrium pH values are consistent with known thermodynamic values for monomeric species and solubility products. The equilibrium pH achieved in a particular solution depends on its initial r_n value. The lower the r_n value, the lower is the equilibrium pH. The rate at which equilibrium is approached depends on the r_n value and also on the rate at which base is added during solution make-up. The lower the r_n value, the slower is the progress toward equilibrium. For example, if r_n value is around 2.8, equilibrium may be achieved in 6 or 8 months (depending on how much Al^b was initially formed during base addition). If r_n value is 1.0 or less, equilibrium probably will not be achieved after several years' aging.

4) The amount of Al^b present in solution at 23 hours' aging depends on rate of base addition during solution preparation. The slower the base is added, the greater the quantity of Al^a plus Al^b formed and the equivalent less Al^c initially formed. Some of the Al^a formed is rapidly converted into Al^b, but this conversion is limited by the particle size solubility effect. Thus, as soon as enough Al^b is formed to bring this effect into operation, the concentration of Al^a remains constant. The amount of Al^b present then decreases as a function of aging time. Over the time interval from about 20–30 days aging to about 120–200 days aging, the disappearance of Al^b in many of the solutions is approximately first order

with respect to concentration of Al^b. As Al^b disappears from the system, an equivalent amount of Al^c is formed. It seems that Al^b particles are coalescing into larger particles as a function of time. At some critical size, the character of the Al^b particles appears to change and they behave as the less reactive Al^c particles. After equilibrium is achieved, the quantity of Al^b present is below the detection limit.

Acknowledgment

This research was performed at the United States Geological Survey's Menlo Park station. The author wishes to thank J. D. Hem of the Geological Survey for providing the original stimulus for this work and for his many direct contributions to it.

Literature Cited

(1) Aveston, J., *J. Chem. Soc. (London)* (1965) **III**, 4438.
(2) Brosset, C., *Acta Chem. Scand.* (1952) **6**, 910.
(3) Brosset, C., Biedermann, G., Sillén, L. G., *Acta Chem. Scand.* (1954) **8**, 1917.
(4) Deltombe, E., Pourbaix, M., *Corrosion* (1956) **14**, 496.
(5) Frink, C. R., Peech, M., *Inorg. Chem.* (1963) **2**, 473.
(6) Fripiat, J. J., Van Cauwelaert, F., Bosmans, H., *J. Phys. Chem.* (1965) **69**, 2458.
(7) Gayer, K. H., Thompson, L. C., Zajicek, O. T., *Can. J. Chem.* (1958) **36**, 1268.
(8) Hem, J. D., Roberson, C. E., *U. S. Geol. Surv. Water Supply Papers* **1827-A** (1967).
(9) Hsu, P. H., *Soil Sci. Soc. Am. Proc.* (1966) **30**, 173.
(10) Hsu, P. H., *Soil Sci.* (1967) **103**, 101.
(11) Hsu, P. H., Bates, T. E., *Soil Sci. Soc. Am. Proc.* (1964) **28**, 763.
(12) Hsu, P. H., Bates, T. E., *Mineral Mag. (London)* (1964) **33**, 749.
(13) Kittrick, J. A., *Soil Sci. Soc. Am. Proc.* (1966) **30**, 595.
(14) Matijevic, E., Janauer, G. E., Kerker, M., *J. Colloid Sci.* (1964) **19**, 333.
(15) Matijevic, E., Mathai, K. G., Ottewill, R. H., Kerker, M., *J. Phys. Chem.* (1961) **65**, 826.
(16) Matijevic, E., Stryker, L. J., *J. Colloid Interface Sci.* (1965) **22**, 68.
(17) Raupach, M., *Australian J. Soil Res.* (1963) **1**, 28.
(18) *Ibid.*, (1963) **1**, 36.
(19) *Ibid.*, (1963) **1**, 55.
(20) Reesman, A. L., Pickett, E. E., Keller, W. D., *Am. J. Sci.* (1969) **267**, 99.
(21) Schofield, R. K., Taylor, A. W., *J. Chem. Soc. (London)* (1954) 4445.
(22) Turner, R. C., *Can. J. Chem.* (1969) **47**, 2521.
(23) Turner, R. C., Ross, G. J., *Can. J. Soil Sci.* (1964) **49**, 389.
(24) Turner, R. C., Ross, G. J., *Can. J. Chem.* (1969) **48**, 723.
(25) Sillén, L. G., *Quart. Rev.* (1959) **13**, 146.
(26) Johansson, G., *Acta Chem. Scand.* (1960) **14**, 771.
(27) Davenport, W. H., Jr., *Anal. Chem.* (1949) **21**, 710.
(28) Rainwater, F. H., Thatcher, L. L., *U.S. Geol. Surv. Water Supply Papers* (1960) **1454**, 97.
(29) Latimer, W. M., "Oxidation Potentials," 2nd ed., p. 282, Prentice-Hall, Englewood Cliffs, N. J., 1952.

(30) Schoen, R., Roberson, C. E., *Am. Mineralogist* (1970) **55**, 43.
(31) Thiessen, P. A., Z. *Elecktrochem.* (1942) **48**, 675.
(32) Thiessen, P. A., Z. *Anorg. Chemie* (1947) **253**, 161.

RECEIVED April 7, 1970.

11

Plutonium in the Water Environment. II. Sorption of Aqueous Plutonium on Silica Surfaces

THOMAS C. ROZZELL and JULIAN B. ANDELMAN

Graduate School of Public Health, University of Pittsburgh, Pittsburgh, Pa. 15213

The sorption and desorption of aqueous plutonium in the range of 10^{-7} to 10^{-8}M was studied on quartz and other silica surfaces. Sorption continued typically for 12 to 15 days before apparent equilibrium was reached, and the distribution of plutonium particle sizes sorbed on the silica was different from that in solution. At pH 7, sorption increased with increasing ionic strength, but decreased when bicarbonate was added. The amount of sorption varied at pH 5 and 7, but differently at high and low ionic strengths, as well as with the age of the solution. Plutonium desorption indicated that there were two basically different sorbed species, and the rate and quantity of desorbed material increased at pH 5 compared with 7 and 9.

In the first paper of this series (*1*), the chemistry of aqueous plutonium was reviewed, with particular reference to the ambient conditions likely to be encountered in natural waters. In addition, experimental work was presented concerning the effects of such variables as pH, plutonium concentration, ionic strength, and the presence of complexing agents on the particle size distribution of aqueous plutonium. It was shown that, at least over an aging period of several days and with pH varied from 5.4 to 8.2, there was an increase in the particle sizes of plutonium in the measured range of 0.07 to 1.2μ (equivalent spherical diameter). Increasing the ionic strength from 0.002 to 0.1M at pH 7 also significantly increased the sizes of the particles. Addition of 10^{-2}M bicarbonate had a generally similar effect, but this was influenced by the ionic strength. The presence of granular silica with particles in the range of 280 to 390μ also significantly affected the plutonium particle size distribution while the solutions were aging for several days at pH 7. At an ionic strength of 0.012M, such additions of silica greatly increased the sizes of the plutonium particles, but decreased them at 0.11M.

This paper is concerned principally with the interaction of aqueous plutonium with silica surfaces. In considering the sorption of plutonium onto a solid surface, such as silica, in an aqueous solution, consideration must be given to the mechanism of interaction. In the silica–Pu(IV) system studied here, the interaction is essentially that of small negatively charged colloidal size particles with a large negative planar surface, as well as ion exchange of positive ions and low-molecular-weight hydrolysis products. In order for the various species to interact with the silica particle surface, they must reach this surface by diffusion through the liquid film adhering to the silica particle. The rate of sorption then should involve and may be controlled by mass transfer through the solution up to the surface.

There have been several laboratory and field studies concerned with the uptake of aqueous plutonium by plants, marine biota, soils, minerals, and glass. These have been discussed in the first paper of this series (1), which shows that several solution variables, as they influence the particle size distribution of the aqueous plutonium, greatly affect its interaction with silica surfaces. Studies were conducted to determine the rates, equilibria, and mechanism of the sorption and desorption of aqueous, colloidal plutonium-239 onto the surfaces of quartz silica. The orientation of these studies is the understanding of the likely behavior and fate of plutonium in environmental waters, particularly as related to its interaction with suspended and bottom sediments.

Experimental

All chemicals, such as sodium hydroxide, nitric acid, sodium perchlorate, and those comprising the buffers, were certified reagent grade. The water used in preparation of the chemicals was demineralized and distilled by passing it through a mixed-bed ion exchanger. Water used to prepare the plutonium solutions was also filtered through a fine-porosity Whatman filter to remove any foreign matter such as dust particles.

The source and purity of the plutonium (IV) stock solution and the method of dilution and pH adjustment were described in a previous paper (1). The composition of the plutonium solutions was such that approximately 95.4% of all alpha activity was owing to ^{239}Pu. The only other significant alpha-active isotope present was ^{240}Pu which contributed approximately 3.5% of the alpha activity. ^{238}Pu and ^{241}Am contributed less than 0.6% each to the total alpha activity.

The plutonium concentration in the ambient sorbing solutions was determined by alpha counting an evaporated aliquot in a gas flow internal proportional counter, also previously described (1). The amount of plutonium sorbed by silica was determined by dissolving it from a silica aliquot (∼0.2 gram) with 1.0 ml of concentrated nitric acid. The silica was first washed with a small amount of the blank ambient solution to remove plutonium in the adhering liquid film. The washing was carried out on a paper filter in a filter chimney with vacuum applied. Experimentation showed that the washing time and volume was low enough that there was no loss of sorbed Pu during the procedure. After drying, an aliquot of the silica was weighed and placed in a small beaker to which the nitric acid was added. The milliliter of nitric acid removed 100% of the Pu sorbed to the silica grains. Two planchets were prepared for each sample, and each received a few drops of distilled water and 0.25 ml of the nitric acid containing the dissolved plutonium from the sample. The planchet was then dried and alpha counted. In this manner, a quantitative determination of the amount of plutonium sorbed per gram of silica could be calculated. The plutonium activity per square centimeter was calculated from the specific surface area of the silica.

The precision of the plutonium analysis was determined and the over-all error at 2σ was ±4.5% for 10 determinations. The 2σ error owing to the counting alone was 0.93%.

Granular silica (crystalline quartz) and transparent silica plates (1 × 3 inches) were used for the sorption studies. The granular silica was supplied by the Ottawa Silica Co. and the Fisher Scientific Co. The silica was sized by dry-sieving with U. S. Standard sieves. In most cases, as narrow a range as possible was chosen, and the mean diameters, except where size was a variable, ranged from 275 to 387 microns. The transparent silica plates were fused silica with a smooth amorphous surface.

The granular silica was first washed several times with demineralized water to remove unwanted fine sizes and debris, and then oven-dried at approximately 120°C. A weighed portion, usually 4 grams, was placed in a 125-ml flask to which was added 80 ml of the plutonium solution being studied. This gave a constant volume-to-mass ratio of 20:1. The flasks were stoppered and placed on reciprocating shakers. Studies of the effect of the shaking rate showed that above 125 opm (oscillations per minute) sorption increased sharply until a speed of 200 opm was attained, above which there was a marked decrease in sorption. A 200-opm operating speed was chosen for all experiments.

All plutonium solutions, except where pH was a variable, were buffered at pH 7 with phosphate. The final concentration of the phosphate, as KH_2PO_4, was typically 0.001M. Experiments on the effect of phosphate on plutonium sorption indicated that there should be no increase in sorption resulting from formation of a plutonium–phosphate complex at this concentration. All experiments, except where temperature was a variable, were carried out at room temperature, which was usually between 24° and 26°C.

In one sorption experiment, the particle size distribution of the aqueous plutonium was determined by the centrifugation technique previously described (1). Simultaneously, a study was made of the size distribution of plutonium sorbed onto silica plates by an autoradiographic method (1, 2, 3). After the plutonium sorbed on the silica plates, the latter were dried and clamped for 2 to 4 days to glass plates coated with Kodak NTA nuclear emulsion. After developing the emulsion, the resulting alpha tracks were counted microscopically at 400× magnification. The number of centers with 1 to 28 tracks were tabulated for several 0.25-cm² areas. Also tabulated as a group were centers having more than 28 tracks and transparent cores, and those with opaque cores having a very large number (>300) tracks. There was a tabulation as well of nonspherical aggregates which, in most cases, had too many overlapping tracks to count. The size of each sorbed plutonium particle was determined by the count of the number of tracks emitted from a common center in the manner developed by Leary (2). The method takes into account the exposure time, and corrections were made for counting statistics, the details being reported elsewhere (4).

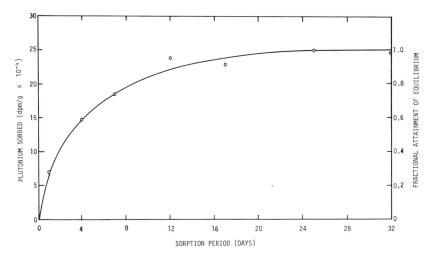

Figure 1.   Rate of sorption of aged 4 × 10⁻⁷ plutonium on 230-μ silica grains at pH 7 and 0.1M ionic strength

Results

Sorption Rate. The rate of sorption of the Pu radiocolloid is typified by results shown in Figure 1. In this case, a $4 \times 10^{-7}M$ Pu solution at a pH of 7 and ionic strength of 0.1M was sorbed onto silica particles having a nominal diameter of 230 microns. Characteristically, the time required to attain 50% of equilibrium was 2 to 5 days. In all experiments, the Pu concentration in the solution was monitored, as well as that sorbed.

The plutonium colloidal aggregates vary in size with age, becoming larger, and thus more filterable, with stirring and age (5). The effect of aging of the plutonium colloid was, therefore, studied in relation to its sorption onto silica. A large master solution of plutonium ($10^{-7}M$) with a pH of 7 was prepared. At times ranging from 0–16 days, aliquots were :aken and added to silica for sorption. The master solution was stirred vhile aging. The amount of sorbed plutonium (in dpm per gram of

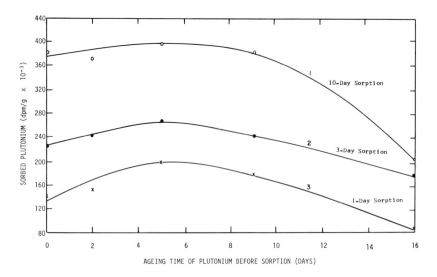

*Figure 2.   Sorption of 10⁻⁷M plutonium at pH 7 on granular silica as a
function of solution age prior to sorption*

Curve 1, 10-day sorption; Curve 2, 3-day sorption; Curve 3, 1-day sorption

silica) is shown in Figure 2 for 1, 3, and 10 days of sorption for solutions which had aged for the times shown (*i.e.*, 0, 2, 5, 9, and 16 days)
before the silica was introduced. The rate of sorption increased with
age of the solution (prior to sorption) up to 5 days and decreased there-
after.

Effect of Ionic Strength and pH on Sorption. The major contribution to the ionic strength, at low values of the latter, came from the
sodium hydroxide used to titrate the acid stock solutions, as well as the
monopotassium phosphate and sodium hydroxide in the buffer used to
maintain the pH at 7. In order to study the effect of ionic strength on
Pu sorption, three pH 7 solutions were prepared and the ionic strengths
adjusted to 0.002, 0.01, and 0.1M. The latter two solutions contained
added $NaClO_4$, which was used to maintain the ionic strength and was
the major contributor to the ionic strength. Each of these solutions was
aged for 1 day ("fresh" solution) or 58 days ("aged" solution) prior to
sorption onto silica for 21 to 24 days, the results being shown in Table I.

The effect of pH on sorption was studied over the pH range of 3 to
8.7 and at an ionic strength of 0.23M. In this case, there was a general
increase in sorption with increasing pH. On the other hand, when the
ionic strength was kept low ($\mu = 0.01M$), there was a decrease in sorption
with increasing pH. The details of this phenomenon were studied further at pH 5 and 7.1 at ionic strengths of 0.004 and 0.23M. Sorption onto
silica grains was carried out for a period of 6 days. The results are shown

Table I.   Plutonium Sorbed and K_s Values for Fresh and Aged Solutions of Varying Ionic Strength at pH 7

Ionic Strength (μ), M	Fresh (1-Day) Solution[a]		Aged (58 Days) Solution[b]	
	Pu Surface Activity, dpm/g, $\times$ 10^{-4}	K_s,[c] Cm	Pu Surface Activity, dpm/g, $\times$ 10^{-4}	K_s,[c] Cm
0.002	8.0	0.04	8.2	0.04
0.01	16.5	0.11	9.4	0.07
0.1	43.5	0.67	28.6	0.15

[a] Values based on 24-day sorption period.
[b] Values based on 21-day sorption period.
[c] K_s is the equilibrium concentration of sorbed Pu in dpm/cm² divided by the equilibrium concentration in the liquid in dpm/cc.

in Table II. Indeed, at the lower ionic strength ($\mu = 0.004M$), there was considerably greater sorption at pH 5 than at pH 7.1. At $\mu = 0.23M$, the opposite occurred.

The Effect of Silica Particle Size. As the particle size of the mineral is decreased, the surface area per unit weight varies inversely with the diameter. Generally, it is expected that sorption will be in direct proportion to the available surface area (6). Using four different size fractions of the same silica, with nominal diameters, $\bar{d}$, ranging from 1.4 $\times$ 10^{-2} cm, the sorption rates and equilibria were studied for a fresh 5.3 $\times$ $10^{-7}M$ Pu solution. The ionic strength of the solution was 0.1M and the pH was 7. Four grams of silica was used in each case. The rate data for the different size fractions are shown in Table III. Even though the total silica surface area available for sorption was quite different for each size fraction, the total amount of Pu removed from the solution by sorption onto the silica was not very different for each fraction. Similarly, the amount remaining in solution at equilibrium was essentially the same in each case.

In this and in all sorption studies utilizing a containing vessel, consideration must be given to sorption onto the wall of the vessel. The interior surface area of the 125-ml flasks used in these studies was approximately 113 cm². This area need not be considered in experiments where the surface area of the silica is constant, for then the ratio of the area of the wall to the area of the silica is constant. In this case, how-

Table II. Sorption of Plutonium onto Silica in High and Low Ionic Strength Solutions at pH 5 and 7.1

pH	Pu Sorption (dpm/g) $\times$ 10^{-4}		K_s, Cm	
	$\mu = 0.004M$	$\mu = 0.23M$	$\mu = 0.004M$	$\mu = 0.23M$
5.0	7.6	0.5	0.2	0.05
7.1	2.4	11.0	0.02	0.52

Table III. Sorption of Plutonium by Different Size Silica Particles

Sorption Period, Days	Pu Sorption, $dpm/g \times 10^{-5}$			
	A	B	C	D
1	2.27	1.90	1.24	1.11
3	2.39	2.10	1.85	1.48
7	2.81	2.93	2.60	2.36
12	3.23	3.45	3.12	2.83
19	3.21	3.64	3.24	2.94
$\bar{d}$, cm	1.4×10^{-2}	1.9×10^{-2}	2.7×10^{-2}	3.9×10^{-2}
Specific surface area, cm^2/g	165	118	82	56
Equilibrium amt. of Pu in soln., cpm/ml	1900	2600	2000	2100
$\dfrac{\text{Wall area}}{\text{Silica area}}$	0.17	0.24	0.35	0.5

ever, this ratio varied, and as the particle size of the silica increased, the fraction of the total surface represented by the wall increased. The ratio of wall area to total silica area (specific surface area times 4 grams) is given in Table III. There was apparently no significant difference in the amount of influence exerted by the wall area in each of the four flasks.

Effect of Temperature. The effect of the temperature of the ambient solution on the rate of Pu sorption onto silica was studied by sorbing for 1.5 hours at temperatures of 5°, 26.5°, and 40°C. The initial concentration of the Pu solution was $4.9 \times 10^{-7}M$, with an ionic strength of $0.1M$ and a pH of 7. The quantity of Pu sorbed was 2.6×10^4, 3.1×10^4, and 2.8×10^4 dpm/gram at 5°, 26°, and 40°C, respectively.

Effect of Pu Concentration on Sorption. The dependence of sorption on the Pu equilibrium concentration was studied for fresh and aged (7 days) solutions. Six solutions of different concentrations were used for each study. All solutions were at pH 7, had ionic strengths of $0.1M$, and were maintained at 25° ± 1°C. The relationships between the Pu equilibrium concentration at 18 days and the amount of Pu sorbed are shown in Figure 3.

Many empirical equations have been proposed to describe sorption behavior of aqueous species on solid surfaces. The most conventional methods of plotting sorption data are the Freundlich and Langmuir isotherms (7). Both isotherms were developed for the sorption of gases by solid surfaces. However, they often, but not always, describe the sorption of nonelectrolytes and other species from liquid solution onto solid surfaces. The attempts to fit the data for the fresh and aged solutions to the Freundlich isotherm did not result in straight lines, and it was concluded that, within the concentration range studied, this isotherm does not adequately describe the Pu–silica system.

The Langmuir isotherm may be expressed in the form

$$A = KC/(1 + K'C) \qquad (1)$$

where A is the amount of Pu sorbed per gram of silica, and C is the concentration of the Pu in the liquid (in moles/liter) at equilibrium. K and K' are constants for the particular system under study. For the sorption systems shown in Figure 3, Langmuir plots were made using Equation 1; these are shown in Figure 4. Sorption from the fresh solution followed the Langmuir isotherm quite well. However, for the aged solution, the plot was linear only in the region shown. For the lower concentrations (larger values of $1/C$), the deviation was considerable.

Effect of Bicarbonate on Sorption. The bicarbonate and carbonate ions are common species in natural waters and have the ability to form complexes with Pu(IV). Starik (8) mentions that in an investigation of the adsorption of uranium there was a decrease in the adsorption after reaching a maximum, which was explained by the formation of negative carbonate complexes. Kurbatov and coworkers (9) found that increasing the bicarbonate ion concentration in UX_1 (thorium) solution decreased that amount of thorium which formed a colloid and became

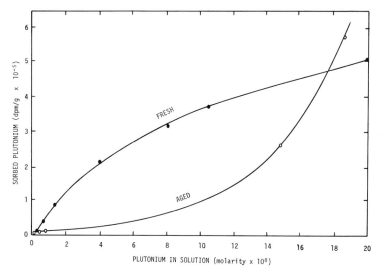

Figure 3.   Sorption isotherms on granular silica for fresh and aged (7 days) plutonium solutions at pH 7 and 0.1M ionic strength

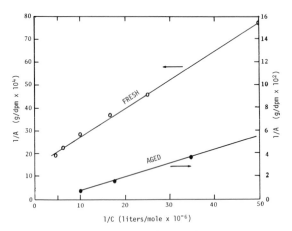

Figure 4. Langmuir isotherm plots for sorption of fresh and aged plutonium solutions; data correspond to Figure 3

filterable. This was attributed to the formation of a soluble complex with the bicarbonate.

In order to determine if bicarbonate may similarly affect colloidal plutonium, a comparison was made of sorption from solutions containing bicarbonate only from atmospheric CO_2 and from those with added $10^{-2}M$ bicarbonate. Two different ionic strengths were used in order to distinguish adequately between and rule out the possibility of an ionic strength effect. The results of the study are given in Table IV. When the bicarbonate concentration was increased from the equilibrium atmospheric value to $10^{-2}M$, the sorption coefficient greatly decreased at both ionic strengths.

Plutonium Particle Size Distributions. Pu particle size distributions were determined for an aqueous system sorbing onto silica plates. The initial Pu concentration was $1.5 \times 10^{-7}M$, with a pH of 7.1 and ionic strength of $0.03M$. Three silica plates were immersed in the solution and removed after sorption periods of 1, 5, and 17 days, respectively. When each plate was removed, aliquots of the solution were taken. Pu particle

Table IV. Effect of Bicarbonate on the Sorption of Plutonium[a]

Ionic Strength (μ), M	HCO_3^- Conc.	Pu Surface Act., dpm/cm^2, $\times 10^{-3}$	K_s, cm
0.112	Atmospheric	6.0	1.17
0.112	$10^{-2}M$	4.4	0.44
0.012	Atmospheric	1.5	0.08
0.012	$10^{-2}M$	0.8	0.03

[a] 13-Day sorption period; pH $= 7$.

size distributions were determined for these solutions by the centrifugation technique previously described (*1*), and for the Pu sorbed on the
plates by autoradiography. The results of these Pu size distributions are
plotted in Figure 5. It is apparent that the distributions in solution and
on the plates are quite different.

Effect of Phosphate. The pH 7 buffer solution used to maintain the
pH of the Pu(IV) solutions contained KH_2PO_4. Because phosphate ions
are known to complex Pu(IV) (*10*), a study was conducted to determine
if adjustment of the phosphate concentration would alter the nature of
the colloidal species and thereby affect the sorption onto silica.

Varying amounts of KH_2PO_4 were added to a fresh Pu(IV) solution,
and sorption was carried out with 4 grams of silica in the usual manner.
At the end of the sorption period, the Pu(IV) concentration on the silica
was determined. The same procedure was used for a Pu(IV) solution
which was aged for 17 days prior to the addition of the phosphate and
the beginning of sorption. The results of both experiments are shown in
Table V. The ionic strengths of the solutions varied as the phosphate
concentration was increased, as shown in the table.

When KH_2PO_4 was added to a fresh Pu(IV) solution just prior to
sorption, the sorption increased as the KH_2PO_4 concentration increased
up to 0.1*M* (Table V). At 0.5*M* KH_2PO_4, there was a decrease in the
amount of Pu(IV) sorbed. According to Denotkina *et al.* (*11*), phosphate complexes should form in the order $PuHPO_4^{2+}$, $Pu(HPO_4)_2$, and

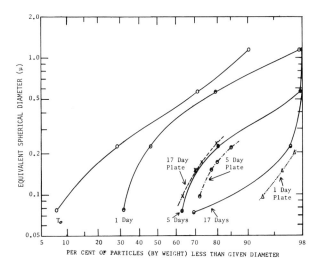

*Figure 5.   Particle-size distributions for plutonium re-
maining in solution and that sorbed on silica plates at
pH 7 and 0.03M ionic strength. The solid curves refer
to the solution distributions.*

Table V.   Effect of Phosphate on Pu(IV) Sorption

Conc. of KH_2PO_4, M/l	Ionic Strength, μ	Fresh Pu(IV) Solution,[a] K_s, Cm	Aged Pu(IV) Solution,[b] k_s, Cm
0.001	0.1	0.1	0.5
0.01	0.2	0.3	1.2
0.1	0.8	1.3	3.6
0.5	3.6	0.7	3.9

[a] 14-Day sorption period.
[b] 18-Day sorption period.

Pu(HPO_4)$_4^{4-}$ in 2M nitric acid. The formation of the phosphate complex, in effect, increases the average negative charge of the ionic or colloidal plutonium species. As in the case of the bicarbonate complex, this should reduce sorption onto the negatively charged silica surface. However, as is shown in Table V, the sorption increased. In another experiment in which the phosphate concentration was varied between zero and 1.25 × $10^{-2}M$, there was no significant difference in the sorption constants.

In this study, the ionic strength of the solutions increased with the phosphate concentration. Therefore, the increase in sorption that occurred is likely caused by the increase in ionic strength. The formation of the phosphate complex is not expected to take place readily at pH 7 owing to the lack of plutonium ions. When complexation does occur, it is a function of phosphate ion concentration (*11*), and the reduced sorption in the fresh solution at 0.5M KH_2PO_4 is probably an indication that the formation of the complex was significant at that concentration.

The effect was essentially the same in the aged Pu(IV) solution. There should have been a larger difference in the sorption of the 0.5M and the 0.1M KH_2PO_4 solution on the basis of the ionic strength effect. The fact that there wasn't leads to the conclusion that complexation did take place. The amount of complex formation would be expected to be less than in the fresh solution because of the removal of ionic species by coagulation during the aging period.

In the majority of the sorption experiments reported herein, the concentration of the phosphate was 0.001M. It is not expected then that the effect of the phosphate was significant in these studies.

Desorption of Plutonium from Silica. A study was made of the effect of pH on the rate of desorption by first sorbing Pu onto silica at pH 7, then desorbing into an "infinite" volume solution at pH's 5, 7, and 9. The "infinite" volume solution condition was approximated by removing the supernatant liquid at each sampling time and replenishing with fresh solution of the same pH. Thus, there was no significant buildup of the Pu concentration in the solution. The results, plotted in Figure 6, indicate that there is a significant increase in the rate of desorption at pH 5, compared with 7 and 9. It was previously observed that there can be significantly lower sorption at pH 5.

A second study dealt with the difference in the rates of desorption of Pu which was fresh and that which was aged when sorbed. Pu solutions of different ages sorb onto silica at different rates, as indicated in Figure 2. However, the rates of desorption of Pu which was fresh when sorbed and that which was 34 days old were not very different, as shown in Figure 7. The approximately exponential desorption curves may be rather precisely represented as reflecting two simultaneous first order desorption processes, probably indicating the presence of two different desorbing species. The curves can be divided into two components by extrapolating the nearly horizontal portions (which represent slow desorption of a tightly-held species) back to t_0 and subtracting this quantity so as to determine the rate of desorption of the 1st (or loosely-held) species. The equation for the 1st component of the curves is then

$$A_{t(1)} = A_{0(1)} \, e^{-\delta_1 t} \tag{2}$$

and that of the 2nd component

$$A_{t(2)} = A_{0(2)} \, e^{-\delta_2 t} \tag{3}$$

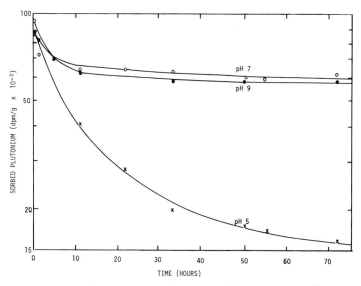

Figure 6.   Effect of pH on the rate of desorption of plutonium from granular silica into an "infinite" solution at an ionic strength of 0.01M

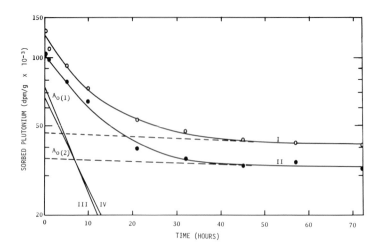

Figure 7. Rate of desorption of plutonium from silica into an "infinite" solution at pH 7 and 0.01M ionic strength

Curve 1, after sorption from fresh solution; Curve 2, after sorption from aged solution; Curves 3 and 4, desorption of "loosely" held plutonium from fresh and aged systems, respectively. See text for definitions of $A_{0(1)}$ and $A_{0(2)}$.

where $A_{t(1)}$ and $A_{t(2)}$ are the Pu activities on the silica surface (in dpm/gram) at time t. $A_{0(1)}$ and $A_{0(2)}$ are the activities on the silica surface at t_0 for species 1 and 2, respectively. δ_1 and δ_2 are the desorption rate constants for the two species in hr^{-1}. The equation for the general curve is then

$$A_t = A_{0(1)} e^{-\delta_1 t} + A_{0(2)} e^{-\delta_2 t} \qquad (4)$$

The half-time ($t_{1/2}$) for the 1st species desorbing is between 6.5 and 7.5 hours. While it is not possible accurately to determine $t_{1/2}$ for the second species from the graph, it may be calculated from the relation for a first order decay process

$$t_{1/2} = 0.693/\delta_2 \qquad (5)$$

and is equal to approximately 350 days. For both the fresh and aged Pu, the ratios of $A_{0(2)}$ to the initial total sorbed Pu are 0.3 and 0.35, respectively. Thus, approximately one-third of the sorbed Pu is strongly held on the silica and desorbs at a very slow rate.

Discussion

In considering the sorption of plutonium onto a solid surface, such as silica, in an aqueous solution, the mechanism of interaction is of particular interest. In the system studied here, the interactions are complex and include those of charged colloidal and larger size particles, low-molecular-weight plutonium hydrolysis products, and ion exchange of plutonium ions with negatively charged, heterogeneous silica surfaces.

The plutonium size distribution measurements made on aging solutions in the absence of silica reported in the previous paper (*1*) indicate that the plutonium hydrolysis products undergo slow coagulation as the suspension ages. In a slowly coagulating suspension, there is a continuous decrease in the fraction of primary particles, as well as changes in the relative amounts of larger particles. This phenomenon adds considerable complexity to the otherwise simple picture of a colloidal size particle diffusing to a solid surface. The 12 or more days required for sorption "equilibrium" to be reached, as shown in Figure 1, is in contrast to the considerably shorter periods of the order of several minutes required for ionic species to equilibrate with the surfaces of nonporous solids, as well as a few hours at most for monodispersed colloids, such as 0.3-mμ particles sorbing on similar planar surfaces (*12*). This slow "equilibration" rate for the plutonium system results from the fact that particulate species of varying size, as well as ionic and low-molecular-weight polymeric hydrolysis products, are diffusing to and sorbing on the surface. At the same time, the size distribution of such species in the ambient solution is changing. This is clearly shown in Figure 5, which indicates that as sorption proceeds an increasing fraction of the remaining plutonium species consists of smaller particles, while simultaneously increasingly larger particles are sorbed on the silica surfaces. Similarly, the effect of aging time prior to sorption, as shown in Figure 2, is indicative of the same kinds of effects. Here the rate of sorption varied with the age of the solution in a nonpredictable manner.

The sorption process and the attainment of apparent equilibrium may be regarded then as involving essentially two kinds of sorbing species. There are a very small number of ionic plutonium species, including monomeric and low-molecular-weight polymeric hydrolysis products (*1*) which sorb relatively quickly and perhaps are involved in a true equilibrium, such as by ion exchange with silanol sites at the silica surface. There is evidence of such sorption of various types of univalent and multivalent cations on silica, and both chemisorption and physical adsorption processes have been deduced (*13, 14, 15*). Filtration of the desorbing plutonium with a 15–40-micron porous silica disc indicated that the very first material to desorb was essentially small, unfilterable Pu(IV).

A second type of sorbing species is that which is essentially colloidal or even a somewhat larger size charged particle, whose sorption on the silica surface may be considered primarily in terms of a charged particle interacting with an essentially planar charged surface. As these particulate plutonium species age and increase in size, their charge-to-mass ratio may also decrease and thereby influence their sorption characteristics (5).

In order for the various species to interact with the silica particle surface, they must reach it by diffusion through the unstirred liquid film at the surface. The rate of sorption then should involve and may be controlled by mass transfer through the solution up to the surface (16). According to Vold (17), the major forces between colloid particles and a surface are long-range London–van der Waals forces of attraction and electrical double-layer interactions, the latter being repulsive for like sign and attractive for unlike sign. However, as Bierman (18) has demonstrated, even if the colloidal particle and the sorbing surface have the same charge, attraction can occur if the interacting substances are nonidentical. The more nonidentical, the greater will be the attractive force.

Kitchener and coworkers (12, 16) have shown that when colloidal size particles with the same charge sign as that of the planar surface diffuse to and sorb on the latter, the rate of uptake may be considerably less than that predicted by simple Fickian diffusion across the unstirred liquid film, principally because of the double layer repulsive forces, as treated in the Derjagin–Landau–Verwey–Overbeek theory of colloid stability. It was found in some systems, particularly at high ionic strength, that the sorption rate decreased with time, eventually becoming negligible.

In the plutonium–silica systems considered here, several sorption mechanisms are plausible and may in fact occur simultaneously. There is no doubt that the granular quartz silica has surface heterogeneities which could result in "favorable" sorbing sites, eventually becoming saturated as the plutonium ages and sorbs. At the same time, ionic and smaller colloid species are decreasing in number. The desorption experiments shown in Figure 7 which are indicative of two different types of sorbed plutonium, loosely and tightly held species, are an additional indication of different types of sorbing mechanisms.

This dual process for sorption of the colloidal species may reasonably account for the very slow attainment of what appears as equilibrium, as well as the different isotherms attained for the fresh and aged sorbing systems shown in Figure 3. In addition, ionic or charged low-molecular-weight species may sorb and equilibrate quickly, also contributing to the complexity of the system. The successful adherence of the fresh system to a Langmuir plot, as shown in Figure 4, might then be fortuitous, or in any event difficult to correlate with a theoretical model.

When the ionic strength of the plutonium solutions was increased by the addition of sodium perchlorate, there was an increase in the degree of aggregation and formation of large particles (1). At the same time it was found, as shown in Table I, that for both fresh and aged solutions at pH 7, increasing the ionic strength over the range of 0.002 to 0.1M greatly increased the sorption. Van Olphen (19) and others point out that increased coagulation results from compression of the electrical double-layer, which leads to a reduction in the range of the repulsive forces between two approaching colloidal particles. The increased aggregation which occurred when the ionic strength was increased should then have caused a decrease in sorption, if the only effect was a reduction in the number of small-size species, and these were the principal sorbing ones. However, the increased ionic strength will similarly reduce the range of the repulsive forces between the surface and sorbing particles. Thus the potential energy curves for the interaction of both large and small colloid species are probably affected, and both such species may then have increased their sorption capabilities. A similar increase in sorption with ionic strength was noted by Kitchener and coworkers for relatively monodispersed sorbing colloids (12, 16), as well as by Egerov *et al.* (20).

As shown in Table II, there was an increase in sorption with increasing ionic strength at pH 7.1, the opposite being obtained at pH 5. Such phenomena may be explained in terms of the types of sorbing species postulated earlier. It has been shown (20, 21, 22) that the nature of the sorption of radiocolloids from aqueous solutions differs from ionic sorption. For example, Schubert and Conn (23) found that the sorption of colloidal zirconium and niobium on a cation exchanger increased with increasing electrolyte concentration, while addition of electrolyte effectively competed with and reduced sorption of the ionic species.

There are ion exchange sites on silica surfaces for which protons and other cations compete on a stoichiometric basis (24). Thus at the lower pH, increasing the ionic strength may have resulted in an increased competition with sorbing ionic and other small charged species, thereby reducing their sorption. At the same time, as shown by Tadros and Lyklema (14), increasing the ionic strength with a uni-univalent electrolyte slightly increases the negative surface charge of silica at pH 5. The simultaneous reduction of the double-layer thickness would be expected to promote sorption of the colloidal plutonium which, however, may make a smaller contribution at pH 5 to the total sorption than that of the low-molecular-weight hydrolysis products and ionic species. The net result would then be the decrease in sorption that was obtained. At the higher pH of 7.1, the increase in sorption as the ionic strength was raised may be regarded as owing primarily to a reduction in double-layer repulsion between the silica surfaces and sorbing colloid species, resulting in a change in the plutonium size distribution, perhaps forming more readily sorbed species, as well as increasing the ability of such species to sorb.

It is also interesting to note from Table II that, at the lower ionic strength of 0.004M, increasing the pH reduced the sorption. This general effect was also obtained by Samartseva (25), although the ionic strength was not stated. This was explained by him as arising from the increased negative charge on the colloidal plutonium as the pH was raised, resulting in increased repulsion by the negatively charged silica surface. Davydov has shown that the point of zero charge of the colloid at a total plutonium concentration of approximately $7 \times 10^{-7}M$ is in the vicinity of pH 3 (26), the particles becoming increasingly negatively charged as the pH is raised.

In contrast, at an ionic strength of 0.23M, Table II indicates that as pH was raised, sorption increased. In this instance, it may be assumed that this is caused primarily by a change in the sorption of the colloidal species, since at this ionic strength ion exchange of both plutonium ions and other small charged species on the silica surface is inhibited. Also at the higher pH, their relative quantities in solution are reduced. At this ionic strength, the plutonium colloidal species were generally larger at pH 7 as compared with 5, probably with more sorbable species being formed (4), thereby accounting for the increase in sorption.

The results of the studies on the effect of silica particle size on sorption indicate that, in the range of silica diameter of 1.4×10^{-2} to 3.9×10^{-2} cm, there is very little difference in the final amount of plutonium sorbed, as shown in Table III, the total weight of silica being the same for each size fraction. The only differences lie in the surface density of the sorbed species and the rate at which the equilibrium was approached. As the surface area decreased (*i.e.*, as the silica size increased), there was more plutonium deposited per unit area of surface. Also, the smaller size particles with the higher surface areas sorbed the plutonium at a faster rate. These results are difficult to interpret, except with the model of a limited number of sorbable species, all of which are taken up eventually by the silica. They would not, however, be consistent with the model of a limited number of sorbing sites.

The addition of $10^{-2}M$ bicarbonate to the plutonium–silica system resulted in a large decrease in sorption compared with that in a solution in equilibrium with atmospheric carbon dioxide, as shown in Table IV, the effect being essentially the same at two values of ionic strength. The previous paper (1) showed that the addition of such bicarbonate had some effect on plutonium particle size distributions, but to different extents at the two ionic strengths. Qualitatively, then, the effect of bicarbonate addition may have been through a complexing effect, both with small plutonium species and those at the surface of the colloid, in both cases resulting in additional negative charge, thereby decreasing sorption.

As noted earlier, similar effects have been found for uranium and thorium sorption (8, 9).

A comparison of the desorption rates at pH 7, shown in Figure 7 for the plutonium sorbed from fresh and aged solutions, indicates that the total desorption curve may be interpreted in terms of two different sorbed species. This is expressed in Equations 2, 3, and 4 as two first order processes. For both the fresh and aged systems, the relative quantities of the $A_{0(1)}$ or loosely-held species were almost identical, as were their desorption rate constants. It is likely that the $A_{0(2)}$ or tightly-held species were colloidal in size, since irreversibility is a widely known characteristic of colloid sorption. This was found to apply, for example, in the case of the sorption of colloidal americium on quartz (27).

The increase in the rate, as well as the quantity of desorbing plutonium at pH 5, compared with that at pH 7 and 9 as shown in Figure 6, may be interpreted partially in terms of ionic and small charged plutonium species. Thus, at the lower pH there is more effective competition by protons for ion exchange sites on the silica surface, thereby increasing the removal of the small ionic and other species. However, the larger or colloidal species may be similarly affected, since their interaction with the silica are also influenced by pH, as noted above.

Although the mechanism of the interaction of aqueous plutonium with quartz silica has been the principal focus of this discussion, the implications of these interactions for the behavior of such species in natural water environments are also of interest. For example, ionic strength exerts a major influence on sorption characteristics. Thus, suspended quartz will sorb plutonium differently in fresh water and estuaries. Since aging affects sorption, a neutralized fresh plutonium effluent immediately released will sorb differently from one held for several days prior to dilution in a stream. Both sorption and desorption were shown to be greatly influenced by pH, but with further differences as the ionic strength was varied. Such variations occur in natural waters and would, therefore, be expected to influence plutonium interactions with suspended and bottom sediments. Since only some 60% of the plutonium at pH 7 was readily desorbed, the remainder could represent a possible cumulative hazard when sorbed onto bottom sediments in a stream. Lowering the pH increased the rate and quantity of plutonium desorbed, and thus also is a significant effect that could be encountered in environmental waters. Finally, the decreased sorption with increased bicarbonate concentration is possibly an important environmental effect because of its variable concentration in natural waters. All of these relationships are useful in predicting the fate of plutonium accidentally released to natural waters.

Literature Cited

(1) Andelman, J. B., Rozzell, T. C., Advan. Chem. Ser. (1970) **93**, 118.
(2) Leary, J. A., *Anal. Chem.* (1951) **23**, 850.
(3) Moss, W. D., Hyatt, E. C., Schulte, H. F., *Health Phys.* (1961) **5**, 212.
(4) Rozzell, T. C., Sc.D. thesis, Graduate School of Public Health, University of Pittsburgh, 1968.
(5) Ockenden, D. W., Welch, G. A., *J. Chem. Soc.* (1956) **653**, 3358.
(6) Sayre, W. W., Guy, H. P., Chamberlain, A. R., *U.S. Geol. Surv. Profess. Papers* (1963) **433-A.**
(7) Wayman, C. H., "Principles and Applications of Water Chemistry," p. 127, S. D. Faust and J. V. Hunter, Eds., Wiley, New York, 1967.
(8) Starik, I. Ye., *et al.*, *Radiokhimiya* (1967) **9**, 105.
(9) Kurbatov, M. H., Webster, H. B., Kurbatov, J. D., *J. Phys. Colloid Chem.* (1950) **54**, 1239.
(10) Cleveland, J. M., "Plutonium Handbook," p. 403, O. J. Wick, Ed., Gordon and Beach, New York, 1967.
(11) Denotkina, R. G., Moskvin, A. I., Schevchenko, V. E., *Russ. J. Inorg. Chem. (Eng. Trans.)* (1960) **5**, 731.
(12) Hull, M., Kitchener, J. A., *Trans. Faraday Soc.* (1969) **65**, 3093.
(13) Allen, L. H., Matijevic, E., *J. Colloid Interface Sci.* (1969) **31**, 287.
(14) Tadros, T. F., Lyklema, J., *J. Electroanal. Chem.* (1968) **17**, 267.
(15) *Ibid.*, (1969) **22**, 1.
(16) Marshall, J. K., Kitchener, J. A., *J. Colloid Interface Sci.* (1966) **22**, 342.
(17) Vold, M. J., *J. Colloid Sci.* (1961) **16**, 1.
(18) Bierman, A., *J. Colloid Sci.* (1955) **10**, 231.
(19) Van Olphen, H., "An Introduction to Clay Colloid Chemistry," p. 10–12, Interscience, New York, 1963.
(20) Egorov, Y. V., Lyubimov, A. S., Khrustalev, B. N., *Radiokhimiya* (1965) **7**, 386.
(21) Egorov, Y. V., Nikolaev, V. M., *Radiokhimiya* (1965) **7**, 273.
(22) Egorov, Y. V., Nikolaev, V. M., Lyubimov, A. S., *Radiokhimiya* (1966) **8**, 8.
(23) Schubert, J., Conn, E. E., *Nucleonics* (1949) **4**, 2.
(24) Vydra, F., Markova, V., *J. Inorg. Nucl. Chem.* (1964) **26**, 1319.
(25) Samartseva, A. G., *Radiokhimiya* (1962) **4**, 647.
(26) Davydov, Yu. P., *Radiokhimiya* (1967) **9**, 94.
(27) Starik, I. Ye., Ginzberg, F. L., *Radiokhimiya* (1960) **1**, 215.

Identification of Manganese in Water Solutions by Electron Spin Resonance

E. E. ANGINO, L. R. HATHAWAY, and T. WORMAN

University of Kansas, Lawrence Kan. 66044; Baker University, Baldwin, Kan. 66006; University of Kansas, Lawrence, Kan. 66044

The use of electron spin resonance spectroscopy to determine the presence and concentration of equilibrium and nonequilibrium species of Mn in natural water systems has not been adequately investigated. The presence of Mn^{2+} in a stream water with pH 8.5 suggests that while the Mn^{2+} was in an equilibrium state with regard to the solubility product of $Mn(OH)_2$ and $MnCO_3$, it was not in equilibrium relative to the oxidation–reduction conditions of the environment. Investigation of signal strength vs. Mn^{2+} concentration as a function of pH and O_2, N_2, and CO_2 saturation suggests that $Mn(H_2O)_6^{2+}$ is the major species present over the pH range 2–6.3. A mixed precipitate of Mn(III) compounds was obtained at pH's above 8.0.

Ο ne of the common problems encountered in studies of aqueous geochemistry and water pollution is proper identification of a particular species of an element or compound that may be present in the system. The use of electron spin resonance (ESR) spectroscopy to determine the presence and concentration of equilibrium and/or nonequilibrium metal species in natural water systems has not been adequately investigated. Coincidentally, Mn^{2+}, one of the easiest elemental species to detect by ESR, is also one of the dissolved species of considerable concern in problems related to heavy metal pollution and aqueous geochemistry. Furthermore, with proper design there exists the possibility of using electron spin resonance as the basis of a remote monitoring system for the detection of appropriate heavy metals in natural water systems.

A theoretical discussion of the chemical and physical principles involved in electron spin resonance can be found in appropriate texts (1, 2) and need not be repeated here.

Procedure

A study of the response of ESR signal *vs.* concentration was made by making up a set of $MnSO_4$ solutions adjusted to a desired pH. Signal response for Mn^{2+} concentrations of 10^{-1}, 10^{-2}, 10^{-3}, 10^{-4}, and 10^{-5} M were studied at different pH values. All solutions were dilutions of 10^{-1} M $MnSO_4$ using ion-free water. Adjustments of pH were made with NaOH and HCl by calculating the amount of one or the other needed to obtain a certain pH at a 100-ml volume, mixing that amount of acid or base with the $MnSO_4$ to be diluted, adjusting volume, and measuring the exact pH obtained. The pH was checked by meter. Determinations of the influence of dissolved N_2, O_2, and CO_2 were made at pH levels of 2, 3, 5, and 6.3 over the 10^{-1} to 10^{-5} M range. Each solution was purged first for 20–30 minutes with N_2 to remove other dissolved gases. The N_2 samples then were used as controls. Subsequently, appropriate solutions were saturated with O_2 and CO_2 and run as previously described. Experiments also were run using 10^{-3} M Mn^{2+} in HCO_3^{-}–CO_3^{2-} buffered solutions at pH values of about 8. Lower detection limits for Mn(II) was about 10^{-6} M. Signal values of several individual samples of different Mn^{2+} concentrations were reproducible to within 5%.

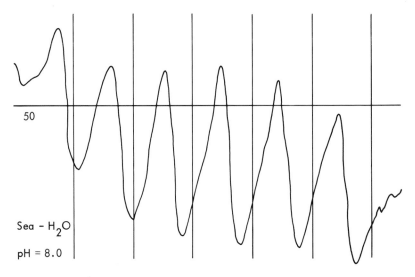

50

Sea - H_2O

pH = 8.0

Figure 1.   ESR spectrum of Mn(II) in artificial sea water

Results and Discussion

Using first derivative curves of signal as a function of field strength at a frequency of 100 K Hertz, Mn^{2+} was positively identified in our work by the $^{55}Mn^{2+}$ nuclear moment. Identification of Mn^{2+} in all cases was made by the characteristic and well-known six-line band of Mn^{2+} shown in Figure 1. To simplify the comparison of the six-line band, all

Table I. Mn²⁺ Equilibrium Data

Equilibria	log K	Ref.
$Mn(OH)_2(s) = Mn^{2+} + 2OH^-$	-12.96	(3)
$Mn^{2+} + H_2O = MnOH^+ + H^+$	-10.59	(3)
$MnCO_3(s) = Mn^{2+} + CO_3^{2-}$	-10.41	(3)
$Mn^{2+} + HCO_3^- = MnHCO_3^+$	$+1.95$	(3)

curves obtained from the solutions were converted to one modulation amplitude and signal level. A mean value was computed for each curve by averaging the six-line amplitudes and plotting the resulting value. A Varian V-4502 EPR spectrometer equipped with a field modulation and control unit (V-4560), EPR control unit (V-4500-10A), X-band microwave bridge (V-4500-41A), selector panel (V-4595), and an output control unit (V-4270) was used in the study.

An extensive survey of the thermodynamic and kinetic properties of manganese in natural aqueous systems has been presented by Morgan (3). From a thermodynamic standpoint, Mn(II) is unstable with respect to oxidation in natural waters. The kinetics of the oxidation reactions are sufficiently slow so that Mn(II) can exist as a metastable species in natural waters. The solubility of Mn(II) in most natural systems probably is limited by the solubility of $MnCO_3$. Soluble complexes such as $MnHCO_3^+$ make varying contributions to the total soluble Mn(II) species in natural waters. Some of the equilibria which are relevant to this study are listed in Table I.

Figures 2–4 represent results of ESR signal *vs.* concentration of Mn(II) for solutions of different pH values which have been saturated with N_2, CO_2, or O_2, respectively. The slopes of all the curves are, as noted, essentially the same within the 5% reproducibility limit. The ordinate intercept values are very similar for all curves. This suggests that the manganese species giving rise to the ESR signals in the three sets of experiments are very similar in nature.

In the solutions saturated with N_2, the solubility of Mn(II) is governed by the solubility product of $Mn(OH)_2$. $Mn(H_2O)_6^{2+}$ is expected to be the major species in solution over the pH range of 2 to 6.3 used in this study. $MnOH^+$ is not expected to make an important contribution to the Mn(II) species in solution until pH values of about 9 are reached (Figure 5A). Over the Mn(II) concentration range used (10^{-1} to 10^{-5} M), it was difficult to maintain a constant pH without occurrence of precipitation above a pH of about 7 in the unbuffered solutions.

The true pH range for the solutions saturated with CO_2 probably is near 2–4 rather than 2–6.3. This reflects the saturating of unbuffered solutions of slight acidity. In these solutions, the solubility product of $MnCO_3$ controls the over-all solubility of the Mn(II) species. At pH

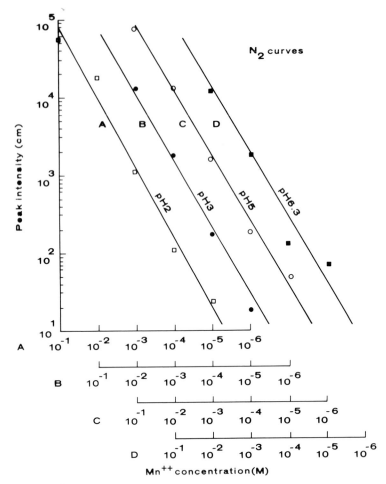

Figure 2. Signal vs. concentration curves for N_2-saturated solutions at different pH values

values of about 6, the $MnHCO_3^+$ species begins to become relatively important. However, under the experimental conditions of this study, it is expected that $Mn(H_2O)_6^{2+}$ is the major species in solution in the CO_2-saturated solutions (Figure 5B).

The data from the O_2-saturated series indicate that oxidation over the pH range and Mn(II) concentration range is a slow process. At pH values of 8–9 in $HCO_3 - CO_3^{2-}$ buffered solutions, loss of ESR signal plus formation of a precipitate is observed from aerated 10^{-3} M Mn(II) samples.

Precipitation is complete at pH 10. An x-ray diffraction pattern of the precipitate is shown in Figure 6. The peaks marked + were identified as those belonging to λ-Mn_2O_3. Those unmarked could not be related in any way to any manganese compounds listed on the ASTM reference cards. We feel they probably represent a poorly organized hydrated manganese oxide compound of open structure. Some substantiation for this view is indicated in Figure 7. This is a diffraction pattern of the sample shown in Figure 6 after heating to 450°C for 6 hours. Note that no peaks are evident below 30°2θ and that a compound identified only as Mn(III) oxide was present. The other peaks present but not marked were assigned similarly to different manganese oxide compounds. What was obtained appeared to be a mixture of Mn^{3+} compounds. No signal for Mn^{4+} compounds was noted.

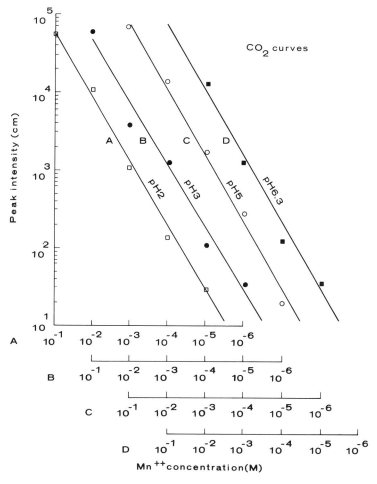

Figure 3.   Signal vs. concentration curves for CO_2-saturated solutions at different pH values

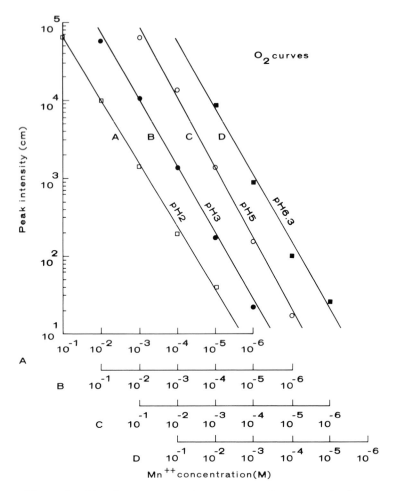

*Figure 4.   Signal vs. concentration curves for O₂-saturated solutions
at different pH values*

Our limited work on the Cl⁻ effects yields a trend toward a decrease of signal strength with saturated NaCl solutions of Mn^{2+}. This signal decrease is probably caused by complexing with Cl⁻. Effect of ions such as Cl⁻, SO_4^{2-}, PO_4^{3-}, etc., have been reported (4, 5) but were not investigated in detail in this study.

While studying the chemical characteristics of a local drainage system, samples were taken and analyzed by ESR. The Mn^{2+} concentration in the water system at pH 8.5 was estimated to be about 10^{-6} M from the experimentally derived plot of Mn^{2+} concentration vs. ESR signal. The signal obtained is shown in Figure 8. Confirmation of the approximate Mn^{2+} concentration level was obtained by atomic absorption analysis of the same sample. The discrepancy between the two analyses was a factor of 2 at the 10^{-6} M level. Other than the Mn^{2+} presence, nothing about the water analysis was unusual.

To find Mn^{2+} present in water of such a pH and low dissolved ion content was a bit unusual. The concentration of Mn^{2+} in this and other

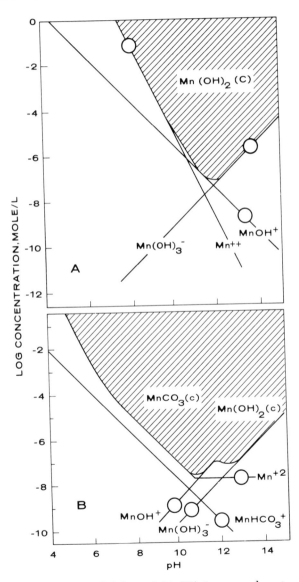

Figure 5. Solubility of Mn(II) in noncarbonate (A) and carbonate containing (B) solutions. Modified after Morgan (3).

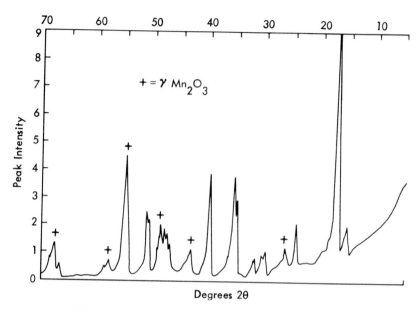

Figure 6. X-ray diffraction pattern of λ-Mn₂O₃

similar samples appears to be well within the solubility equilibrium ranges for $Mn(OH)_2$ and $MnCO_3$ in waters of the pH found and exposed to a normal atmospheric partial pressure of CO_2. However, Mn^{2+} is not in equilibrium with the redox potential of the natural system.

Investigation of ESR signal *vs.* pH at a Mn^{2+} concentration of 10^{-5} M shows a behavior more complex than that expected for the solubility equilibria of $Mn(OH)_2$. The cause of the increased complexity of the line spectra is not known at this time.

To test further for the common occurrence of Mn^{2+} in natural waters at pH levels above 7 and to eliminate especially the possibility of organic complexing, a solution of artificial sea water ("Instant Ocean") was made up according to directions and an ESR determination run on the resulting solution. The pattern obtained at a solution pH of 8.0 is shown in Figure 1. Clearly, the presence of Mn^{2+} is indicated. Furthermore, the shape of the curve is somewhat suggestive of the possible presence of Fe^{3+} in the solution with the actual signal observed being two signals superimposed one on the other, as suggested by Heise (6). A similar signal was obtained with some of the river samples examined.

In an attempt further to test the potential of the method, a leaching of carbonate-rich marine sediments with 20% H_2SO_4 gave a solution which after filtering yielded the well-defined Mn^{2+} ESR spectrum. The filtrant from the H_2SO_4 was subjected to stirring with a 20% HCl solution. Again after filtering, the solution gave the characteristic Mn^{2+} spectrum. A third portion was stirred for several hours in artificial sea water (Mn total = 2.3 ppm) with similar results. The presence of Mn^{2+} in these leached solutions suggests the presence of Mn^{2+} adsorbed on the carbonates from sea water or acid-soluble Mn^{2+} minerals in the sediments. However, the presence of finely divided MnO_2 cannot be eliminated from consideration. We lean toward the adsorbed hypothesis.

As noted, we have previously demonstrated the presence of Mn^{2+} in other Kansas stream samples at pH 8–8.5. This leads us to believe that the presence of the disequilibrium state of Mn^{2+} in natural water is much more common than presently suspected. The divalent manganese, as ESR spectroscopy indicates, is of special interest because it is the most chemically reactive form in natural water systems.

Investigation of what we believe to be weak signals of other appropriate transition metals either as free ions or complexes in polluted and nonpolluted natural water systems is expected.

Summary

The possibility of using ESR spectroscopy to identify the presence of Mn^{2+} and selected transition metal ions (*e.g.*, Cr, V, Mo, and Cu) as

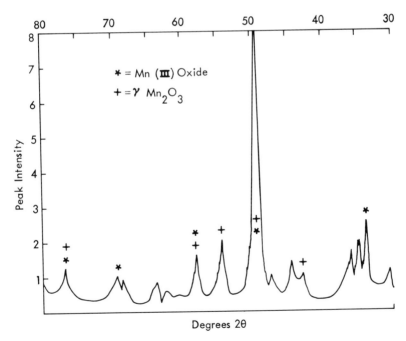

Figure 7.   X-ray diffraction pattern of heated sample shown in Figure 6 indicating peaks of several Mn(III) compounds

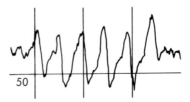

Waste Water

Figure 8. Typical stream or waste water, ESR pattern showing presence of low levels of Mn^{2+} at pH 8.5

free ions or complexes in natural water systems in equilibrium or non-equilibrium situations deserves further study. ESR may also provide qualitative information on the concentration levels of these same species in water and in specific problems involving polluted water. Data from controlled laboratory experiments are consistent with the known chemical behavior of Mn(II) in aqueous solutions. There also exists the possibility of using ESR as basis of a remote monitoring system for the detection of Mn and other appropriate heavy metals in natural water systems.

Acknowledgment

Publication authorized by the Director, State Geological Survey of Kansas.

Literature Cited

(1) Assenheim, H. M., "Introduction to Electron Spin Resonance," Plenum, New York, 1966.
(2) Squires, T. L., "An Introduction to Electron Spin Resonance," Academic, New York, 1964, 140 pp.
(3) Morgan, J. J., "Principles and Applications of Water Chemistry," S. D. Faust and J. V. Hunter, Eds., pp. 561–624, Wiley, New York, 1967.
(4) Ermakov, V. I., Zagorets, P. A., Grunau, A. P., Orlov, V. V., *Zh. Fiz. Khim.* (1967) **41**, 1669.
(5) Iordanov, N. D., *Compt. Rend. Acad. Bulgare Sci.* (1968) **21**, 111.
(6) Heise, J. J., *Marine Sci. Instrumentation* (1968) **4**, 25.

RECEIVED May 27, 1970.

13

Kinetics of the Nonbiological Decomposition and Racemization of Amino Acids in Natural Waters

JEFFREY L. BADA[1]

Department of Geological Sciences, Hoffman Laboratory, Harvard University, Cambridge, Mass. 02138

Amino acids dissolved in natural waters can undergo a variety of nonbiological decomposition reactions. The rates of most of these reactions are very slow at neutral pH and 0°C. The fastest decomposition reaction appears to be a metal ion catalyzed oxidation; even this reaction is fairly slow, indicating that amino acids dissolved in natural waters would undergo little nonbiological decomposition even over a period of a thousand years. Since the amino acids in natural waters are of biological origin, they should be largely of the L-configuration. However, amino acids are slowly racemized at neutral pH. The slow racemization of amino acids in the sea and other natural waters is discussed, and it is suggested that the racemization of amino acids might be used to calculate residence time or "age" of the dissolved amino acids.

Most of the amino acids which occur in the proteins of organisms, and also a few nonprotein amino acids, have been found dissolved in natural waters (*1, 2, 3, 4, 5, 6, 7, 8, 9*). The concentrations range from about 1 to 20 μg per liter. These dissolved amino acids differ from the amino acids associated with the particulate organic matter in natural waters in that they are free and not bound in a peptide or polymeric linkage. Table I summarizes some of the amino acids and concentrations which have been reported. Amino acids not included in the table but which may be present in low concentrations in natural waters include

---

[1] Present address: Scripps Institution of Oceanography, University of California at San Diego, La Jolla, Calif. 92037.

Table I.   Amino Acid Concentrations

Location (Ref.)	Ala[a]	Asp	Glu	Gly	Leu + Iso
Irish Sea (2) surface water 54° 00′N 3° 20′W	6	6	2	4	2
Pacific Ocean (4) 32° 32′ N 120° 30′W					
depth = 200m	3	2	1	12	<1
depth = 1000m	5	<1	2	10	3
depth = 3120m	6	7	1	17	3
York River estuary, Virginia (5)	1.5	1.9	1.0	16.9	iso = 0.6 leu = 0.6
Petroleum Brine Waters (3) Uinta Basin, Utah Geological age of formation, tertiary	9	8	1	12	7
Black Sea (9) surface	4.5	3.0	Tr.	7.8	Tr.
150 meters	2.4	Tr.	–	7.8	Tr.

[a] Abbrevations used: Ala = α-alanine; Asp = aspartic acid; Glu = glutamic acid; Gly = glycine; Leu = leucine; Iso = isoleucine; Lys = lysine; Met = methionine; Phe = phenylalanine; Ser = serine; Thr = threonine; Tyr = tyrosine; Val = valine; Orn = ornithine.

histidine, proline, and β-alanine. Great care must be used in the isolation of amino acids from natural waters to avoid contamination since several of the reagents used in the isolation procedure contain minute amounts of amino acids (10, 11, 12). The use of distilled and freshly prepared reagents substantially reduces these amounts. Analytical blanks should always be reported along with the amino acid analyses. Unfortunately, this is not usually the case, and as a result most of the present available data should be considered as only preliminary.

One of the sources of the amino acids dissolved in natural waters is probably the excreta of living organisms (13, 14, 15, 16, 17); a sizeable fraction likely comes from the decomposition of the proteins of dead organisms. Rain (12) and the discharge from rivers (7) probably are contributing sources in oceans and lakes. In recent times, the effluent from sewage treatment plants (18) has become an increasingly significant source in coastal and inland waters. In oceans and lakes, the amino acids

Found in Various Natural Waters (μg/liter)

Lys	Met	Phe	Ser	Thr	Tyr	Val	Orn	Total
4	4	6	4	2	–	4	–	44
<1	–	<1	8	2	<1	<1	2	~34
1	–	2	21	3	1	<1	3	~52
2	–	2	22	4	1	2	11	~78
1.0	1.3	0.2	4.9	1.5	0.8	0.9	3.2	36.3
–	–	4	15	7	–	2	8	~72
Tr.	–	Tr.	8.8	2.4	Tr.	Tr.	–	36.9
–	–	Tr.	Tr.	–	–	Tr.	–	15.2

are likely introduced into the water-mass in the largest quantities in surface and shore waters. The rate at which they are mixed into the water column depends on the mixing time of the particular water-mass, which is on the order of 500 to 1000 years for oceanic water-masses, but is substantially shorter in lakes. In the sea, dissolved amino acids are more or less uniformly distributed in the water column below a depth of a few hundred meters (4, 6). This homogeneous distribution of amino acids in the sea is similar to the total dissolved organic matter which also has a relatively uniform concentration in the oceanic water column below 300 meters (19, 20, 21, 22), a situation which has been said to imply that the dissolved organic matter in the deep sea is stable with respect to chemical and microbial degradation (23, 24, 25). According to Menzel and Ryther, the entire process of synthesis and decomposition of the dissolved organic matter may take place only within the upper 200–300 meters of the sea (22).

Table II.   Approximate Half-Lives (Hours) for the Oxidative Deamination of Several Amino Acids in the Presence of Copper Ions[a]

$$Amino\ acid\ =\ 10^{-3}M;\ CuSO_4\ =\ 10^{-3}M$$
$$Temperature\ =\ 100°C$$

	pH 9.6	pH 4.0
Glycine	5.6	11.2
Alanine	5.6	16.8
Glutamic acid	3.7	11.2
Phenylalanine	5.6	16.8
Tryptophan	6.8	11.2
Histidine	1.3	5.6
Threonine	3.7	7.4

[a] Taken from M. Ikawa and E. E. Snell (28).

The amino acids dissolved in natural waters can undergo a variety of nonbiological reactions. To determine whether these reactions would be important in natural waters, the kinetics of the reactions must be considered. In this paper, the kinetics of several important amino acid reactions are discussed, and these kinetics in turn are used to estimate the importance of the various reactions in natural waters.

Decomposition Reactions

Oxidation. The oxidation of amino acids is probably the principal nonbiological decomposition reaction under aerobic conditions. Except for some preliminary studies, however, there have been few investigations of this reaction. The reaction is likely an oxidative deamination, producing ammonia and the α-keto acid of the corresponding amino acid. The α-keto acid may decarboxylate to give an aldehyde. The over-all reaction sequence can be written as

$$\underset{\underset{{}^{+}NH_3}{|}}{RCHCOO^-} + 1/2\ O_2 \rightarrow R\overset{O}{\overset{||}{C}}COO^- + NH_4^+ \tag{1}$$

$$R\overset{O}{\overset{||}{C}}COO^- \xrightarrow{-CO_2} R\overset{O}{\overset{||}{C}}{}_- \underset{\rightleftarrows}{\overset{H_2O}{}} R\overset{O}{\overset{||}{C}}H + OH^- \tag{2}$$

Oxidative deamination is the major reaction which takes place during the photolysis of aqueous solutions of amino acids under aerobic conditions (26).

Conway and Libby (*27*) estimated a half-life of 10^6 years for the oxidation rate of alanine at 0°C. This slow oxidation rate indicates that even over periods of hundreds of thousands of years dissolved amino acids would undergo only small amounts of nonbiological oxidation. However, metal ions appear to catalyze the oxidation of amino acids (*28, 29*). The apparent function of the metal ions in the reaction is to chelate the amino acid (*29*). Ikawa and Snell (*28*) investigated the oxidative deamination of several amino acids in the presence of Cu^{2+} ions at 100°C and pH 9.6 and 4.0. Some of the results are shown in Table II. The rates are very nearly the same for all of the amino acids studied with the exception of histidine. At pH 9.6, each of the various amino acids would be completly chelated by Cu^{2+}. At pH 4.0, however, the amino acids are not completely chelated, and as a result the oxidation rates are slower. A value of 0.01 yr^{-1} at 0°C can be calculated for the oxidation rate (k_{oxid}) of alanine completely chelated by Cu^{2+} by assuming an Arrhenius activation energy of 25 kcal/mole and extrapolating the pH 9.6 data shown in Table II. In the presence of pyridoxal and metal ions (*28, 30*), amino acids are rapidly oxidized; pyridoxal has not been isolated, however, from any natural waters so this reaction is probably not significant.

It is difficult to estimate what the rate of the metal ion catalyzed oxidative deamination reaction of amino acids would be in natural waters. Hamilton and Revesz (*30*) found that the rate of oxidation of alanine in the presence of pyridoxal and manganese ions was inhibited by EDTA. Since metal ions in natural waters can be complexed by a variety of organic and inorganic compounds, their effectiveness in catalyzing the oxidative deamination of amino acids may be reduced. Also, the fraction of dissolved amino acids which would be complexed by metal ions at the pH and metal ion and amino acid concentrations found in natural waters must be considered. At neutral pH, where the amino group of the amino acid is protonated, the fraction of the amino acid that would be in the form of the metal ion complex depends upon the equilibrium constant for the formation of the complex and the pK of the amino proton of the amino acid. The reactions for the formation of the Cu^{2+}–alanine complexes can be written as

$$Cu^{2+} + \underset{+NH_3}{(CH_3CHCOO^-)} \rightleftarrows \underset{NH_2}{Cu^{2+} (CH_3CHCOO^-)} + H^+ \qquad (3)$$

$$\underset{NH_2}{Cu^{2+} (CH_3CHCOO^-)} + \underset{+NH_3}{CH_3CHCOO^-} \rightleftarrows \underset{NH_2}{Cu^{2+} (CH_3CHCOO^-)_2} + H^+$$
$$(4)$$

The equilibrium constants for Equations 3 and 4 as a function of pH can be written as $K_1 K_a/(H^+)$ and $K_2 K_a/(H^+)$, respectively, where K_1 is the stability constant for mono-alanine complex, K_2 the stability constant for the di-alanine complex, and K_a the equilibrium constant for the ionization of the amino proton of alanine; at $0°C$ (31, 32) $K_1 K_a = 2.6 \times 10^{-2}$ and $K_2 K_a = 6.9 \times 10^{-4}$. The equilibrium constants for the reactions of other amino acids with Cu^{2+} have similar values. In the ocean, $Cu = 2 \times 10^{-7}M$ (33, 34), alanine $= 7 \times 10^{-8}M$ (2, 35), and the pH of deep water is about 7.7 (36). Assuming all the copper is in the form of Cu^{2+} and substituting the above concentrations into the equilibrium expressions for Equations 3 and 4 gives an estimate of about 17% for the amount of alanine that would be in the form of the Cu^{2+} complexes in the deep ocean. Probably not all the dissolved copper is in the form of Cu^{2+}, and also a sizeable fraction of the Cu^{2+} is likely complexed by carbonate ion. Therefore, the actual amount of alanine chelated by Cu^{2+} is probably less than the estimated 17%. For other metal ions present in the sea and other natural waters, the equilibrium constants for reactions similar to the ones given above for Cu^{2+} are several orders of magnitude smaller than those of Cu^{2+}. Although the Mg^{2+} concentration in sea water is $\sim 10^5$ times larger than that of Cu^{2+}, the equilibrium constants for the formation of the Mg^{2+}–amino acid complexes are on the order of 10^6–10^7 smaller than those for the formation of the Cu^{2+}–amino acid complexes. The 17% estimate for the fraction of alanine complexed by Cu^{2+} therefore represents a maximum value. The fraction of amino acids complexed by other metal ions would be much less than 17%, possibly by several orders of magnitude.

The rate constant for the metal ion catalyzed oxidative deamination of an amino acid when only a fraction of the amino acid is chelated is given by

$$k_{\text{oxid}} = k_{\text{oxid}} \text{ (amino acid 100\% chelated)} \cdot \text{(fraction of amino acid chelated)} \tag{5}$$

Substituting into Equation 5 the value for the k_{oxid} of completely chelated alanine estimated previously and 0.17 for the fraction of chelated alanine gives

$$k_{\text{oxid}} = (0.01\text{yr}^{-1})(0.17) = 2 \times 10^{-3}\text{yr}^{-1}$$

for the maximum rate of oxidation of alanine in the deep ocean. This rate corresponds to a half-life of ~ 350 years. Since the oxidation rates are about equal for most chelated amino acids (*see* Table II) the rates of oxidation of other amino acids in the deep ocean are probably close to the value of 2×10^{-3} yr^{-1} estimated for alanine. These calculations indicate that amino acids would be stable with respect to nonbiological oxidation for periods of several hundred years. As stated earlier, the amino acids are homogeneously distributed in the oceanic water column below 200–300 meters. This situation suggests that the residence time of amino acids in the deep ocean must be less than one or two thousand years. With longer residence times, significant amounts of oxidation would take place, and the concentration of amino acids in the deep ocean would become much lower than that of near surface waters, which according to the present available data is not the case. In natural waters where amino acids have very short residence times, only very small and insignificant amounts of nonbiological oxidation would take place.

Deamination. Aspartic acid and asparagine undergo a slow, reversible deamination reaction. The aspartic acid deamination reaction, which produces fumaric acid and ammonia, can be written as

$$^-OOCCH_2CHCOO^- \rightleftarrows {}^-OOCCH{=}CHCOO^- + NH_4^+ \qquad (6)$$
$$\mid$$
$$+NH_3$$

The kinetics of this reaction have been investigated in detail between pH -1 and 13 over the temperature range 60° to 135°C (*37, 38*). Only deamination of aspartic acid was observed; there was less than 0.2% decarboxylation to α- or β-alanine. Thus, the β-alanine reported in sea water and marine sediments (*4*) could not have arisen from the nonbiological decarboxylation of aspartic acid. The estimated half-lives (*39*) for the aspartic acid deamination reaction at pH 7 are 28×10^6 years at 0° and 96,000 years at 25°C; for pH values greater than 10, the values are 330,000 years at 0° and 4100 years at 25°C. The rate of deamination of asparagine (*38*) is 100 to 200 times faster than the aspartic acid deamination rate. Fumaric acid has been found in several natural waters (*40, 41*) but it is unlikely that it could have arisen from the deamination of aspartic acid because the rates of the reaction are so slow. Its likely source is from organisms. Fumaramic acid, the deamination product of asparagine, has not been reported in any natural waters. Since this compound is not known to be metabolized or synthesized by organisms, any fumaramic acid eventually found in natural waters would have come from the deamination of asparagine. The half-life for the asparagine deamination at neutral pH and 0°C is of the order of 100,000 years. In natural waters where the residence time of asparagine might be substantial, small amounts of fumaramic acid could be produced from the asparagine deamination. However, these small quantities would be extremely difficult to detect by the present analytical methods.

Although the aspartic acid deamination reaction is slow and unimportant in present-day natural water systems, the reaction has geochemical implications, since it has been used to estimate the minimum NH_4^+ concentration in the oceans of the primitive earth (*39*). The argument is basically as follows. The kinetics and equilibrium of reactions involving the compounds presumably needed for life to arise should be able to fix the optimum concentrations of molecules present in the atmosphere and oceans of the primitive earth. A relevant reaction of this type is the reversible deamination of aspartic acid (Equation 6) since this reaction involves ammonia, a molecule generally held to have been present on the primitive earth (42). The heterotrophic hypothesis of the origin of life assumes that the basic constituents of the first living organism were available in large quantities in the primitive oceans. Aspartic acid is assumed to be one of these constituents. The equilibrium constant for the aspartic acid deamination reaction can be written as

$$K \frac{\text{(aspartate)}}{\text{(fumarate)}} = (NH_4^+) \qquad (7)$$

Although the concentration of aspartic acid in the primitive ocean cannot be estimated, it is assumed that the ratio of aspartic and fumaric acids did not fall substantially below 1.0. On the basis of these assumptions, it can be said that the minimum concentration of NH_4^+ in the primitive ocean is given by the aspartic acid equilibrium constant (43). At 0° and 25°C, the estimates are $1.0 \times 10^{-3}M$ and $2.7 \times 10^{-3}M$, respectively. The rates of the aspartic acid deamination reactions are fast enough that equilibrium would be attained in the time available, which is less than 10^9 years, but probably several hundred million years, since the earth was formed 4.5×10^9 years ago, and the earliest evidence for life is in rocks 3.5×10^9 years old (44). From these estimated minimum NH_4^+ concentrations in the primitive oceans, the minimum pressure of ammonia and hydrogen in the atmosphere can be calculated. The estimated values at 0°C are $pNH_3 = 2.9 \times 10^{-8}$ atm and $pH_2 = 3.5 \times 10^{-8}$ atm. A maximum NH_4^+ concentration of $0.01M$ can be estimated from the ion exchange of NH_4^+ and K^+ on clay minerals (39). These calculations represent the first attempt to calculate the possible composition of the atmosphere and oceans of the primitive earth. Since we have no geological record of the first billion years of the earth's history, the use of the kinetics and equilibria of reactions involving important prebiological organic compounds could prove to be a powerful method of describing the chemical environment of the primitive earth.

Decarboxylation. Most of the amino acids commonly found in proteins decompose by a slow, irreversible decarboxylation. The kinetics of decarboxylation of several amino acids have been studied by Abelson (45, 46), Conway and Libby (27), and Vallentyne (47, 48). The decarboxylation rates were determined by heating unbuffered aqueous solutions of the amino acids at elevated temperatures (between 100° and

280°C). Table III gives the half-lives for several amino acids at 0° and 25°C estimated from Arrhenius plots of the high-temperature data. The decomposition of serine is complex: in addition to decarboxylation, some deamination takes place (47). The half-lives for alanine and serine seem to be the maximum and minimum values, respectively, since other amino acids which decarboxylate have intermediate half-lives. Glutamic acid, instead of decarboxylating directly, first undergoes a reversible dehydration to form pyroglutamic acid. Pyroglutamic acid irreversibly decarboxylates giving pyrrolidone. At neutral pH and 0°C, the half-life for the over-all reaction is on the order of a million years (49).

The rates of the various decarboxylation reactions are so slow in the temperature range of 0° to 25°C that the reactions must be considered insignificant in all present-day natural waters.

Table III.   Half-Lives (Years) for the Decarboxylation of Several Amino Acids[a]

	$0°C$	$25°C$
Alanine	10^{14}	10^{11} (10^{10}) [b]
Phenylalanine	5×10^{8}	4×10^{6}
Serine	2×10^{6}	17,000

[a] Taken from J. R. Vallentyne (47).
[b] Value from measurements of D. Conway and W. F. Libby (27).

Racemization

With the exception of glycine, the amino acids commonly found in proteins are asymmetric at the α-carbon. The amino acids can therefore have two optical isomers, which are designated as the D and L enantiomers. Isoleucine and threonine, which are also asymmetric at the β-carbon, have four different optical isomers. In living organisms, only L-amino acids are usually found.

Amino acids are racemized by concentrated acid (50, 51, 52) and base (52, 53, 54) at elevated temperatures, and some preliminary experiments have shown that at 116° aspartic acid is racemized slowly at neutral pH values (38, 55). Also, the amino acids in fossil shells are partially racemized, with the amounts of racemization increasing with the age of the shell (56, 57, 58); racemization is essentially complete in shells of Miocene age. Since the kinetics of racemization of amino acids have not been investigated in detail at any pH, I have recently carried out a detailed study of the kinetics of racemization of aspartic acid between pH 0 and 13 and also the kinetics of racemization of phenylalanine, alanine, and isoleucine at pH 7.6. The results of these investigations are reported herein.

Kinetics of Racemization of Amino Acids. Phosphate was used to buffer solutions of the L-amino acids at pH 7.6. Solutions of L-aspartic acid were buffered at the various pH values by either hydrochloric acid, oxalate, succinate, phosphate, borate, or NaOH. The pH values of the buffered solutions at the elevated temperatures were estimated as described previously (38). Sodium chloride was added to the solutions to adjust the final ionic strength to 0.5. The solutions were degassed and sealed under vacuum in borosilicate glass ampules. The ampules were sterilized immediately after being sealed by heating at 100°C for 15 to 20 minutes. The rates of racemization were determined from measurements of the rate of change of optical rotation (α) of the solutions. The measurements were made on a Perkin-Elmer 141 polarimeter at 365 nm. With the exception of the aspartic acid solutions at pH values less than 2 and the phenylalanine solutions, all samples were diluted in $1M$ HCl before measuring the rotation.

The racemization reaction for most amino acids can be written as

$$\text{L–amino acid} \underset{k}{\overset{k}{\rightleftarrows}} \text{D–amino acid} \qquad (8)$$

where k is the first order rate constant for interconversion of the enantiomers. The first order rate constant for racemization equals $2\,k$. At equilibrium, the concentration of L-amino acid = concentration of D-amino acid. The expression for the rate of disappearance of the L-amino acid from the buffered solutions is

$$\frac{-\text{d(L–amino acid)}}{\text{d}t} = k\,(\text{L–amino acid}) - k\,(\text{D–amino acid}) \qquad (9)$$

The various decomposition reactions outlined in the preceeding section would remove amino acid from the buffered solutions, and if these reactions were significant compared with the k values, Equation 9 would have to be modified. Since the racemization was studied under anaerobic conditions, the decarboxylation and deamination reactions are the only important amino acid decomposition reactions. With the exception of the deamination of aspartic acid, the decomposition rates of the other amino acids studied were negligible compared with the racemization rate of the amino acid. Integration of Equation 9 yields

$$\ln\left\{\frac{(\text{L–amino acid})_0}{2\,(\text{L–amino acid})_t - (\text{L–amino acid})_0}\right\} = 2\,k \cdot t \qquad (10)$$

where $(\text{L–amino acid})_0$ and $(\text{L–amino acid})_t$ are the L-amino acid concentrations at time zero and time t, respectively. Equation 10, rewritten in terms of the optical rotation of the solution, is

$$\ln \left\{ \frac{\alpha_0}{\alpha_t} \right\} = 2\,k\cdot t \tag{10a}$$

For the aspartic acid racemization reaction, Equation 9 must be modified to

$$\frac{-d(\text{L–asp})}{dt} = k(\text{L–asp}) - k(\text{D–asp}) + k_{\text{deam}}(\text{L–asp}) \tag{11}$$

where k_{deam} is the rate constant for the deamination of aspartic acid. This equation assumes that the amount of decomposition of the D-aspartic acid is negligible. This assumption is valid since the solutions initially contained only L-aspartic acid, and the racemization reaction was not studied for times greater than one half-life. Integration of Equation 11 yields

$$\ln \left\{ \frac{(\text{L–asp})_0}{2\,(\text{L–asp})_t - (\text{L–asp})_0} \right\} = \ln \frac{\alpha_0}{\alpha_t} = (2k + k_{\text{deam}})\cdot t \tag{12}$$

The racemization kinetics of isoleucine are slightly complicated by the fact that this amino acid is asymetric at both the α- and β-carbons. The racemization reaction for isoleucine can be written as

$$\text{L–isoleucine} \underset{k'}{\overset{k}{\rightleftarrows}} \text{D–alloisoleucine} \tag{13}$$

At equilibrium, the concentration of alloisoleucine does not equal that of isoleucine (*58*). The rate expression for Equation 13 is similar to that given in Equation 9 except k (D-amino acid) is replaced by k' (D-allo-isoleucine). Since the solutions initially contained only L-isoleucine, when the amount of racemization is small, the term k' (D-alloisoleu-cine) can be neglected in the rate expression and in this case the integrated rate equation is

$$\ln \left\{ \frac{(\text{L–isoleucine})_0}{(\text{L–isoleucine})_t} \right\} = \ln \frac{2\alpha_0}{\alpha_t + \alpha_0} = k\cdot t \tag{14}$$

For the isoleucine investigations, the heating times were chosen such that only small amounts of racemization occurred. The rates of racemization of isoleucine were therefore calculated from Equation 14.

The rate of racemization of aspartic acid as a function of pH at 117.2°C is shown in Figure 1. The rates were calculated from Equation 12. The rates of deamination of aspartic acid at each pH value were calculated from the data given by Bada and Miller (*38*). The results in Figure 1 are the average values for samples heated for at least two

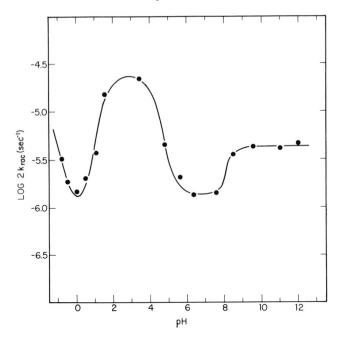

*Figure 1. Log 2k for aspartic acid as a function of pH
at 117.2°C*

different lengths of time. The uncertainty of the measurements is about
±3%. Figure 1 indicates that the rate of racemization of aspartic acid
is independent of pH between 10 and 13, and is approximately first order
in H^+ concentration at pH values less than 0. The maximum rate of
racemization of aspartic acid is at pH 3. The racemization kinetics of
other amino acids have a similar pH dependence except there is no maximum at pH 3; the rate of racemization of valine is independent of pH
between pH 3 and 8. Aspartic acid has a maximum at pH 3 because it is
a dicarboxylic amino acid. The interpretation of the kinetics and the
mechanism of the racemization reaction of amino acids will be discussed
elsewhere. An Arrhenius plot of the rate constants determined at pH 7.6
for aspartic acid, phenylalanine, alanine, and isoleucine is shown in
Figure 2. The values of E_a (kcal mole^{-1}), the Arrhenius activation energy,
calculated for the racemization of phenylalanine, aspartic acid, alanine,
and isoleucine at pH 7.6 are 28.6, 31.1, 31.0, and 31.5, respectively. The
data in Figure 2 were fitted by the method of least squares to give the
following equations for the interconversion rates of the various amino
acids at pH 7.6:

$$\text{isoleucine} \quad \log k \ (\text{yr}^{-1}) = 17.98 - 6853/T \tag{15}$$

$$\text{alanine} \quad \log k \ (\text{yr}^{-1}) = 18.11 - 6750/T \tag{16}$$

$$\text{aspartic acid log } k \text{ (yr}^{-1}) = 18.73 - 6780/T \qquad (17)$$

$$\text{phenylalanine log } k \text{ (yr}^{-1}) = 17.05 - 6208/T \qquad (18)$$

The half-lives (*i.e.*, when D-amino acid = 25%, L-amino acid = 75%) for the racemization reaction can be calculated by substituting (L-amino acid)$_0$ = 1 and (L-amino acid)$_t$ = 0.75 into Equation 10. The resulting relationship is

$$t_{1/2} = \frac{\ln 2}{2k} \qquad (19)$$

The estimated half-lives at pH 7.6 at 0° and 25°C are given in Table IV.

Metal Ion Catalyzed Racemization of Amino Acids. In dilute alkali, metal ions like Cu^{2+} and Al^{3+} catalyze the racemization of amino acids (*59, 60, 61, 62, 63*). The metal ions chelate the amino acid, and in this

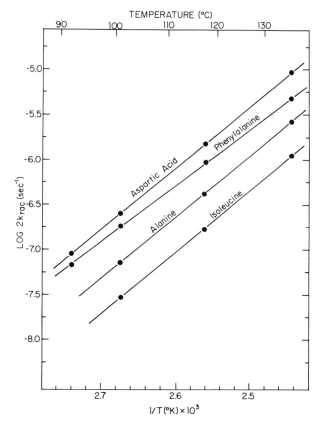

Figure 2.   Log 2k vs. 1/T (°K) for several amino acids at ph 7.6

Table IV. Half-Lives (Years) for Racemization of Amino Acids at pH 7.6 and Ionic Strength of 0.5

	$0°C$	$25°C$
Phenylalanine	160,000	2030
Aspartic acid	420,000	3460
Alanine	1.1×10^6	11,000
Isoleucine[a]	4.4×10^6	34,700

[a] These would be the half-life values if the equilibrium ratio of alloisoleucine to isoleucine was 1.0. At 140° the ratio is 1.25 (56). The temperature variation of the ratio is not, however; a value of 1.0 is therefore assumed for 0° and 25°C. The uncertainty arising from this assumption is probably less than 10%.

complex, the reactivity of the α-proton of the amino acid is greatly increased (59, 63). The kinetics of racemization of valine and alanine in the complex ions Co(ethylenediamine)$_2$amino acid^{2+} have been investigated by Buckingham, Marzilli, and Sargeson (59). The rate of racemization of these chelated amino acids is first order in OH$^-$ concentration. At an ionic strength of 0.5, $k \cong 2 \times 10^{-2}M^{-1}$ sec^{-1} at 34.3° C, and $E_a = $ 17.1 kcal mole^{-1} for the racemization of chelated alanine. Extrapolating these data to pH 7.6 and 0°C gives $k \cong 7 \times 10^{-4}$ yr^{-1} for the racemization of alanine in the complex Co(ethylenediamine)$_2$ alanine^{2+}. This k value is $\sim$ 3000 times that of nonchelated alanine at pH 7.6 and 0°C. In an earlier discussion of the metal ion catalyzed oxidation of amino acids, it was estimated that only about 17% of the alanine dissolved in natural waters would be chelated by Cu^{2+}. This indicates that the value of k for the metal ion catalyzed racemization of alanine in natural waters at pH 7.6 and 0°C should be on the order of 1×10^{-4} yr^{-1}.

It is difficult to determine whether the complex ions studied by Buckingham and coworkers would be good models of the amino acid–metal ion complexes which may be present in natural waters. It is apparent from their data, however, that the chelation of amino acids by metal ions greatly accelerates the rate of racemization of the amino acid. A study of the rate of racemization of amino acids at the metal ion and amino acid concentrations and ionic strengths usually found in natural waters would provide better estimates of the rate of the metal ion catalyzed racemization of amino acids in natural waters. A detailed investigation of these kinetics is in progress in this laboratory.

Racemization of Amino Acids in Natural Waters. Since the amino acids dissolved in natural water are of biological origin, they should be largely of the L-configuration. However, during the residence time of amino acids in natural waters, a small amount of racemization may take place.

The residence time of amino acids dissolved in natural waters is likely governed by the rate at which the amino acids are removed by organisms. Dissolved amino acids can be utilized by a variety of organisms, including invertebrates (*64, 65, 66*), planktonic algae (*17, 67, 68, 69, 70*), and bacteria (*71, 72, 73, 74, 75*). Of these various organisms, the heterotrophic bacteria probably consume the largest quantity of amino acids. In studies with glucose and acetate, Wright and Hobbie (*76*) found that in Lake Erken, Sweden, the rate of algal uptake was always less than 10% of that of the bacteria. Similar results might be expected for the uptake of amino acids since in growth studies of marine planktonic algae, amino acids in general have been found to be poor sources of nitrogen (*67, 69, 70*). There have been few investigations of the rate of microbial utilization of amino acids in natural waters. From studies of the net zooplankton excretion rates, Webb and Johannes (*17*) calculated a turnover time (time required to release or remove an amount of dissolved amino acids equivalent to that in solution) of 30 days for amino acids in the surface waters of the oceans. In the York River estuary, Hobbie and coworkers (*5*) measured the turnover times of several amino acids and obtained values which ranged from a few hours for methionine to four days for arginine. Both of these rate estimates were determined in regions where there is a very high density of microorganisms. In the deep sea where the density of microorganisms is very low, the turnover rates would be expected to be very much longer. Williams (*77*) has recently observed that while small but detectable amounts of uptake of dissolved amino acids occurred at depths of 400 and 600 meters in the western Mediterranean Sea, no uptake was observed at 2000 meters. Also, Jannasch (*78*) has found that marine bacteria at very low population densities are probably not capable of utilizing the dissolved organic matter in the sea, and in experiments with sea water samples taken from various depths, Skerman (*74*) found essentially no amino acid-requiring bacteria in deep water samples, while in surface waters, a high percentage of the bacteria required amino acids. These studies suggest that in the sea the rate of microbial removal of amino acids is substantial only in the upper few hundred meters and in shallow bays and estuaries. Once the amino acids are circulated out of these areas, they may be fairly stable with respect to microbial breakdown until they are recycled. In shallow lakes where the amino acids (*1*) and the bacteria (*79, 80*) are fairly uniformly distributed in the water column, the turnover rate of amino acids is probably fast at all depths.

The "age" of the dissolved organic carbon in the deep ocean supports the concept that below a depth of a few hundred meters the rate of microbial consumption of dissolved organic compounds is very slow. Based on the oxygen consumption rate in deep waters, Postma (*81*) has calculated a minimum age of 500 years for the dissolved organic matter in the deep sea, while Williams and coworkers (*82*) have recently determined an age of 3400 years from C^{14} measurements. If the dissolved amino acids in the deep ocean have an age near that of the total dissolved organic matter, they should be slightly racemized. Based on the k values at pH 7.6 and 4°C, the amounts of racemization that would be observed for amino acids of various ages are shown in Table V; also included in Table V are calculated amounts of racemization for alanine partly chelated by Cu^{2+}. Since microorganisms can probably also utilize D-amino acids (83, 84), once the amino acids in the deep ocean are recycled into the upper few hundred meters of the sea, both the D- and L-amino acids are likely rapidly consumed by organisms. The process of racemization of amino acids in the sea is shown in Figure 3.

This discussion suggests that the racemization of amino acids might be used to calculate amino acid residence times in the sea. Assuming that only L-amino acids are initially introduced into the oceans and that when these L-amino acids are mixed into deep waters racemization occurs, the observed amount of racemization during a residence time is given by

$$ t = \frac{\ln \left\{ \dfrac{(\text{D--amino acid}) + (\text{L--amino acid})}{(\text{L--amino acid}) - (\text{D--amino acid})} \right\}}{2k} \tag{20} $$

where the parentheses refer to the observed concentrations of the D and L enantiomers of a particular amino acid and k is the first order rate constant for interconversion of the enantiomers of the amino acid in the ocean. This equation assumes that D-amino acids are introduced into the sea only from the racemization of L-amino acids. There may be other contributing sources. D-amino acids have been found in insects and worms (85), in the proteins of some bacteria (86), and possibly in algae (87). Rain (12) has been suggested as a possible source, but this seems unlikely. The discharge from rivers may be a contributing source. These other

Table V. Percent of D-Enantiomer[a] of Several Amino Acids Which Would Be Produced During Various Residence Times

	500 Years	3400 Years
Aspartic acid	0.09%	0.6%
Phenylalanine	0.2%	1.4%
Alanine	0.03 (~5%)[b]	0.02% (~25%)[b]
Isoleucine[c]	0.009%	0.06%

[a] Calculated as $\left\{ \dfrac{\text{D--amino acid}}{\text{D--amino acid} + \text{L--amino acid}} \right\} \times 100.$

[b] Amount of D-alanine based on the estimated value of $k = 10^{-4}$ yr^{-1} for the metal ion catalyzed racemization of alanine at pH 7.6 and 0°C.

[c] Calculated as $\left\{ \dfrac{\text{alloisoleucine}}{\text{alloisoleucine} + \text{isoleucine}} \right\} \times 100$

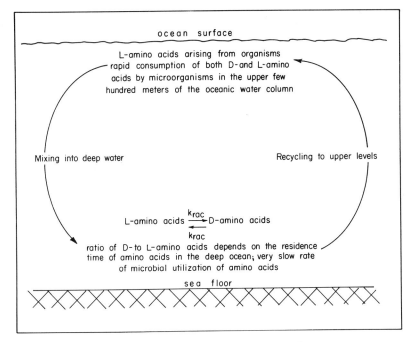

Figure 3. Racemization of amino acids in deep ocean

sources would introduce the D-amino acids mainly into surface waters. To correct for this addition, Equation 20 would be rewritten as

$$t = \frac{\ln \frac{\text{(D–amino acid)} + \text{(L–amino acid)}}{\text{(L–amino acid)} - \text{(L–amino acid)}}_{\text{deep water}} - \ln \frac{\text{(D–amino acid)} + \text{(L–amino acid)}}{\text{(D–amino acid)} - \text{(D–amino acid)}}_{\text{surface}}}{2\,k} \tag{21}$$

In order to calculate amino acid residence times from Equation 21, the ratio of the D to L enantiomers of the various amino acids are required as a function of depth in the oceanic water column. Unfortunately, there have been no investigations of the amino acid enantiomers dissolved in any natural waters. The analyses are difficult because most D- and L-amino acids are not separable by the usual amino acid analytical techniques. One exception is isoleucine, which forms alloisoleucine when it racemizes (Equation 13). Isoleucine and alloisoleucine are separable on the buffered columns of the automatic amino acid analyzer (88). However, as can be seen from Table V, only very small amounts of alloisoleucine would be produced from the racemization of isoleucine, unless the racemization reaction is catalyzed by metal ions. The enantiomers of other amino acids can also be determined on the automatic amino acid analyzer by making diasterisomeric dipeptides obtained by derivitization with an L-amino acid N-carboxyanhydride (50). Gas chromatography can also be used to separate the D- and L-amino acids (89, 90).

Recently, I have been doing some preliminary investigations of the alloisoleucine/isoleucine ratio in a sample taken from a depth of 2500 meters in the Atlantic Ocean. After isolating the amino acids by a procedure similar to that used by Chau and Riley (2), the sample was analyzed for the presence of alloisoleucine on the Beckman–Spinco automatic amino acid analyzer. The results indicate that a small amount of alloisoleucine appears to be present in the sample. It is impossible to make any conclusions from this one experiment, however, since the analysis of a blank which had been carried through the same isolation steps as the sea water sample contained a significant amount of isoleucine. Also, several dipeptides appear at about the same location as does alloisoleucine on the chromatogram from the automatic amino acid analyzer. Many samples from the world's oceans will have to be analyzed before it can be determined whether the racemization of amino acids can be used to calculate amino acid residence times in the sea.

Although only small amounts of racemization of the dissolved amino acids may take place in the open ocean, relatively large amounts may occur in certain natural water masses. For example, it would be interesting to look for D-amino acids in the hot brine pools which have been found in the Red Sea (91). Temperatures of 50°C and higher have been recorded in these areas. If there are dissolved amino acids in these waters, they would be racemized rapidly (for phenylalanine at pH 7.6 and 50°C, $k = 0.7\%$ yr^{-1}). Even if the residence times of the amino acids in the hot brine pools are only a few years, significant amounts of racemization would take place. With long residence times in the hot brine pools, only racemic amino acids would be observed. Another area of interest would be the Black Sea, where organic molecules may have long residence times, and as a result, the dissolved amino acids would be expected to be appreciably racemized. Indeed, in any natural water system characterized by slow mixing rates and/or relatively high temperatures, the dissolved amino acids should be significantly racemized.

Racemization might also be used to estimate the age of amino acids found in ground waters. Degens et al. (3) claim that the amino acids isolated from various petroleum brine waters are of the same age as that of the formation in which they are found. The formations which were investigated ranged in age from a million to a few hundred million years. If the amino acids found in the brine waters are the same age as the formations, they would be expected to be completely racemized. On the other hand, if the amino acids were the result of recent bacterial contamination, only L-amino acids would be found. The ratio of D- to L-amino acids in the brine waters was not investigated by Degens and coworkers. Also, in the carbonate aquifer described by Back and Hanshaw (92) where the ground waters have ages of tens of thousands of years and a temperature of 25°C, the amount of racemization would appear to be an excellent indicator of the age of any dissolved amino acids which may be present.

The slow racemization of amino acids can take place not only in the water column but also in the bottom sediments, and this racemization has important geochemical implications. Most of the amino acids commonly found in the proteins of organisms have been found in small quantities in marine sediments (9, 35, 93, 94). The concentrations near the sediment–sea water interface are on the order of 0.1 to 2 mg per gram of dry sediment. The concentration decreases with increasing depth into the sedimentary column.

Recently, it has been shown that with increasing depth into the sedimentary column, an increasing amount of racemization of isoleucine is observed (95). The core which was studied was taken from the Atlantis fracture zone, about 30 nautical miles west of the crest of the Mid-Atlantic ridge. The sedimentation rate in this general area is on the order of 4–5 mm per 1000 years. As mentioned earlier, when isoleucine is racemized, alloisoleucine is formed, and these two amino acids are separable on the automatic amino acid analyzer. The amino acids were isolated from various sections in the core, and the ratio of alloisoleucine to isoleucine has been determined as a function of depth below the sediment–sea water interface. At 145–155 cm below the sediment–sea water interface, alloisoleucine/isoleucine = 0.055 while at 445–455 cm, the ratio was 0.154. The racemization of isoleucine in the sedimentary column was used to calculate a sedimentation rate of 4.2 mm/1000 yrs and an age of 1.23 million years for the bottom of the core. This estimated sedimentation rate is in close agreement with values determined in the general vicinity by both paleomagnetic and radioactive nuclide decay techniques. Although the racemization of other amino acids were not investigated, as can be seen in Table IV, phenylalanine, aspartic acid, and alanine would be expected to be racemized much more rapidly than isoleucine in the sedimentary column.

These results suggest that the slow racemization of amino acids in the sedimentary column can be used to estimate the sedimentation rate, and in turn the age, of both marine and fresh water sediments. In sediments from areas in the deep ocean where sedimentation rates are very slow (*i.e.*, a few mm per 1000 years), the amount of racemization of isoleucine would be the easiest to determine since the investigations can be carried out on the automatic amino acid analyzer. In coastal areas and in lakes with sedimentation rates on the order of 1–10 cm every 1000 years, the amount of racemization of other amino acids like aspartic acid and phenylalanine would have to be determined in order to observe significant amounts of racemization in a core of reasonable length. In lakes with fast sedimentation rates (*i.e.*, 1 cm every few years) only small amounts of racemization of aspartic acid and phenylalanine would be observed even with long cores unless the bottom water temperature averaged above 15°C.

Summary

The rates of several nonbiological reactions involving the amino acids have been evaluated and these rates used to estimate the importance of the various reactions in natural waters. The fastest reaction under aerobic conditions is a metal ion catalyzed oxidation. In the sea, this reaction has an estimated half-life of $\sim$ 350 years. Under anaerobic conditions, the decomposition rates are very slow, and as a result there would be little nonbiological decomposition of the amino acids dissolved in anoxic natural waters even over periods of hundreds of thousands of years. The decomposition kinetics of the various amino acids indicate that if dissolved amino acids are not consumed by organisms, they would remain in natural waters for periods of thousands of years or longer. Of the various amino acid reactions which may take place in natural waters, the racemization reaction is of particular interest. Small amounts of D-amino acids may be produced in natural water from the slow racemization of the L-amino acids arising from organisms. By measuring the ratio of the D to L enantiomers of a particular amino acid and knowing the rate of interconversion of the enantiomers of the amino acid, it is possible to estimate an age of the amino acids dissolved in natural waters. Also, the racemization of amino acids in the sedimentary column can be used to estimate sedimentation rates and the ages of sediments.

The amino acids make up only a small fraction of the organic matter dissolved in natural waters. By studying the kinetics of reactions involving other dissolved organic constituents, the rate of the nonbiological decomposition and alteration of the dissolved organic matter in natural waters can be determined. These rates would be useful in estimating how long organic molecules would persist in the aquatic environment if they are not degraded by organisms.

Acknowledgment

I thank Werner Stumm and Raymond Siever for helpful discussions. This work was supported by the Committee on Experimental Geology and Geophysics at Harvard University.

Literature Cited

(1) Brehm, J., *Arch. Hydrobiol. Suppl.* (1967) **32**, 313.
(2) Chau, Y. K., Riley, J. P., *Deep-Sea Res.* (1966) **13**, 1115.
(3) Degens, E. T., Hunt, J. M., Reuter, J. H., Reed, W. E., *Sedimentology* (1964) **3**, 199.
(4) Degens, E. T., Reuter, J. H., Shaw, K. N. F., *Geochim. Cosmochim. Acta* (1964) **28**, 45.
(5) Hobbie, J. E., Crawford, C. C., Webb, K. L., *Science* (1968) **159**, 1463.
(6) Park, K., Williams, W. T., Prescott, J. M., Wood, D. W., *Science* (1962) **138**, 531.
(7) Semenov, A. D., Pashanova, A. P., Kishkinova, T. S., Nemtseva, L. I., *Gidrokhim. Materialy* (1966) **42**, 171.
(8) Siegel, A., Degens, E. T., *Science* (1966) **151**, 1098.
(9) Starikova, N. D., Korzhikova, L. I., *Okeanologiya Akad. Nauk. SSSR* (1969) **9**, 625.
(10) Hare, P. E., *Carnegie Inst. Year Book* (1965) **64**, 232.
(11) Schopf, J. W., Kvenvoldern, K. A., Barghoorn, E. S., *Proc. Natl. Acad. Sci. U.S.* (1968) **59**, 639.
(12) Sidle, A. B., *Tellus* (1967) **19**, 128, 132.
(13) Hellebust, J. A., *Limnol. Oceanog.* (1965) **10**, 192.
(14) Jawed, M., *Limnol. Oceanog.* (1969) **14**, 748.
(15) Johannes, R. E., Webb, K. L., *Science* (1965) **150**, 76.
(16) Stewart, W. D. P., *Nature* (1963) **200**, 1020.
(17) Webb, K. L., Johannes, R. E., *Limnol. Oceanog.* (1967) **12**, 376.
(18) Hunter, J. V., Heukelekian, H., *J. Water Pollution Control Federation* (1965) **37**, 1142.
(19) Menzel, D. W., *Deep-sea Res.* (1964) **11**, 757.
(20) Menzel, D. W., *Deep-Sea Res.* (1967) **15**, 228.
(21) Menzel, D. W., Ryther, J. H., *Deep-Sea Res.* (1968) **15**, 327.
(22) Menzel, D. W., Ryther, J. H., *Symp. Organic Matter in Natural Waters, University of Alaska, College, Alaska, September 1968.*
(23) Barber, R. T., *Nature* (1968) **220**, 274.
(24) Williams, P. M., *Nature* (1968) **219**, 152.
(25) Williams, P. M., Gordon, L. I., *Deep-Sea Res.* (1970) **17**, 19.
(26) McLaren, A. D., Shugar, D., "Photochemistry of Proteins and Nucleic Acids," p. 97, Pergamon, New York, 1964.
(27) Conway, D., Libby, W. F., *J. Am. Chem. Soc.* (1958) **80**, 1077.
(28) Ikawa, M., Snell, E. E., *J. Am. Chem. Soc.* (1954) **76**, 4900.
(29) Nyilasi, J. Pomogáts, E., *Acta Chim. Hung. Tomus* (1964) **42**, 27.
(30) Hamilton, G. A., Revesz, A., *J. Am. Chem. Soc.* (1966) **88**, 2069.
(31) Izatt, R. M., Wrathall, J. W., Anderson, K. P., *J. Phys. Chem.* (1961) **65**, 1914.
(32) Sillen, L. G., Martell, A. E., "Stability Constant of Metal–Ion Complexes," p. 398, Spec. Publ. No. 17, The Chemical Society, London.
(33) Leisegang, E. C., Orren, M. J., *Nature* (1966) **211**, 1166.
(34) Slowey, J. F., thesis, Texas A&M University, College Station, Tex., 1966.
(35) Degens, E. T., "Geochemistry of Sediments," p. 206–24, Prentice-Hall, Englewood Cliffs, N. J., 1965.
(36) Park, P. K., *Science* (1968) **162**, 357.
(37) Bada, J. L., Miller, S. L., *J. Am. Chem. Soc.* (1969) **91**, 3946.
(38) Bada, J. L., Miller, S. L., *J. Am. Chem. Soc.* (1970) **92**, 2774.
(39) Bada, J. L., Miller, S. L., *Science* (1968) **159**, 423.
(40) Goncharova, I. A., Khomenko, A. N., Semenov, A. D., *Gidrokhim. Materialy* (1966) **41**, 116.
(41) Lamar, W. L., Goerlitz, D. F., *J. Am. Water Works Assoc.* (1963) **55**, 797.

(42) Miller, S. L., Urey, H. C., *Science* (1959) **130**, 245.
(43) Bada, J. L., Miller, S. L., *Biochemistry* (1969) **7**, 3403.
(44) Barghoorn, E. S., Schopf, J. W., *Science* (1966) **152**, 758.
(45) Abelson, P. H., "Researches in eGochemistry," p. 79, P. H. Abelson, Ed.,
      Wiley, New York, 1959.
(46) Abelson, P. H., *Progr. Chem. Org. Nat. Prod.* (1959) **17**, 379.
(47) Vallentyne, J. R., *Geochim. Cosmochim. Acta* (1964) **28**, 157.
(48) Vallentyne, J. R., *Geochim. Cosmochim. Acta* (1968) **32**, 1353.
(49) Povoledo, D., Valletyne, J. R., *Geochim. Cosmochim. Acta* (1964) **28**,
      731.
(50) Manning, J. M., Moore, S., *J. Biol. Chem.* (1968) **243**, 5591.
(51) Wiltshire, G. H., *Biochem. J.* (1953) **55**, 46.
(52) Neuberger, A., *Advan. Protein Chem.* (1948) **4**, 298.
(53) Crawhall, J. C., Elliott, D. F., *Biochem. J.* (1951) **48**, 237.
(54) Gunness, M., Dwyer, I. M., Stokes, J. L., *J. Biol. Chem.* (1946) **163**, 159.
(55) Bada, J. L., thesis, University of California, San Diego, 1968.
(56) Hare, P. E., Mitterer, R. M., *Carnegie Inst. Year Book* (1967) **65**, 362.
(57) Hare, P. E., Abelson, P. H., *Carnegie Inst. Year Book* (1968) **66**, 526.
(58) Hare, P. E., Mitterer, R. M., *Carnegie Inst. Year Book* (1969) **67**, 205.
(59) Buckingham, D. A., Marzilli, L. G., Sargeson, A. M., *J. Am. Chem. Soc.*
      (1967) **89**, 5133.
(60) Hirota, K., Izumi, Y., *Bull. Chem. Soc. Japan* (1967) **40**, 178.
(61) Olivard, J., Metzler, D. E., Snell, E. E., *J. Biol. Chem.* (1952) **199**, 669.
(62) Toi, K., *Bull. Chem. Soc. Japan* (1963) **36**, 739.
(63) Williams, D. H., Busch, D. H., *J. Am. Chem. Soc.* (1965) **87**, 4644.
(64) Fontaine, A. R., Chia, F. S., *Science* (1968) **161**, 1153.
(65) Stephens, G. C., Schinske, R. A., *Limnol. Oceanog.* (1961) **6**, 175.
(66) Stephens, G. C., "Estuaries," p. 367, G. H. Lauff, Ed., Publication No.
      83, American Association for the Advancement of Science, Washington,
      D. C., 1967.
(67) Guillard, R. R. L., "Marine Microbiology," p. 93, C. H. Oppenheimer,
      Ed., C. C. Thomas, Springfield, Ill., 1963.
(68) North, B. B., Stephens, G. C., *Biol. Bull.* (1967) **133**, 391.
(69) Pinter, I. J., Provasoli, L., "Marine Microbiology," p. 114, C. H. Oppen-
      heimer, Ed., C. C. Thomas, Springfield, Ill., 1963.
(70) Provasoli, L., McLaughlin, "Marine Microbiology," p. 105, C. H. Oppen-
      heimer, Ed., C. C. Thomas, Springfield, Ill., 1963.
(71) Botan, E. A., Miller, J. J., Kleerekoper, H., *Arch. Hydrobiol.* (1960) **56**,
      334.
(72) MacLeod, R. A., Onofrey, E., Norris, N. E., *J. Bacteriol.* (1954) **68**, 680.
(73) Ostroff, R., Henry, B. S., *J. Cellular Comp. Physiol.* (1939) **13**, 353.
(74) Skerman, T. M., "Marine Microbiology," p. 685, C. H. Oppenheimer, Ed.,
      C. C. Thomas, Sprinfield, Ill., 1953.
(75) Kaksman, S. A., Hotchkiss, M., Carey, C. L., Hardman, Y., *J. Bacteriol.*
      (1938) **35**, 477.
(76) Wright, R. T., Hobbie, J. E., *Ecology* (1966) **47**, 447.
(77) Williams, P. J. B., *J. Marine Biol. Assoc. U. K.*, in press.
(78) Jannasch, H. W., *Limnol. Oceanog.* (1967) **12**, 264.
(79) Collins, V. G., *Proc. Water Treat. Exam.* (1963) **12**, 40.
(80) Potter, L. E., Baker, G. E., *Ecology* (1961) **42**, 338.
(81) Postma, H., "Advances in Organic Geochemistry," p. 47, P. A. Schenck
      and I. Havenaar, Eds., Pergamon, Braunschweig, Germany, 1969.
(82) Williams, P. M., Aeschgr, H., Kinney, P., *Nature* (1969) **224**, 256.
(83) LaRue, T. A., Spencer, J. F. T., *Can. J. Botany* (1966) **44**, 1222.
(84) Rydon, H. N., *Biochem. Soc. Symp.* (1948) **1**, 40.
(85) Corrigan, J. J., *Science* (1962) **164**, 142.

(86) Meister, A., "Biochemistry of the Amino Acids," p. 113–8, Academic, New York, 1965.
(87) Wagner, M., *Zentra. Bakteriol. Parasitenk.* (1962) **115**, 66.
(88) Spackman, D. H., Moore, S., Stein, W. H., *Anal. Chem.* (1958) **30**, 1185, 1190.
(89) Gil-Av, E., Charles-Sigler, R., Fischer, G., Nurok, D., *J. Gas Chromatog.* (1966) **4**, 51.
(90) Pollock, G. E., Oyama, V. I., *J. Gas Chromatog.* (1966) **4**, 126.
(91) Degens, E. T., Ross, D. A., "Hot Brines and Recent Heavy Metal Deposits in the Red Sea," Springer-Verlag, New York, 1969.
(92) Back, W., Hanshaw, B. B., ADVAN. CHEM. SER. (1971) **106**, 77.
(93) Clarke, R. H., *Nature* (1967) **213**, 1003.
(94) Palacas, J. G., Swanson, V. E., Moore, G. W., *U.S. Geol. Surv. Prof. Paper* **550C**, p. C102.
(95) Bada, J. L., Luyendyke, B., Maynard, J. B., *Science* (1970) **170**, 730.

RECEIVED June 1, 1970.

INDEX

INDEX

335